PERGAMON INTERNATIONAL LIBRARY
of Science, Technology, Engineering and Social Studies
The 1000 volume original paperback library in aid of
education, industrial training and the enjoyment of leisure
Publisher: Robert Maxwell, M.C.

INTRODUCTION TO LABORATORY ANIMAL SCIENCE AND TECHNOLOGY

THE PERGAMON TEXTBOOK INSPECTION COPY SERVICE

Other titles of interest

ANDERSON:
Nutrition of the Dog and Cat

CHRISTOPH:
Diseases of Dogs

LANE:
Jones's Animal Nursing, 3rd Edition

PARKER:
Health and Disease in Farm Animals, 2nd Edition

ROBINSON:
Genetics for Cat Breeders, 2nd Edition

By the same author

INGLIS:
A textbook of Human Biology, 2nd Edition

LEE and INGLIS
Science for Hairdressing Students, 2nd Edition

Introduction to Laboratory Animal Science and Technology

J. K. INGLIS, B.Sc., B.A., Dip.Ed., M.I.Biol.

Section Leader—Life Sciences,
College of Further Education, Oxford

Sometime
Lecturer Anatomy and Physiology,
College of Lake County,
Grayslake, Illinois, USA

PERGAMON PRESS

OXFORD ● NEW YORK ● TORONTO ● SYDNEY ● PARIS ● FRANKFURT

U.K.	Pergamon Press Ltd., Headington Hill Hall, Oxford OX3 0BW, England
U.S.A.	Pergamon Press Inc., Maxwell House, Fairview Park, Elmsford, New York 10523, U.S.A.
CANADA	Pergamon of Canada, Suite 104, 150 Consumers Road, Willowdale, Ontario M2J 1P9, Canada
AUSTRALIA	Pergamon Press (Aust.) Pty. Ltd., P.O. Box 544, Potts Point, N.S.W. 2011, Australia
FRANCE	Pergamon Press SARL, 24 rue des Ecoles, 75240 Paris, Cedex 05, France
FEDERAL REPUBLIC OF GERMANY	Pergamon Press GmbH, 6242 Kronberg-Taunus, Hammerweg 6, Federal Republic of Germany

First edition 1980

British Library Cataloguing in Publication Data

Inglis, John Kenneth
Introduction to laboratory animal science and technology. (Pergamon international library).
1. Laboratory animals
I. Title
636.08′85   SF406   79-41552

ISBN 0-08-023772-X hardcover
ISBN 0-08-023771-1 flexicover

Printed in Great Britain by A. Wheaton & Co., Ltd., Exeter

Acknowledgements

FEW works of non-fiction can claim to be entirely original. The author is influenced strongly by his own reading and practical experience. This book is the result of just such influences.

The Bibliography pays tribute to some of the authors who have influenced the writer of this book.

The students that have attended the College in Oxford are nameless, but represent a strong influence on any teacher's manner of presentation, inclusion or exclusion of irrelevancies, and motivation.

The employing authorities and their skilled technicians are particularly important influences because it is they that have made it possible for the author to be "at the scene of the work" so that which is written is not dated nor irrelevant. It is these "experts" that have been tolerant of my continuous pestering for assistance, information, or time. I name some of those that have over the years influenced the contents of this book.

Mr. Bagnall	Department of Physiology, Oxford University.
Mr. Buckland	Medical Research Council, Harwell.
Mr. Coughlin	Department of Psychology, Oxford University.
Mr. Davys	OLAC 1976 Ltd., Bicester, Oxon.
Mr. Dear	Department of Biochemistry, Oxford University.
Mr. Elvidge	Medical Research Institute, Oxford.
Mr. Hovell	Veterinary Officer, Oxford University.
Mr. Kent	Department of Pathology, Oxford University.
Mr. King	Manager, Park Farm, Oxford University.
Mr. Manual	Department of Zoology, Oxford University.
Mr. Millican	Manager, Animal House, John Radcliffe Hospital, Oxford
Mr. Pettigrew	Nuffield Department of Medicine, Oxford University.
Mr. Turner	Department of Psychology, Oxford University.
Mr. Small	Department of Zoology, Oxford University
Mr. Weston	Veterinary Science Laboratory, Oxford University.

The commercial organizations, as listed opposite, that supplied the figures are also acknowledged. The publishers who gave permission to use their illustrations are acknowledged in the text. My technicians for their work over the years in maintaining a comparatively large stock of laboratory animals in less than adequate circumstances. Mr. Ross McKay, Senior Technician in the Life Sciences at the College for his art work, as acknowledged in the text. The invertebrate illustrations are adapted from *General Zoology*, by Tracey Storer, McGraw Hill (1943). Particular thanks are due to my wife, Ulrike, who gave up many hours of her free time to type and retype the manuscript.

Despite the many influences, any errors in fact, or poverty of style belong solely to the author. I would appreciate any constructive correspondence that may contribute to improvements and updatings.

Commercial Organizations

All-Type Tools (Woolwich) Ltd., Purland Road, Woolwich Industrial Estate, London SE28 0AS, UK. Telephone: 01-310 5376.

Associated Crates (Fabrications) Ltd., Coronation Street, Stockport, Cheshire SK5 7PL, UK. Telephone: 061-480 3016.

Bantin and Kingman Ltd., Laboratory Animal Consultants, The Field Station, Grimston, Aldbrough, Hull HU11 4QE, UK. Telephone: 04017-555.

Better Built Machinery Corp, 441 Market Street, Saddle Brook, New Jersey 07662, USA. Telephone: (201) 843-1010.

John Burge (Equipment) Ltd., 35 Furze Platt Road, Maidenhead, Berkshire SL6 7NE, UK. Telephone: Maidenhead 27840.

Elliott Fox (Elstree) Ltd., Home Farm, Aldenham Road, Elstree, Hertfordshire WD6 3AY, UK. Telephone: 01-953 5322.

Forth Tech Services Ltd., Mayfield, Dalkeith, Midlothian EH22 4AQ, UK. Telephone: 031–663 4474.

Th. Goldschmidt Ltd., Chemical Products, Initial House, 150 Field End Road, Eastcote, Middlesex HA5 1SA, UK. Telephone: 01-868 1331.

North Kent Plastic Cages Ltd., Home Gardens, Dartford, Kent, UK. Telephone: Dartford 21488.

R. B. Radley and Co. Ltd., Metabolism Cages, London Road, Sawbridgeworth, Hertfordshire CM21 9JH, UK. Telephone: 0279 722661. Telex: 817074 Radley G.

Oxford Laboratory Animals Centre (OLAC 1976), Shaw's Farm, Blackthorn, Bicester, Oxon. OX6 0TP, UK. Telephone: 08692 43241-2-3 Telex: 83683 (OLAC UK).

Preface

Student audiences

"Laboratory Animal Science and Technology" is directed at the following categories of reader:

The technician trained, in training and "on the job".
The student or pupil using animals.
The teacher or researcher keeping animals.
The interested public who are concerned for animals and their use in laboratory work.

The contents have been assembled and presented in such a manner as to be suitable and relevant for students working on introductory animal science and technology courses in both the UK and USA. The spelling conforms to the American style throughout.

The academic bodies conducting programs of study which have syllabus cover in this book are as shown below:

Technician Education Council—(TEC).
Institute of Animal Technology—(IAT).
Registered Animal Nursing Auxiliary—(RANA).
City and Guilds of London Institute—Craft Course (CGLI).
American Association for Laboratory Animal Science—(AALAS).

Subject-matter and the objectives

The subject-matter is animal care and welfare. The animals considered are those most frequently encountered in the laboratory situation. The greater proportion of the study being confined to the more common laboratory mammals.

The primary objective is to discuss the principles involved in the healthy maintenance of animals in the laboratory or animal house. Further objectives are to present factual information about the physical requirements of animals, physiological data, and techniques of husbandry. Much of these introductory data are reduced to tabulations for ease of quick reference.

The book is recommended to those with a minimum background in science although much that is within the book will be of benefit to those in research or education.

Introduction

Organization

The book is presented as six units of study followed by summary data capsules and recommended further reading. Practical work is suggested for each study unit and a selection of self-testing exercises are provided for the reader to monitor his progress in understanding the text. For ease of cross-reference the Glossary and Index is indexed to page numbers, as is the Bibliography.

Study breakdown

Pre-study Unit

0.1. Introduction to laboratory animals

"One hundred million animals die every year in the world's laboratories—many in excrutiating agony"; "Beagles were force fed weedkiller and after days of agony they died"; "The eyes of kittens sewn shut in order to study their response to a permanently dark world".

Above are the sorts of statements familiar to most when involved in discussion about the use of animals for experiment. (See Ryder, R. D., *Victims of Science*, Davis-Poynter Ltd., 20 Garrick Street, London WC2E 9BJ.)

It is these extreme cases of research and drug testing that rightly attract public attention, but it may be true to say that millions of living organisms are used for experimental purposes from primary school through to graduate level, without attracting comment.

In the United Kingdom it has been estimated that over forty species of organisms are used in the primary school and more than one hundred are used at the secondary school level. The educational use to which these organisms are put have been grouped as follows:

(*a*) Biological work and enquiry.
(*b*) Centers of interest and activity (e.g. in creative work).
(*c*) Associated work in other subjects (e.g. in mathematics or geography).
(*d*) Remedial and beneficial uses (e.g. with deprived or maladjusted pupils).

Extensive information concerning the educational use of living organisms in schools in the United Kingdom is to be found in the publication of the *Schools Council* (*Educational Use of Living Organisms Project*).

The use of living animals in American schools has different dimensions, as can be seen from the many laboratory manuals written for the secondary school student. The experimental tasks outlined in some textbooks would not be appropriate, nor perhaps be permitted in schools in Great Britain. American educators are generally recommended to follow the *Guiding Principles in the Use of Animals by Secondary School Students and Science Club Members* (Institute of Laboratory Animal Resources, National Academy of Sciences, National Research Council, Washington, DC). The American Cancer Society has published *Biology Experiments for High School Students* (1964), which presents a valuable link between the school laboratory and an area of research with international importance. Some of the suggested experimental work in this book would not be attempted in schools in the United Kingdom.

Within the educational context, animals kept for instruction purposes demand extra work from the educator and the technical staff. Living organisms require proper accommodation and daily attention. No living organism should be brought into the school or college if there is no individual who is prepared to give up time to supervize the welfare of the organisms. Animals have no regard for week-ends, holidays, or vacations and so strict routines must be drawn up and adhered to. (See Regan, T., and Singer, P. (1976), *Animal Rights and Human Obligations*, Prentice-Hall, Englewood Cliffs, New Jersey.)

No wild mammals or birds (dead or alive) should be brought into educational quarters. Both may carry disease that could be communicable to man or to other animals being kept for instruction purposes. (See Scott, W. M., "Diseases of Animals Communicable to Man", *Biology and Human Affairs*, vol. 32, no. 2, 1967; "Hazards of Animal Maintenance", *School Science Review*, vol. 50, no. 172, March 1969.)

Animals should be housed in such a manner as to prevent them having any contact with wild species of mammal or bird. The housing should be escape proof and should be purchased from specialist manufacturers (p. vii), not from pet shops as their cages are generally inadequate and difficult to clean. Mammals and birds should be purchased from accredited dealers, details of which can be obtained from the Laboratory Animals Centre (p. 304).

Non-mammals, such as the reptilian, tortoises, and terrapins are known carriers of food poisoning bacteria. The terrapins are particularly dirty creatures in their feeding habits and not recommended for educational use. Budgerigars and other members of the parrot family are susceptible to psittacosis which can be transmitted to man and is fatal. They should never be allowed to fly free or come in contact with wild birds.

Under no circumstances should primates (monkey family) be kept on educational premises. Monkeys are carriers of hookworm, Marburg virus, and the B virus infection that is fatal to man. There is no known antidote to this infection. (See *Safety in Science Laboratories*, DES Safety Series No. 2, HMSO.)

The previously mentioned animals have backbones. Fish also have backbones but tend to attract less sympathy and consideration, perhaps because they are aquatic and make no meaningful gestures or noises. Cruelty to animals, in legislation, *seems* to be limited to those animals with backbones that are warm blooded. This should not limit the attentions of those concerned for animal welfare, just because legislation stops short of frogs and fish in some aspects (i.e. pithing). All animals require equal consideration with regard to their biological and physical needs. This philosophy should extend into the maintenance of non-backboned animals, however small. If we wish to successfully culture water fleas, earthworms, locusts, or protozoans, then the same attitudes that we apply to the larger mammals need to be applied.

Animals used for research or testing could be thought of as being used for educational or instructional purposes, but there are major differences. Investigations into cancer producing agents, the effects of shampoos on the eyeball, the results of inhaling poisonous fumes, like burnt tobacco plant leaves, all involve possible physiological damage to the animal. It is in these areas of important research and drug testing that the public conscience is exercised. Animal husbandry considerations in the commercial and research institutions are even more important than for the smaller school or college animal facility.

Animals used for research may need to be inbred or random bred, they may need to be "optimum breeders" to make them an economic proposition in terms of the expensive outlay on housing, food, and services. They may need to be animals that have spontaneous tumours or be spontaneously hypertensive, or hair free. It is in this area of commercial breeding for particular strains or sub-strains that the science of animal technology becomes most apparent. In the pages that follow, the elements of laboratory animal science and technology are outlined.

0.2. Introduction to laboratory mammals

0.2.1. MICE

These are perhaps the most numerous of all laboratory mammals. Their popularity lies in their prolific reproduction and comparative ease of management in domestication.

The ancestor of the laboratory mouse is the wild mouse (*Mus musculus*) which has an agouti coat coloration. The laboratory mouse is generally albino, but there are a whole range of coat colors produced by selective breeding.

There are large numbers of inbred strains of mice developed as a result of many years of research into genetics and cancer. Random bred strains are used for a variety of routine-testing procedures in the drug industry. The strains, and sub-strains of inbred mice, may number in excess of 200. The precise characteristics of these different types needs to be known to the experimenter and so a system of coding has had to be introduced. Some of the strains are designated by capital letters, others are given names. For example, CF No. 1 strain originated at Carworth (New York), while ICR originated at the Institute of Cancer Research.

Documentation upon the many strains is to be found in the *Mouse News Letter*, and the *International Index of Laboratory Animals*.

0.2.2. RATS

These small mammals have been used for many years as a laboratory species. They originate from the wild brown Norwegian rat (*Rattus norvegicus*). These animals themselves may perhaps have originated in eastern Asia, reaching the UK and USA by way of ships in the 1700s. The brown rat took over the dominant position held by the already existing black rat (*Rattus rattus*).

The laboratory rat is an albino strain of the Norway rat. It has many characteristics that make it useful for research work. There are three well-known strains employed by laboratory workers. (More information available in the *Rat News Letter*.)

(i) *Sprague–Dawley* (albino). This strain had its start at Sprague–Dawley Farm, Madison, Wisconsin. It is a longer animal with a narrower head. The long tail is a noticeable feature. It tends to develop less respiratory infections than some other strains.

There are sub-strains developed from this animal such as the CFE produced at Carworth, New York.

(ii) *Wistar* (albino). This strain originated at the Wistar Institute, Philadelphia, Pennsylvania. This animal has longer ears and a wider head. Its tail is not equal to the length of its body as is the Sprague–Dawley. It develops less spontaneous tumors than some other strains.

There are also many sub-strains of Wistar rat such as the CFN produced at Carworth, New York.

(iii) *Long Evans* (hooded). This strain has colored heads and shoulders and sometimes dorsal markings. The coloration can vary from black to cream. The animal is smaller than the two previous strains.

0.2.3. GUINEA-PIGS

These docile mammals are popular as pets as well as laboratory animals. They originate from South American burrowing guinea-pigs that still inhabit Peru.

Guinea-pigs (*Cavia porcellus*) are vegetarian animals that generally live in groups or colonies in the wild. There are three main strains of animal, distinguished by their hair lengths and its direction of growth.

(i) *English types* have short hair with a wide variety of colors, or combination of colors. It is the more popular choice for experimental work. There are many sub-strains of this animal that may be checked out by making reference to standard references (e.g. Dunkin–Hartley albino).

(ii) *Abyssinian types* have short hair, which is rough, and radiates from a number of centers.

(iii) *Peruvian types* have long, rough hair.

0.2.4. HAMSTERS

The wild hamster is a solitary animal with territorial behavior. It burrows and hoards plant materials and seeds. They are active during the evening and night, sleeping for much of the day. They would seem to be quite unsuitable as pets for this reason.

There is a wide range of hamsters in the wild, some more suitable for domestication than others. The two most commonly reported species being mentioned below.

(i) *Syrian golden hamsters* (*Mesocricetus auratus*) are comparatively recent newcomers to the research laboratory. They were first used in the early 1930s. They are of a light brown coloration, but their variations in color make the title "golden" less than appropriate.

(ii) *Chinese* (*grey*) *hamsters* (*Cricetulus griseus*) are smaller than the previous mentioned and have a darker dorsal surface and a grey to white undersurface.

(iii) *European* (*black*) *hamsters* (*Cricetus cricetus*) have thicker light brown and white coats. They are slightly larger than the Syrian type. This species is less commonly kept for laboratory work.

0.2.5. MONGOLIAN GERBIL

This desert animal (*Meriones unguiculatus*) was first domesticated in the early 1930s. In the wild it lives in large colonies, demarcating its territory by means of scent released by the male. This species of gerbil is widely distributed in China and Mongolia where temperatures range from winters of $-10°C$ (14°F) to summers of 30°C

(86°F). Despite these cold conditions there appears to be no clear evidence that the gerbil hibernates.

0.2.6. RABBIT

The Old World rabbit (*Oryctolagus cuniculus*) is classified together with the following species.

Lepus species are the hares and Jack "rabbits" which mostly occur in Eurasia, as well as Africa, North America, and Australia.

Sylvilagus species are the cottontail rabbits of North America and parts of South America.

Pronolagus species are the red and rock hares of central and southern Africa.

The varieties of rabbit differ in size and weight from 1 kg (2 lbs) Netherland Dwarf, to 7 kg (15 lbs) Flemish Giant.

(i) *Fancy varieties*: Angora, the Dutch, the English, the Himalayan, the Polish and the Netherland Dwarf. These types are bred for particular hair lengths and color patterns, or for distinct body shape or size.

(ii) *Normal fur varieties*: the Chinchilla, the New Zealand White. Their coats have a short undergrowth with projecting guard hairs.

(iii) *Rex fur varieties*: self-colored, tans, and agoutis. They have the short undercoat hairs and guard hairs of the same length.

(iv) *Satin fur varieties*: self colored, tans, and agouti. These types have flatter shaped hairs producing a characteristic sheen to the coat.

0.2.7. FERRET

This mammal (*Mustela putorius furo*) is not recommended for use in educational institutions with young people. The animal has a reputation for aggressive behavior, and being carnivorous is thought to be a danger. It is not always true, it depends upon the way in which the animal has been reared.

Ferrets are in the same classificatory group as the weasel, mink, and otter.

0.2.8. CAT

The European domestic cat (*Felis species*) is thought to be the result of a cross between two wild types. The tabby cat (*Felis catus*) is a comparatively small member of the cat family, which include the lions and tigers. There is a wide range of varieties of cat; the results of selective breeding.

0.2.9. DOG

The dog (*Canis familiaris*) used in laboratory work can be the medium-sized mongrel or the beagle. Dogs range from small to very large and placid to excitable. There are some breeds that are used as standards for a particular program of work. The beagle is a popular choice.

0.2.10. PRIMATES

Monkeys have been used for anatomical studies back in the second century A.D. Experimental work on live monkeys may date back to 1873.

A popular monkey used in laboratory studies is the Rhesus (*Macaca mulatta*). These "old world" monkeys are grouped as *macaques* to distinguish them from others in the same classification group, such as the baboons, the patas, and the green monkeys. This super family of monkeys, in common with most old world monkeys, have storage cheek pouches and tails that are incapable of being used for grasping (not prehensile).

A few other laboratory primates are listed below:

Papio (Baboon)	*Cebus* (pauchin ringtail)
Callithrix (Marmoset)	*Cercopithecus* (African Green)
Ateles (Spider monkeys)	*Cercocebus* (Mangabeys)
Saimiri (Squirrel monkeys)	*Pan* (Chimpanzee)

0.3 Introduction to laboratory non-mammals

Non-mammals make up the majority of the animals in the world. Professional animal technicians are usually knowledgeable of the needs of a restricted group of animals (the mammals) because they are more commonly employed in research as "human models" in experiments. Rarely is an individual employed to cater to the needs of the full range of animals reviewed in this book. For this reason, the space given over to mammals is at a premium.

Non-mammals are usually employed in educational institutions for demonstrations within the disciplines of Zoology and Physiology. Investigations into the effects of drugs on the heart-beat may employ tortoises or frogs. Studies in photoperiodicity, or behavior, may be carried out using birds. Examples of parthenogenesis (virgin birth) may be demonstrated using stick insects or water fleas. Asexual reproduction can be seen in the more lowly forms such as Hydra and Amoeba.

Culturing or rearing non-mammals in the laboratory requires quite exacting routines, especially with the microscopic animals. A training in several laboratory skills becomes a necessity for those maintaining many of these animals. Some of the skills, such as the use of microscopes and the making up of culture solutions, are described later.

STUDY OBJECTIVES

UNIT 1:

Animal Accommodation

(*a*) Describes various forms of animal housing.
(*b*) Describes the importance of designs, materials, and traffic within the animal house.
(*c*) Describes different types of animal room.
(*d*) Describes the importance of design, materials, and facilities within the animal rooms.
(*e*) Describes the important requirements of cage construction for different animal species.
(*f*) Describes the various methods of shelving and racking of cages.
(*g*) Describes different types of animal pen.
(*h*) Describes aquaria and terraria used in laboratories for holding small animals.
(*i*) Suggests project work.

UNIT 1:

Animal Accommodation

ANIMALS are kept in captivity by man for a variety of reasons. The reasons may be one of the following:

> Recreational.
> Educational.
> Experimental.
> Production/breeding.

The accommodations provided for animals kept in the laboratory situation are to be examined under the following sub-headings:

1.1. Animal houses.
1.2. Animal house rooms.
1.3. Animal cages.
1.4. Animal pens.
1.5. Aquaria.
1.6. Terraria (vivaria).
1.7. Project program (construction materials).

The choice of suitable accommodation will depend upon several factors, such as:

(*a*) The species and size of animal.
(*b*) The number of animals.
(*c*) The reason for confining the animal.
(*d*) The space and finance available.

This study unit, in common with all others, is only an introduction. Further, more detailed information may be obtained by turning up the references mentioned in the text and in the bibliography.

1.1. Animal houses

It can be said that the animal house is where laboratory animals live and where animal technicians work, but this is not the complete picture. The student or researcher has a partial interest in the animal house. Without the "user" the animal house would have no reason to exist, the technician no employment.

The animal house may be a small unit within the teaching facilities of a school or college. It may be a larger complex within a commercial organization. Whatever the size, some of the basic requirements will be similar and will be taken into account when considering the design and the layout and so forth.

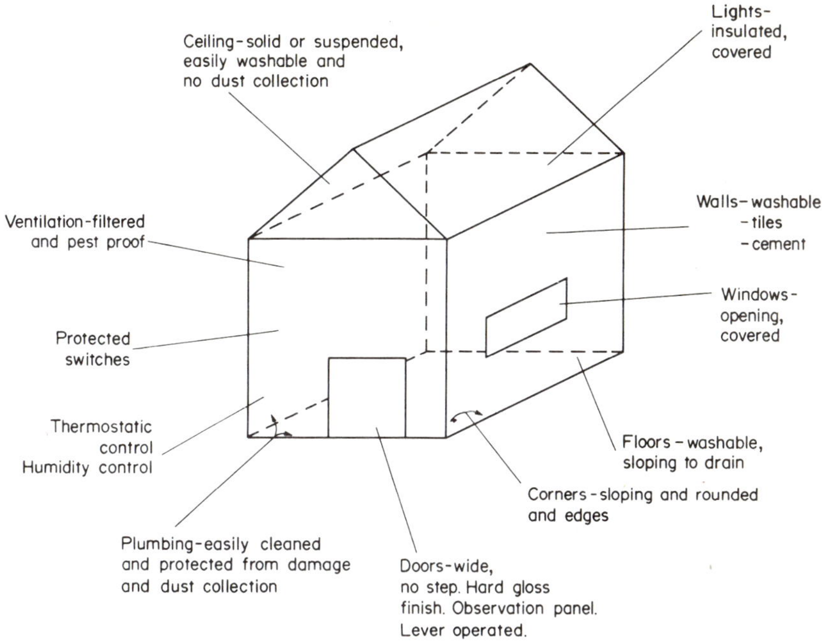

FIG. 1. Points requiring attention in the animal house

The species of animals that are to be housed may not be the same year after year, and so it is as well to provide facilities that are as flexible as possible. For instance, mobile cage units and even moveable slide-back walls are preferable to fixed units and walls that are expensive to modify if a different species of animals need to be accommodated.

Whilst attempting to be as flexible as possible in designing the layout of the animal house, one should not become too remote from the practical issues such as bringing in water, gas, and electricity, getting rid of wastes, maintaining a degree of isolation from possible sources of infection or infestation, or nuisance. Whatever scheme of layout is proposed, it is less than useful if it does not function efficiently.

1.1.1. ANIMAL HOUSE FACILITIES

Some of the facilities to be found in the larger animal house may be listed as below:

1. Reception area for receiving incoming materials such as animals, bedding, feed, and machinery.
2. Quarantine room(s) in which incoming animals may be held for a period of time in order to observe whether any disease symptoms make themselves obvious.
3. Administrative offices where records of materials and stock are kept.
4. Locker room with facilities for personnel, including lavatories, showers, and a refreshment area.
5. Store-rooms for bedding, feed, cleansing materials, chemicals, instruments, and caging materials not in use. Stores should be classed as "dirty" or as "clean".

6. Engineers room, or a space in which all the mechanical plant is located, such as switches, furnace, air-conditioning, and fuse boxes.
7. Cleansing area where soiled cages and equipment may be sterilized. This area requires a fair amount of space in order to accommodate those dirty materials awaiting steaming, etc. A disposal unit or incinerator may be provided in this area for getting rid of soiled bedding, carcasses, etc.
8. Surgery/X-ray room may be an added facility to the animal house where anesthetics can be administered, and blood or other tissue samples taken. A post-mortem room may also be required.
9. Treatment room where sick animals may be examined and treated, if and when necessary.
10. Animal breeding room, where animals are being mated.
11. Animal "growing room", where offspring are held until reaching a suitable size/age/weight.
12. Animal stock room, where animals are held until such time as they are required for use.
13. Animal experiment room, where animals undergoing some form of experimental procedure are held.
14. Isolation room(s), where animals may be kept in a "germ-free" or relatively germ-free environment.
15. Corridors, linking up the functional areas of the animal house as just described.

From what has just been said above, it seems that the animals occupy only about half of the animal house floor-space. That may well be the case as the back-up services for animal breeding are quite considerable. Corridors alone are of great importance, as they are the means of communication between all the animal support areas and their construction and layout must be carefully considered. One way of displaying the relative areas occupied by the facilities just mentioned is shown in Fig. 2.0 adapted from an article by Lane Petter (p. 301). Does this representation give a true picture of your animal house?

For the purposes of design and convenience, we may suggest that there are two major sub-divisions to the animal house; "clean" and "dirty". The traffic between these two areas should always be from the clean to the dirty, and never the reverse. Between the clean and dirty areas there should be some kind of barrier, physical or behavioral, to prevent pests or parasites reaching the animals. It should be remembered that these distinctions are no more than conveniences. All areas in the animal house should be clean, but there are grades of cleanliness perhaps in reality.

1.1.2. ANIMAL HOUSE FUNCTIONAL AREAS

If the animal house is to be considered as having "clean" and "dirty" parts, then all rooms of the animal house will fall within one or other of these sub-divisions.

"Clean" areas   Animal rooms.
                    Stores (food, bedding, cages and equipment).
                    Corridors (clean side).

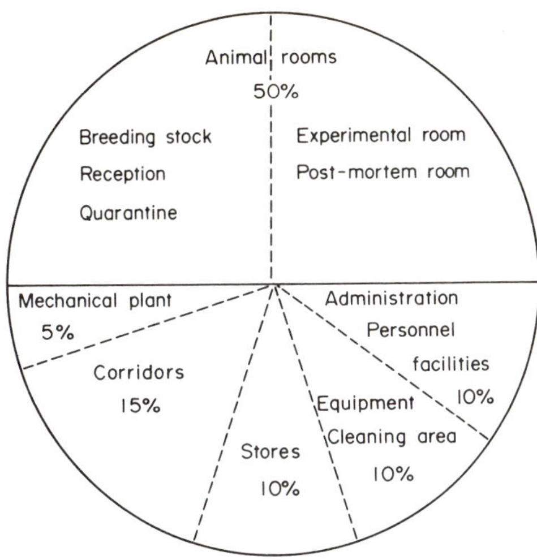

FIG. 2. Approximate relative areas of the subdivisions of an animal house (adapted with permission from *UFAW Handbook*, Churchill Livingstone)

"Dirty" areas   Cleansing rooms.
Stores (soiled cages, equipment, chemicals, bulk food and bedding).
Engineers room.
Administration room.
Locker room.
Corridors (dirty side).

Looking at these two lists it seems that some definition of "clean" and "dirty" needs to be made. This is not really necessary when breeding "conventional" animals, but it is of some importance in SPF (specific pathogen free) units, which will be studied later. In the "normal" animal house "dirty" and "clean" are defined by the administrators. In different animal houses the definitions will differ. Food pellets trodden under foot and returned to the food hopper in one animal house may be no problem, but in others this behavior could not be tolerated. Visitors walking around the animal rooms may be permissible in some animal houses, but in others they would be forbidden, or must at least shower and wear special clothing before entry.

1.1.3. ANIMAL HOUSE TRAFFIC—CORRIDOR USE

The functional areas of the animal house are usually linked together by corridors that permit wheeled vehicles carrying large and awkward-shaped objects. Any feature of the corridor that impedes the efficient movement of these materials is not only a bad design, but also slows down the animal house routines. Steps should always be avoided where possible in the design of an animal house. For this reason corridors should feature early in the planning. Some of the factors to take into account when

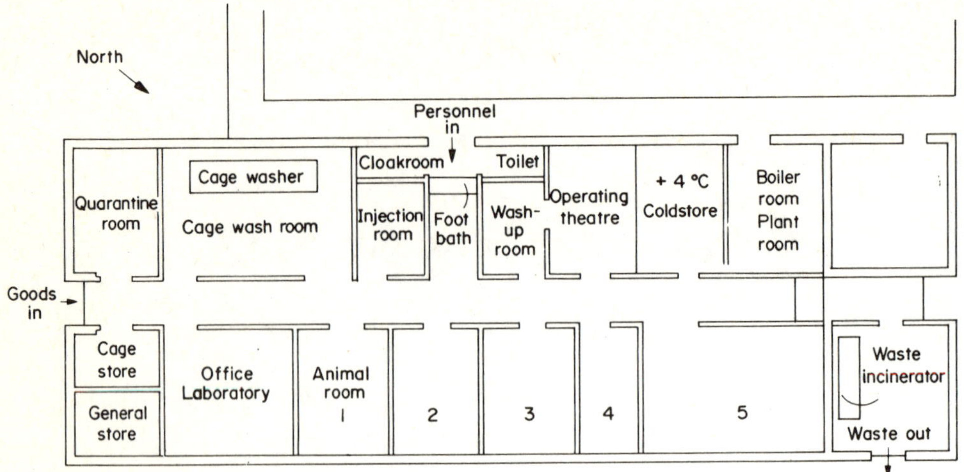

Fig. 3. Plan of a conventional animal house (adapted with permission from *IAT Journal*, vol. 22, no. 1, March 1971)

considering the design and layout of corridors may be summarized as follows:

1. Which rooms are in greatest need of communication?
2. What is the volume of traffic likely to be along any communicating corridor?
3. What type of objects are transported along such a corridor?

With the answers to these questions available a corridor can be designed. A short corridor is the ideal if the traffic is great and the materials carried bulky. There seems little logic in constructing a short corridor for little traffic, and a lengthy one for the busiest link. This type of information will dictate to us the positioning of the rooms in our overall plan.

1.1.4. CORRIDOR DESIGN

Constant traffic along corridors will make it necessary for them to be constructed of materials resistant to wear. There should be as few corners as possible and any protruding edges should have protective guards fitted to prevent damage to the walls. If doors are needed along the length of a corridor they will need to be made of a robust material, with a viewing window to prevent collisions. They would ideally be self-opening and closing, or at least capable of being pushed open and self-closing. These doors would need to be draught-proof.

Doors giving access to animal rooms should ideally be of the double type to reduce the amount of temperature change as doors are frequently opened and closed.

The width of the corridors will need to be carefully calculated because too much width represents wasted space. If a wide corridor is needed then some areas can be specially designed as parking areas, or laybys for wheeled vehicles.

The materials used for the walls and floors will be considered later, under the general heading of construction materials (p. 34).

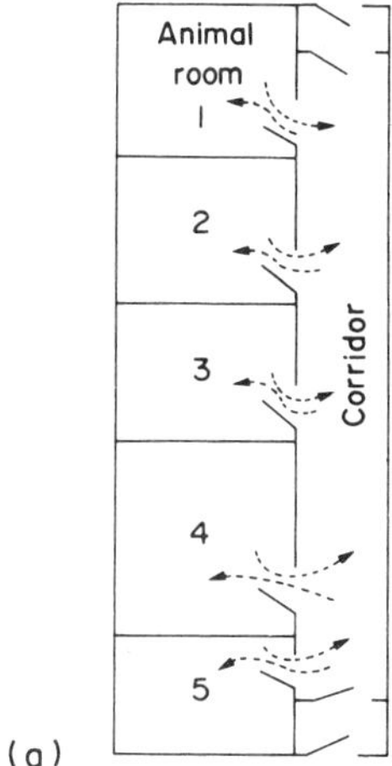

FIG. 4*a*. One-corridor system

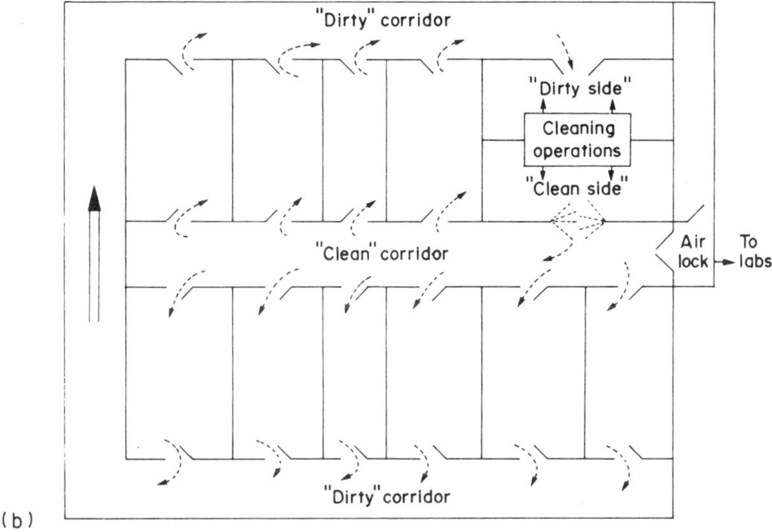

FIG. 4*b*. Two-corridor system

FIG. 4. Corridor systems (adapted from *Manual of Laboratory Animal Technicians*, AALAS, 1967)

1.1.5. CORRIDOR LAYOUT

The distribution of corridors within the animal house are very important because it is around these corridors that the specialist rooms are to be arranged.

The decision to be made at the planning stage is whether to have a *one-corridor system* or to have a *two-corridor system*.

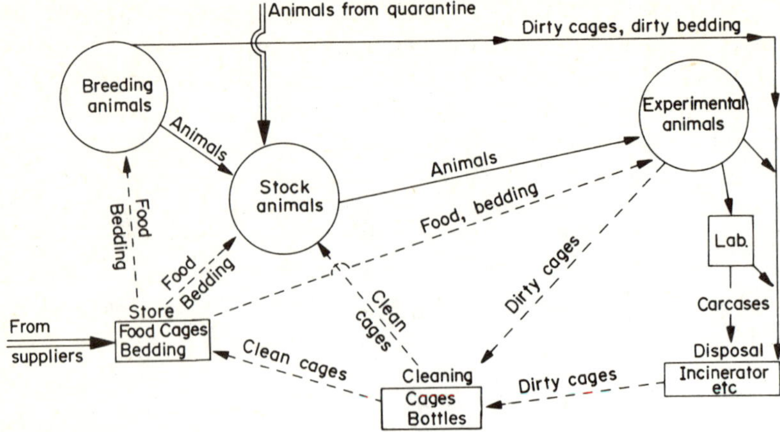

FIG. 5. Traffic between rooms of an animal housing unit (adapted with permission from *UFAW Handbook*, Churchill Livingstone)

The two corridor system is naturally more expensive. Cross contamination can still occur despite this financial outlay and the idea of two doors to each animal room may be impractical in any case, because the animal rooms may be too small.

1.1.6. TRAFFIC FLOW

The movements of materials, animals, and staff within the animal house should ideally be regulated to conform to the principle of "clean to dirty", and not the reverse. In the smaller animal houses, conforming to this ideal (in the strictest sense) could be very limiting to the staff in terms of efficiency and speed. However, as a general rule, the circulation within the animal house may approach the sort of ideal flow shown in the diagram above. Use this adopted figure as a discussion topic and attempt to draw out a flow diagram that represents your work situation.

1.2. Animal house rooms

The animal house facilities briefly described earlier are located in specialist rooms. It is the characteristics of the animal holding rooms only that will be considered here.

Those rooms used for holding animals will have some, or all, of the following requirements and characteristics:

1.2.1. FLOORS

These will need to be constructed of materials that are not liable to crack or be easily damaged by frequent washing with detergents and disinfectants. The floor cover should be non-slip, strong enough to withstand heavy traffic and it should ideally have few junctions, such as may be found on tiled floors. Such junctions are convenient traps for dust and other particles. There seems to be no ideal floor material. A cement floor may produce a great deal of dust but it has many advantages in terms of strength and durability, provided it has a covering of varnish or sealer.

The floors of animal rooms will need to be cleaned frequently. Some designers make the washing down process somewhat easier by having the whole floor sloping towards a drainage gulley. This slope can have disadvantages when wheeled cage racks are in use. A level floor will need mop-washing but does not have the disadvantages associated with the floor slope or drainage system; such as blockages and potential sources of infection remaining within the gulleys.

A common feature of animal room floors is that the floor slopes upwards at the point of contact with the wall. This curved edge prevents the accumulation of waste in awkward floor-wall angles.

1.2.2. WALLS AND CEILINGS

The walls, like the floors, need to be able to withstand frequent washing with detergents and disinfectants. The finish applied to the walls in order to give a resistance to the atmosphere of an animal room needs to be reasonably cheap and easily renewable. Tiles are very good but they are expensive to renew and they do have junctions that may harbour unwanted materials such as bacteria or dust.

A good emulsion paint applied to the walls is washable and reasonably cheap to renew. The construction of internal walls needs to be fairly substantial in the event of cages being stored on wall fixed racks. This has to be taken into account when thinking of designing a flexible animal room.

The solid ceiling is easier to maintain in terms of repainting and the prevention of dust accumulation. The suspended ceiling is better for hiding the piped services, but more of a problem in terms of concealed dust and pest control.

1.2.3. DOORS AND WINDOWS

A door giving access to an animal room is to be preferred if it is animal-proof when closed. The hinge-type door is the best in this respect, but it can be a nuisance in terms of daily work because it needs opening space. A sliding door is more practical from the work viewpoint, but it can be more easily penetrated by a mouse or an unwanted pest. Wooden doors and their frames need attention from time to time if the paint becomes damaged and the wood attacked by airborne chemicals.

Windows are to be found in some animal rooms. The value of windows is debatable. They may be of more value to the technician working in the room than they are of value to the animals. It is advisable to have any such windows permanently locked because they can be an additional entry point for pests. The atmospherics within the animal room may damage window frames, more so perhaps if they are wooden. Metal window frames are more durable.

1.2.4. SERVICES

The animal holding-rooms are supplied with various services, such as all or some of these now mentioned.

Water may be let into an animal room for the automatic drink apparatus, if it exists, or for hose-pipe connection when washing down floors or for bottle filling and handwashing.

Electricity outlets may be available to give an energy source for special instruments used in an animal room. An experimental room, for instance, may employ recording equipment measuring animal activity; this will require an electric source. The electrical sockets and light switches are best enclosed within a purpose-built housing, or enclosed by a plastic cover to prevent corrosion or water contact. The conduit-tubing leading electricity through the animal room should be reduced to the minimum, or enclosed in some manner so that dust collection is at a minimum. All lighting, and any time switches should also be enclosed.

Gas, air and vacuum may be needed in some animal rooms.

Heating and ventilation is a very important service provided for an animal-holding room. The species of animal being held will determine the temperature and humidity range required in the animal house. The equipment installations will normally be located in the plant, or engineer's room, but the controlling switches may be situated in the animal room or in a control panel outside in the corridor. In many parts of the world it is necessary to equip the animal rooms with air-conditioning. This is an expensive facility but is nevertheless vital when the climate is very hot and humid. The air-ducting used for ventilation should be installed in such a way as to be escape-proof and pest-proof. Both incoming and outgoing air may or may not be filtered.

The problems associated with maintaining an optimum temperature and humidity are considered below.

1.2.5. HEATING AND VENTILATION

This topic can be approached on two fronts, i.e. the problems, and the correction of the problems associated with the atmosphere in the animal room. At this level of study we will mainly consider the problems because the corrections would need more detail than is appropriate here. This is one way in which the heating and ventilation engineer can come to the aid of the laboratory scientist.

The problems confronting those setting up animal rooms in terms of atmospheric conditions are outlined below.

(a) A specific temperature range needs to be maintained for the rearing of given types of animals. This necessitates thermostatic control.

(b) Air temperatures must be evenly distributed avoiding "hot spots" in cages or beneath shelving, and "cold spots" elsewhere.

(c) A required relative humidity needs to be maintained. This necessitates de-humidifying and humidifying equipment, depending upon the seasons or climatic conditions.

(d) Fresh air must be supplied to the room either by pulling in filtered outside air, or by recycling some of the room air after it has been filtered. This latter method has certain disadvantages.

(e) In some rooms a positive air pressure needs to be maintained so that air is not drawn in beneath doors from corridors and such like. In the case of rooms that are "dirty" or infective they need to have a negative internal pressure to reduce the risk of airborne pathogens moving out of the room.

The way in which these problems or requirements are attended to depends upon the amount of cash available.

Space heating and cooling can probably be achieved most efficiently by the installation of air-conditioning. The expense of installing such a unit would be inappropriate for smaller animal rearing organizations but advisable, if not essential, for large-scale work. If full air-conditioning is not possible, then the air may need to be warmed by some means to keep it within a comfort range. As an extra to this thermostatically controlled background heating, individual rooms may need booster heating devices to keep them at a higher required temperature.

On the other hand, there are times and places where the concern is to reduce the temperatures. This requires that the animal house and individual rooms are fitted with thermostatically controlled cooling devices.

The methods of warming the air within the animal house are numerous, some less suitable than others. The problem of uneven heat distribution is a major reason why radiating or convecting heaters are undesirable. A desirable method is the feeding of warmed air into the room by way of well-insulated ducting. The air is heated by an electrical or burning device. It is important that the warm air input is filtered to remove particulate and biological matter. The entry point of the air to the animal room has to be correctly positioned to avoid draughts.

Air in-put to the animal room will be such that it raises or lowers the atmospheric temperature as required. The filtered air will need to be of a required humidity, bearing in mind that the room air is already receiving a fair amount of water vapor from the animals themselves, as well as from their drinking vessels. This humidity level can normally be controlled by adjusting the numbers of air changes per hour, but in some climates it is important and necessary to have dehumidifying or humidifying accessories.

Ventilating the animal rooms adequately will maintain the temperature and humidity at the required values providing an even distribution of the air to all parts. The air inlets to the room must be located correctly to give adequate distribution.

The fresh-air intakes need to be located outside in the atmosphere in such a position that they do not pull in dirty or offensive air. Polluted air will put an unnecessary load on the filtering devices which will require more frequent cleaning. Recycling the air within the animal house may seem to be an economic thing to do because fresh air needs warming to bring it up to a required temperature whereas recycled air does not. Recycled air may need more effective filtering because it contains considerable air-borne particles or even pathogens. The method is, however, recommended provided adequate high-grade filters are used.

The outlets for stale air should be situated in positions such that no draughts are caused and offensive air is quickly dispersed thus reducing airborne dust and odors. It should be filtered before being returned to the atmosphere.

Filters may in time become ineffective because of trapped particles and thus need to be cleaned or changed. This failing function can be monitored by taking air-pressure readings either side of the filters.

Filters are constructed in such a way as to trap airborne particles of different sizes. The rough filters hold back the larger particles whereas the ultra high efficiency filters will hold back sub-microscopic organisms as small as 5 μm down to 0.1 μm (see p. 71). Filters as fine as this are necessary if animals are to be reared in "germ-free" conditions. The efficiency of such a filter is described as being about 99.99%. It will even hold back the smallest of animal pathogens, the viruses.

On the average, the air in an animal room is changed fifteen times per hour. Variations on this figure are sometimes necessary for climatic reasons or because of conditions within the animal room.

1.3. Animal Cages

Animal caging will vary according to the animal species held, their numbers, the length of time they are held, and the purpose for which they are held. The design, the materials, and the accessories will depend upon some or all of the forementioned.

1.3.1. CAGE REQUIREMENTS

The design of the caging will need to conform to certain requirements.

(a) *Security.* The caging is required to be escape-proof. The doors and the locks should be of such a design that no animal can tamper with them and escape. The bars and wire mesh should be constructed in such a way as to prevent the escape of the smallest animals, the infants. The materials from which the caging is built should be indestructible by the animal species being housed. It should be resistant to gnawing and pulling, and not subject to excessive corrosion.

(b) *Accommodation.* The caging is required to accommodate the animals in sufficient space for comfort and health. To this end there should be at least enough space for individual animals to stand up, turn around and lie down. Specific requirements will be mentioned later for individual species. The accommodation is required to be supplied with food and water on a fairly regular basis and without being spilt and fouled by the inhabitants. Some provision is required to minimize the contact of urine and feces with the animals and their bedding. This may necessitate a slip-tray holding an absorbent litter beneath a grid floor. The accommodation is required to be easily cleaned and free of injurious sharp projections. Adequate ventilation and light is required.

(c) *Economy.* The initial cost of the caging should not be too expensive and may have to be within a stated budget. Caging which is easily replaced with easily available accessories will argue in favor of some kind of standardization. There seems to be little in favor of repairing worn-out units, or buying-in cheaper non-standard units. Much time and money can be saved if the parts of the caging are interchangeable.

(d) *Function.* The caging must serve the function intended, i.e. breeding, isolation, experiment, etc. The caging available commercially can be seen in the trade catalogues. There is a wide variety of design to choose from. The caging chosen will depend upon personal preference, determined by the function to be served. Some popular caging is shown in the illustration (Fig. 6).

Caging for use in some experiments may have to be custom made. They may not conform to the requirements stated earlier and so the animals will in general not be confined to such cages for any great length of time.

1.3.2. CAGE SIZES

The space provisions of a cage will depend upon the species of animal, the numbers of animals, and the purpose for which they are being confined. The space provided should permit the animals to live in comfort and health. The authorities differ as to what is the "optimal" size for confining different animal species, but guidelines are offered in the chart below.

TABLE 1.

Space recommendations for laboratory animals (the figures should be regarded as minimal)

Animal type	Occupants	Floor area	Cage sizes
Mouse	Breeding group	310–390 cm² (48–60 sq in)	30 cm × 13 cm × 15 cm (12″ × 5″ × 6″)
Rat	Breeding pair	1077 cm² (168 sq in)	36 cm × 30 cm × 26 cm (14″ × 12″ × 10″)
Guinea-pig	Sow and litter	960 cm² (150 sq in)	A wide variety including floor pens
Hamster	Female and litter	310–390 cm² (40–60 sq in)	30 cm × 13 cm × 15 cm (12″ × 5″ × 6″)
Rabbit	Doe and litter	0.38–0.56 m² (4–6 sq ft)	1.20 m × 0.46 m × 0.46 m (4′ × 1.5′ × 1.5′)
Cat	Queen and litter	0.56 m² (6 sq ft)	0.91 m × 0.61 m × 0.46 m (3′ × 2′ × 1.5′)
Ferret	Gill and litter	0.39–0.56 m² (4–6 sq ft)	0.91 m × 0.61 m × 0.30 m (3′ × 2′ × 1′)
Dog	Individuals 15–30 kg	Pen or run 1.12 m² (12.0 sq ft) Cage 1.12 m² (12.0 sq ft)	Varies with the breed

[Adapted from *Guide for the Care and Use of Laboratory Animals*, ILAR, USA, and IAT Manual.]

1.3.4. CAGE TYPES

Cages differ in several ways, in design and size. The floors and walls may be solid or grid. The door openings may be in the walls or on the cage top. The water bottles and pellet feeders may be presented to the animals in different ways.

The type of cage chosen will be a matter of professional judgement, taking into account cost and usage. Some cages are described below.

Shoebox-type cage. This type of cage is in common use. It is the rectangular cage with the grid top or lid serving as the opening. The water bottle and food hopper may be inserted through or hung from this lid. The walls of this type of cage may be plastic or metal. In the case of the metal grid-floor type, a solid slip-tray is inserted beneath the floor in order to catch the feces and other wastes.

The metal grid-floor types are unsatisfactory if bedding for the animals is to be placed directly upon the floor. If this is the case, a solid floor type is more suitable.

This sort of cage is used for small mammals, like rodents.

Front opening-type cages. This type of cage has metal grid walls and floors. The opening is a hinged wall on the longer or shorter side of the rectangle. The hinge can be at the top or the bottom of the cage. The hinge position is a matter of choice and depends upon the species being held. The grid-floor type needs a slip tray to catch the wastes. This side opening type of cage is used for guinea-pigs, dogs, rabbits, cats, and monkeys.

Suspension-type cage. Some rectangular cages may be without a top because they are suspended within a rack and the slip-tray of the cage above acts as the top of the cage below.

This type of cage is suspended within a mobile rack unit. All parts of this caging are best if detachable and easily sterilised.

Specialist cages. Within this category of caging are included the following: *The metabolism cage* is designed to collect feces and urine for the purposes of biological assay. Some of these cages are so designed as to separate the feces from the urine (see Fig. 52). *The mechanical exercise cage* is provided with a rotating exercise wheel. The animal may use this wheel voluntarily or in some cases it may be motorized and the animal is obliged to exercise. The wheel is often attached to a recording device.

1.3.5. CAGE RACKS AND SHELVING

The most economic way in which to store animal cages is to shelve or rack them. This shelving is best kept clear of the floor to permit easier cleaning. The highest shelf should not really be above the level of observation by the technician. With these points in mind, it seems that a maximum of about four tiers of cages can be stored on a fixed or mobile rack.

The racks or shelves can be fixed to the walls or suspended from the ceiling. The fixed variety will be more permanent and demands particularly tough wall and/or ceiling structures in order to carry the weight of cages and shelves. This increases the constructional costs as well as making floor and wall cleaning behind the racks and shelves more laborious.

The mobile racks on wheels are a practical answer to both the previous problems. These storage units can be conveniently moved away whilst floor and wall cleaning operations are in progress.

The stacking of animal cages can produce a ventilation problem. The still air held between shelving is best prevented by some kind of air circulatory device (without causing draughts). There are proprietary devices advertised in the literature (filter racks). The materials employed for racking, be it fixed or mobile, should resist attack by urine, feces, and detergents.

The shelving used to hold cages can be slatted to allow debris to fall through, or it can be solid. These both have disadvantages. The slats permit bedding and debris to fall into cages beneath, but they do permit air circulation. The solid shelving restricts air circulation.

1.3.6. CAGE WATERING EQUIPMENT

There are two ways in which water may be supplied to cages; by water bottle, or by some kind of automatic watering device with valves that are activated by the animals.

Most animals will take water from a stainless-steel drinking tube supplying water from a suspended bottle. Open pans or dishes of water on the floor of the cage are

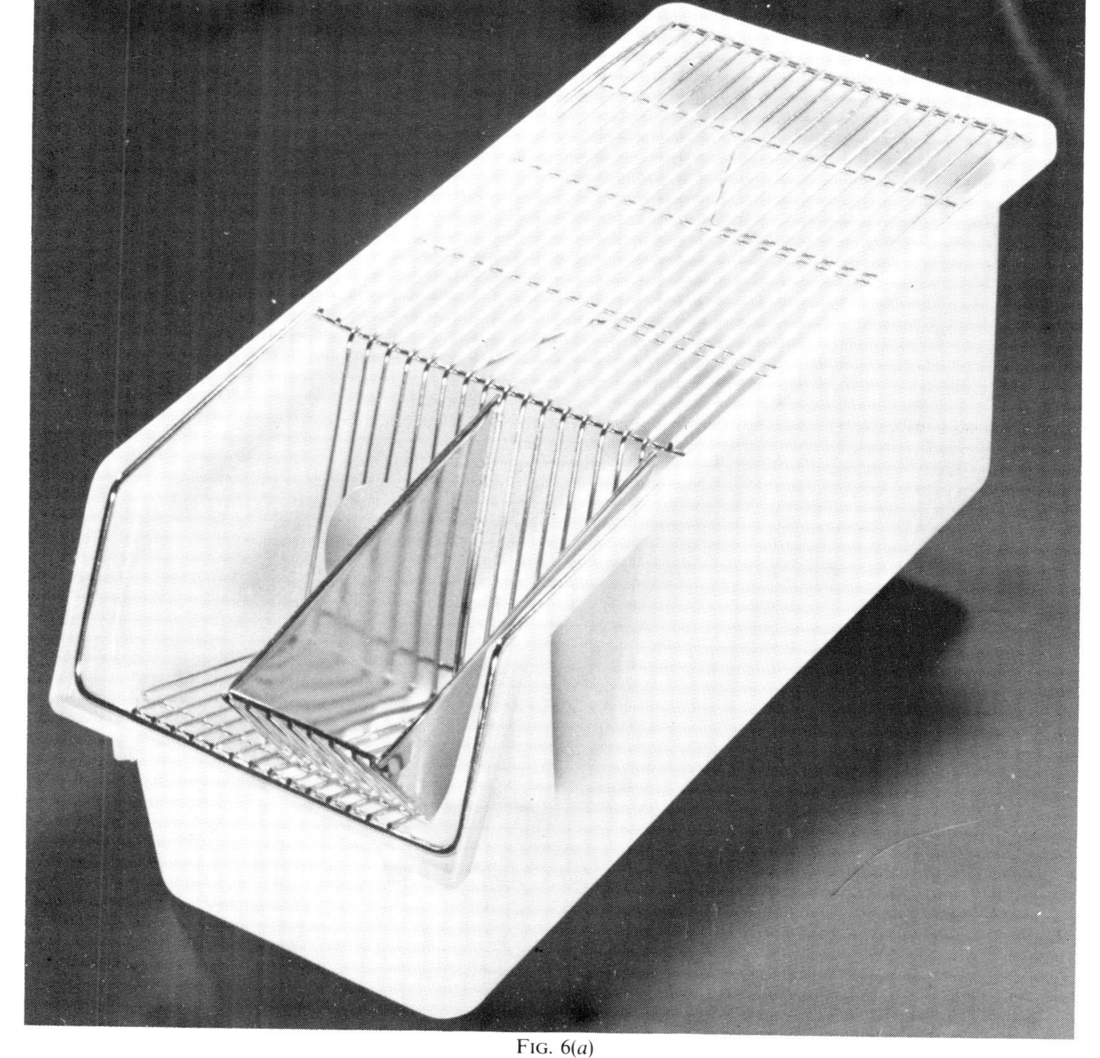

Fig. 6(a)

Fig. 6. Types of animal caging (photographs supplied by North Kent Plastics). (a) Mouse cage molded in polypropylene, wire work polished in stainless steel, 33 cm × 15 cm × 13 cm (13 in. × 6 in. × 5 in.).

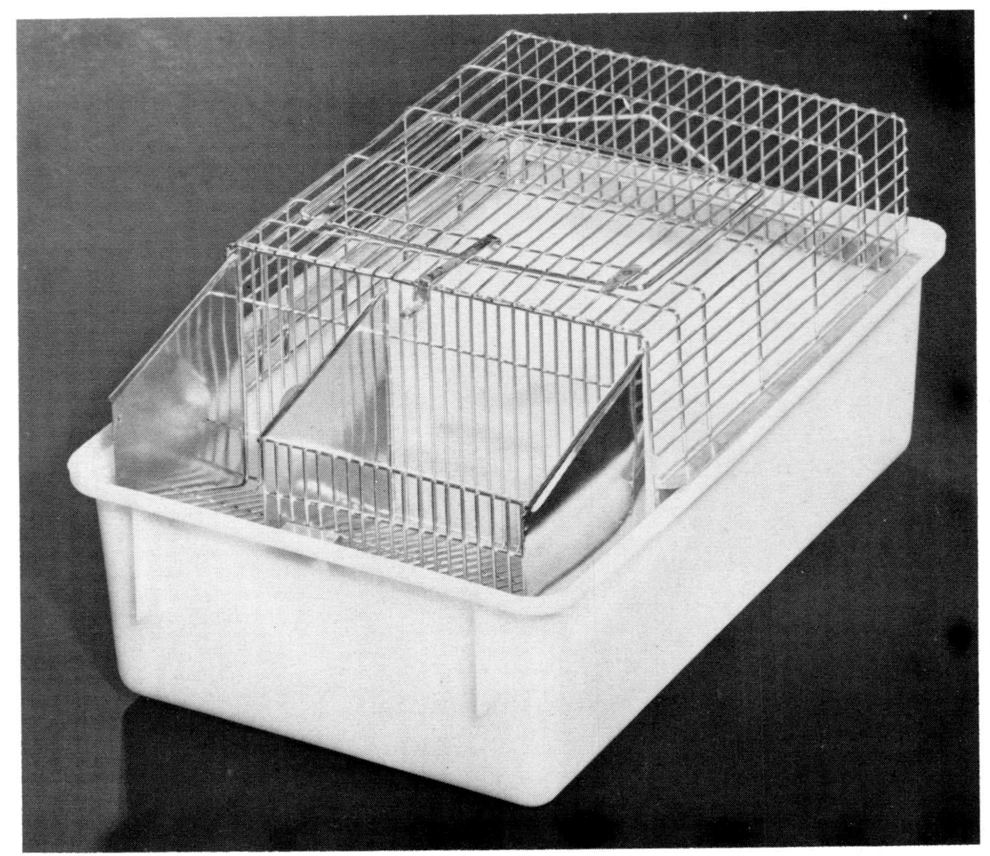

Fig. 6(b). Rat, hamster, gerbil caging, 45 cm × 28 cm × 20 cm (17½ in. × 11 in. × 8 in.).

FIG. 6(c). Rat, guinea-pig caging, 56 cm × 38 cm × 25 cm (22 in. × 15 in. × 10 in.).

FIG. 6(*d*). Rabbit caging in anodized aluminum alloy; perforated or wire aluminum floor; stainless-steel wire work and hopper, 61 cm × 46 cm × 46 cm. (24 in. × 18 in. × 18 in.).

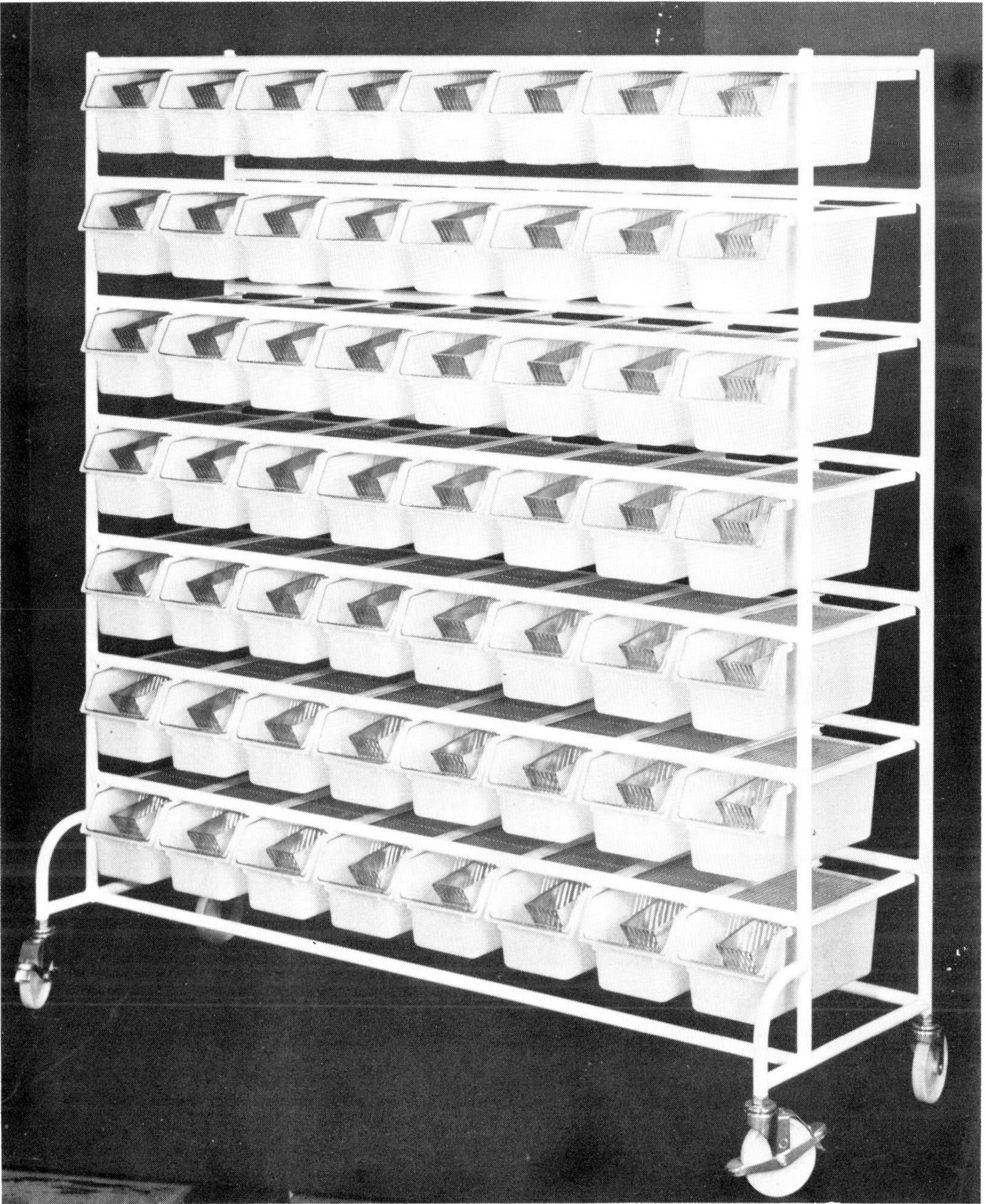

FIG. 7. Racking systems. (a) Mouse cage unit. Steel tubular rack white coated or aluminum; mobility on castors with brakes on two castors.

Fig. 7(b). Rat, guinea-pig cage racking.

FIG. 7(*c*). Rabbit cage racking.

Fig. 7(d). Adjustable wall racking. Steel brackets, white coated; aluminum tubes to form shelves; galvanized steel wall uprights (shelf width, 43 cm ($16\frac{3}{4}$ in.).

Fig. 8(a)

Fig. 8. Food hoppers and drinking tubes. (a) Interchangeable rat pellet food hoppers.

FIG. 8(*b*). (*left*) Interchangeable anodized aluminum food hopper for guinea-pig cage. (*right*) Powder food hopper.

FIG. 8(c). Stainless-steel ball-tip nozzles with rubber bungs.

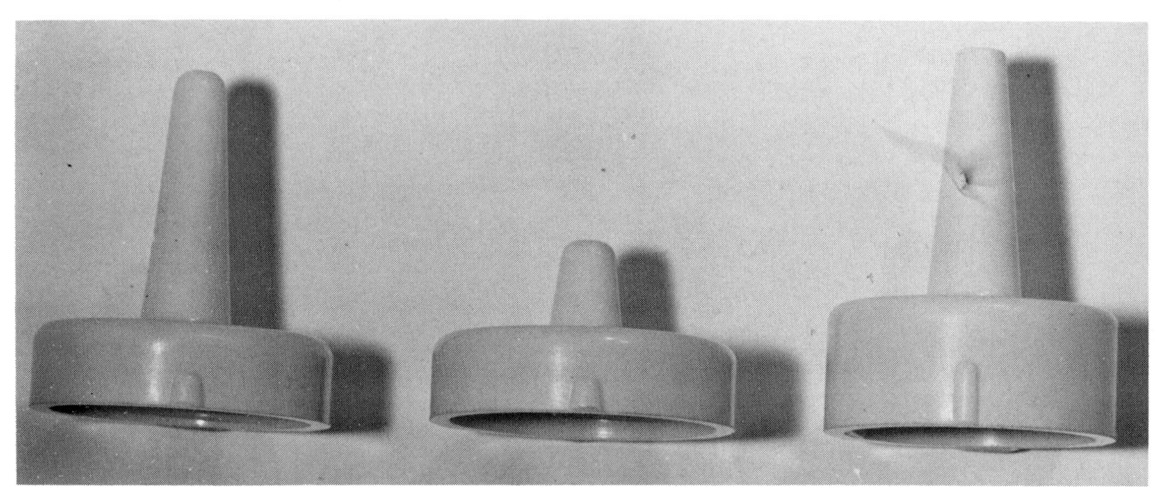

FIG. 8(*d*). Melamine nozzles for use with glass or polythene drinking bottles.

clearly not always suitable because of spillage and contamination. The problems associated with bottles are as follows:

(a) Supplies a limited quantity of water because over a certain volume (about 500 cm^3) spontaneous emptying may take place if the spout is maltreated (i.e. grit pushed into the mechanism by the animal).

(b) Leakage and blockage may occur for a variety of reasons; flooding the cage or killing the animals over a week-end.

(c) Bottles need to be changed frequently and cleaned to prevent the green growths of algae.

There are a large variety of drinking bottles, plastic or glass, but they all need to be of an optimal size in order to permit the steady flow of water. The choice of bottle material will be a personal preference depending upon the facilities and bottle-cleaning equipment.

The spout is held in the bottle by means of a bung or by a screw top. The diameter of this drinking tube is critical in order to prevent leakage, but it should permit a waterflow at a steady rate when the spout tip is touched. For this reason the heaping up of sawdust or bedding against water spouts should be avoided to prevent any leakage.

The drinking tube is best made from stainless steel as it is easier to clean and not easy to chew.

The water bottles are suspended at an angle on the top or side of the cage. Since the individual bottles rarely exceed half a liter, more than one bottle may need to be supplied. These bottles are usually suspended on the outside of cages to permit speedier replacement and to prevent interference by the cage inhabitants.

If a water bottle is functioning properly, an air bubble will move up the bottle as water is withdrawn. This check should be made before leaving animals for any length of time.

Automatic water supplies to animal cages can be a great asset when they are functioning without problems. When problems develop, those problems are magnified well above those of an individual leaky bottle.

The expense of putting in an automatic watering device has been considered justified for larger animals such as dogs and monkeys but for smaller animals like mice and rats problems have been described. One problem is that licking animals, such as the latter, can push grit into the valves on the drinking spout causing them to stop open and flood the cage. The automatic devices may produce more headaches than actually laboring daily to provide fresh bottle water.

1.3.7. CAGE FEEDING EQUIPMENT

In general, laboratory animals are supplied with proprietary dried food pellets. These pellets are presented to the animals in food hoppers suspended above cage floor level with the opening at convenient feeding height. The opening through which the food is presented should not be large enough to permit the escape of the smallest cage inhabitant. The food should be presented in the hopper in such a manner as to prevent animals standing on, or in the food, and fouling it.

The species of animal that appear to feed well from slotted hoppers (wire baskets) are few. Rats, mice, hamsters, and gerbils feed well, but wider openings are more suitable for the larger species, such as guinea pigs, rabbits, cats, and dogs.

If wet foods are to be provided in bowls, these are best made of stainless steel rather than any other material.

1.4. Animal pens

Keeping animals on the floor, enclosed within partitions, is only suitable for some species. It is a popular method for animals like guinea-pigs, goats, chickens, ducks, and pigs. This system can be very efficient for cleaning purposes as the detachable partitions can be laid out in such a way that animals need only be driven from one pen to another whilst the floors are being cleaned up and fresh bedding provided. Routine observations are much easier with this type of enclosure.

The pen system of housing can have disadvantages, such as expense. To provide adequate accommodation considerable floor space is needed and this may be regarded as wasteful. If the pens are indoors, then heating and building costs may be substantial. Disease can spread more easily throughout the floor pens, and the floors may be cold unless correctly insulated.

The pen system can be used in a variety of ways. Some institutions will cover a large percentage of the available floor space with animals. Other institutions will use only half the floor space each day, the other half being prepared for occupation the next day by the same animals.

1.5. Aquaria

The types of aquaria set up for use in animal rooms and laboratories can be grouped as follows:

Fresh water—cold or warm.
Sea water—cold or warm.

The aquaria tanks available from suppliers may be one of the three described below.

(i) *Plastic tanks.* These tanks are not very practical for a busy laboratory. The only advantages would seem to be that they are less easy to break, and that they are less heavy than glass. They do have the serious disadvantage of being easily scratched and they eventually deteriorate in optical quality. If they do become damaged, the whole thing has to be disposed of.

(ii) *Whole glass tanks.* These tanks are best used for small culture work, rather than larger display work. The problem with this type of tank is that if it should get cracked, it is virtually useless. They cannot easily be repaired. The sizes available from the suppliers range through 25 cm × 17.5 cm × 17.5 cm to 35 cm × 22.5 cm × 22.5 cm.

(iii) *Metal angle-iron and glass tanks.* These tanks are often described as "angle-iron" even though the metal is not necessarily iron. The framework can be chromium plated metal, or stainless steel for greater resistance to corrosion. The iron-type frame needs to be covered in tough plastic, or a non-toxic paint.

The metal frame of these rectangular tanks can be mounted on a floor stand, or located on table surfaces. The smaller tanks (that is those below a meter in length) can range through 45 cm × 25 cm × 25 cm to 95 cm × 37.5 cm × 37.5 cm.

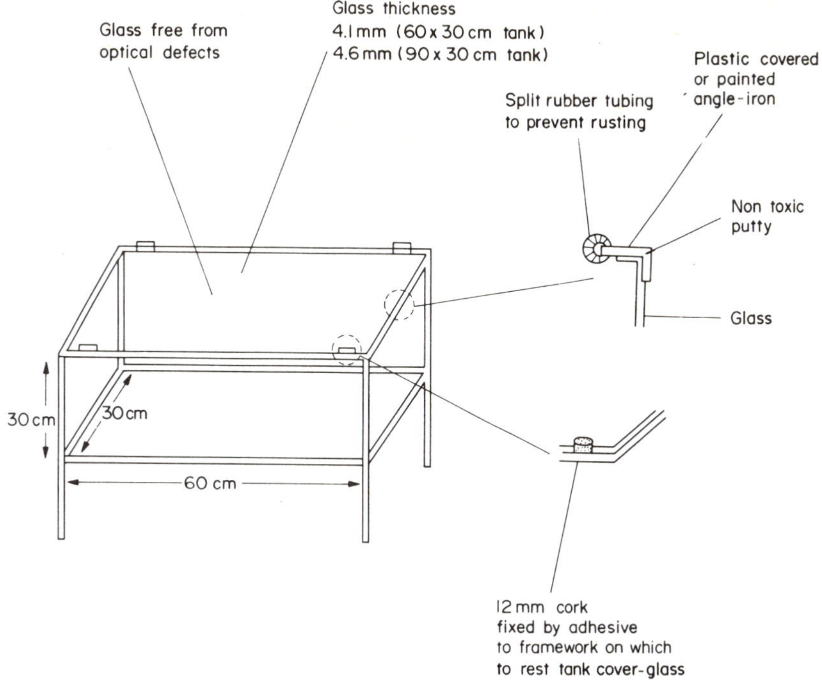

Fig. 9. Angle-iron aquarium

The glass sides of this type of aquarium need to be accurately located into the angle of the metal and fixed there by means of a non-toxic glass putty. This type of putty is available from aquarist suppliers. It does not set hard and should not permit leaks. Any tank should be tested for leaks before it is set up and stocked. Leave some water in the tank overnight and track down any leaks the next day. Carry out a sealing procedure using proprietary sealants, if necessary.

The advantage of this type of aquarium is that it can be repaired by replacing any broken glass panel. This sounds easier than it turns out to be in practice, particularly if one has to replace more than one piece of glass. Accurate measurements and planning in this repair work is most important. If glazing the whole tank, follow this procedure:

(*a*) Secure the bottom glass.
(*b*) Secure the back and front glass.
(*c*) Secure the side glass.

Remember the glass must be tough enough to hold several kilograms of water. The framework of the tank should never be allowed to corrode, if it does, prepare the metal for painting. Rub down the metal to remove all rust and then give at least two coats of aluminum paint. Apply an undercoat on top of this aluminum and then follow by two or three coats of enamel paint.

Before briefly considering the aquaria accessories, it should be mentioned that a sheet of thinner glass will be needed to rest over the top of the tank in order to cut down water loss by evaporation. Some tanks are supplied with opening lids and built in strip-lighting.

1.5.1. AQUARIA ACCESSORIES

The aquarium needs the water oxygenating and the water filtered, if it is to support life successfully. In the case of warm water aquaria, the temperature of the water may need to be raised and regulated thermostatically. Light is essential for some part of the day for all aquaria.

(i) *Lighting* is important for an aquarium because the water plants need it in order to manufacture their carbohydrates. There is a problem with too much light in that it encourages green algal growths on the glass and elsewhere. Artificial lighting may generate unwanted heat and so it needs to be carefully checked before use. There are proprietory fluorescent tinted strip-lights (15 watt) which generate little heat and yet are satisfactory for plant growth. In practice, it is best to keep the tank illuminated at least 8 hours a day. This could be regulated by a time switch. With experience one begins to estimate the amount of daily illumination required to keep the tank balanced and healthy. The lighting should come from above the tank and all wiring checked out for insulation and safety. A tank of about 60 cm (or approx. 24 in.) in length will require two 40 watt lamps illuminating for 8 hours, if no other light source is available.

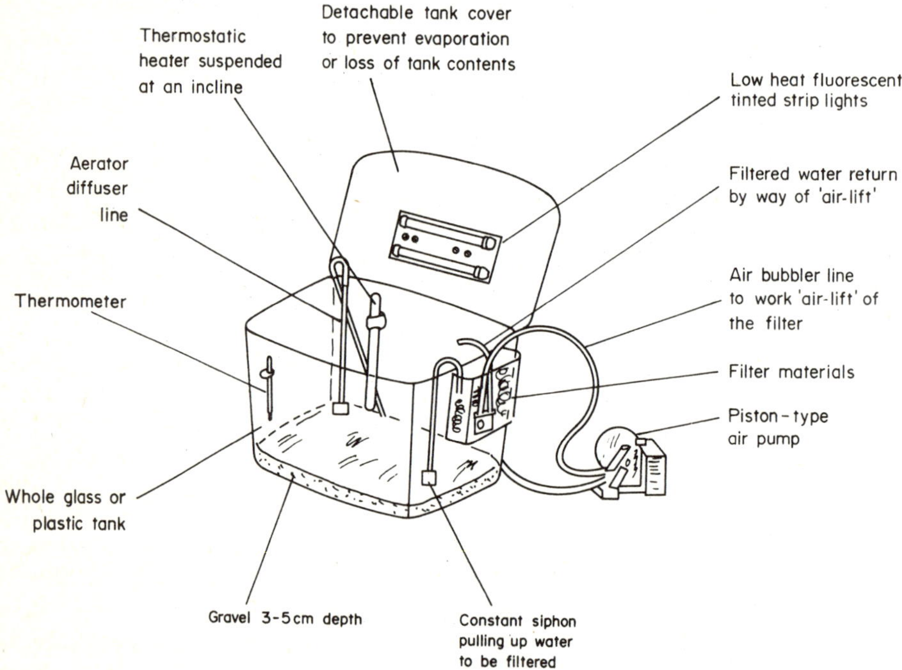

Fig. 10. Aquarium set up with accessories (adapted with permission from Griffen & George Ltd.)

(ii) *Aeration* is not always necessary for a well-balanced cold water tank. If the aquarium has the necessary population of healthy plants, they should oxygenate the water sufficiently. During the day, plants photosynthesize and release oxygen, but they also respire which uses up oxygen. At night time, it is more reasonable to run an aerator when the plants are producing no oxygen, but are using it.

The warm water and marine aquaria generally require constant aeration. The aerator pumps, when used, push air into the water and release it at the bottom of the tank through a stone-type diffuser. These are attached to plastic, or some non-metal tubing. The released air bubbles break up the water and aid its circulation. They disturb the tank water surface, which permits an increased degree of gas exchange there. A very rapid stream of air bubbles rising to the surface has dubious value as an efficient source of oxygen.

(iii) *Aquarium pumps* used for aeration, or for running a filtration system can be described as cheap and less cheap.

Electric vibrator or diaphragm pumps are cheaper, but they can be noisy and less powerful than others. Their life span is not too great.

Electrical piston pumps are more expensive, but they are quieter and their life span is considerable. The writer has employed one such pump for 10 years with no more attention than occasional oiling.

(iv) *Filtration* of the tank water is perhaps a more relevant use for the pumps previously described. A crowded aquarium may need aerating, but all aquaria need the water filtering over a period of time. Accumulations of toxic products can destroy the natural balance. Filtration equipment may be inside or outside the tank.

The bottom located filters are for most purposes perfectly suitable, provided they are given routine attention. This type of undergravel filter is used with an aerator pump. The suspended particles are drawn down into the gravel where they are acted upon by bacteria. The water is kept clear, and there is comparatively little work to be done, except very occasional cleaning of the filter unit.

There are other more complex filters, which are hooked over the tank rim. They contain layers of filtering material through which the water passes before being returned to the tank. With this type of filter it is necessary to change the filter materials fairly frequently.

The principle behind these "air lift" filters is that air is pumped in beneath the filter material, and as it pushes upwards it pulls filtered water to the surface. The water is continuously being replaced from the tank by the constant level siphon.

The filter materials can be layers of glass, or polymer wool which removes the suspended solids; and activated charcoal which removes dissolved gases. Remember not to use this type of filter if small creatures, such as young fish fry are present. They may be siphoned off!

(v) *Heating the water* is necessary if a tropical tank is to be set up. The heating is provided by thermostatically regulated electric elements. There is a wide range of heating equipment available commercially.

The first question to be asked is, "what size of heater is necessary?" Can we for instance use the same heater for the following tanks? A $36 \times 16 \times 12$ in. tank holding 23 gallons and a $90 \times 38 \times 30$ cm tank holding 104 liters?

The power of the immersion heater will depend upon the amount of water to be heated and the prevailing temperatures surrounding the tank. The average water

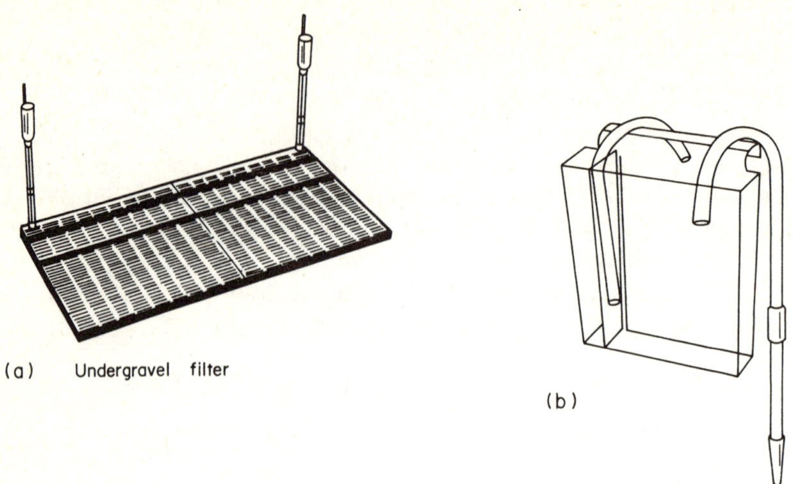

FIG. 11. Aquarium filters (drawn from Gallenkamp trade literature). (*a*) Undergravel type. (*b*) Siphon type

temperature required for tropical fish is 24°C (75°F). The power (wattage) of the electric heater required can be calculated by simply applying the formulae below.

Metric units

$$\text{Wattage needed} \rightarrow \frac{\text{Volume of water}}{\text{(liters)}} \times \frac{\text{Temperature difference}}{\text{(Celsius)}} \times 0.2$$

Example:
 Tank size, 60 cm × 30 cm × 30 cm.
 Capacity, 50 liters (approx.)
 Temperature, 25°C required; 15°C expected lowest; 10°C difference.
 Power requirement, 0.2 watt per liter per degree C.

Calculation:
 50 × 10 × 0.2 = 100 watt heater.
 For a tank of 90 cm × 30 cm × 30 cm the heater size needed is 150 watt.

Imperial units

$$\text{Wattage needed} \rightarrow \frac{\text{Temperature difference (Fahrenheit)}}{10} \times \frac{\text{Volume of Water}}{\text{(gallons)}} \times 2.5$$

This calculation is based upon the fact that 2.5 watt will raise the temperature of 1 gallon of water by 10°F. A 24-in tank requires about a 60-watt heater.

Some commercially available tanks in Imperial measurements are listed below, with useful associated data.

Measurement in inches	Water surface area (sq. in.)	Capacity in gallons
48 × 15 × 15	720	40
30 × 12 × 15	360	20
24 × 12 × 15	288	15
24 × 12 × 12	288	$14\frac{1}{2}$ approx.
18 × 10 × 10	180	6
12 × 6 × 6	72	$1\frac{1}{2}$

(Remember 1 gallon equals 4.5 liters, 1 inch equals 2.5 cm)

The tubular immersion heaters must be installed in the tank so that they are inclined, not horizontal, vertical, or buried in the gravel. These heaters can be supplied as a unit with combined thermostat, or singly with separate thermostat. The first mentioned are more convenient, but they do have the disadvantage of needing replacement totally should they become faulty. The separate outside thermostat clipped on the tank side is more convenient for checking and adjusting.

To heat the tank safety precautions should be exercised. The circuit should include a pilot-light to indicate when the heater is functioning. All wires should be well insulated. The heaters should be of the sort that blow the fuse in the event of water entering a cracked heater tube. For safety, the tank metal frame should be earthed.

1.5.2. COLD FRESHWATER AQUARIA

This type of aquarium is commonly used to keep pond, river, or lake water organisms in the laboratory or animal room. If a large number of aquaria are in use for display purposes, space can be a problem. The manner of display can save space. The aquaria can be single, in line, or displayed in tiers with one above the other. This latter method makes it difficult to attend to the top tank. Whatever method is employed, it is well to bear in mind (before filling up the tank) that an aquarium full of water is extremely heavy and difficult to move (1 liter weighs 1 kg). The dimensions of the aquarium should ideally be in the ratio indicated below:

Length   15   (1.5 m).
Breadth 10   (1.0 m).
Depth   6   (0.6 m).

This kind of ratio provides a bulk of water with a good surface area exposed to the air for oxygenation. The depth of water in any case must be sufficient to permit active fish movements.

1.5.3. SETTING UP A FRESH-WATER AQUARIUM

The operations involved in setting up a fresh-water aquarium may be summarized as follows:

1. Fill up the selected aquarium and leave it to stand overnight. Check for leakages.
2. Empty the aquarium and wash it thoroughly after removing any grease and dust. Running water through the tank many times is a useful way to clean away any detergent that may have been used in the washing.
3. Select some appropriate sand, gravel, or rocks and wash them to remove any loose or soluble materials. Be careful not to use any materials that may contain poisonous or soluble materials. Now put in the filter devices.
4. Add the sand, gravel, and rocks to a depth of about 30 to 50 mm attempting to slope the material towards the front of the tank. This makes it easier to remove debris that falls towards the front of the tank. As a guide to the quantities of sand and gravel needed for a particular tank, refer to the following calculation.

   For every 10 liter of water use 1.5 kg of gravel. A tank of 60 cm × 30 cm × 30 cm holds about 50 liters of water and so will require 7.5 kg of gravel.
5. It is now possible to add the water which should be suitable "natural" water from a stream or a lake. If tap water is used it must be allowed to stand for at least 24 hours before use in order that it becomes dechlorinated.

   The water is poured into the tank slowly onto a piece of plastic sheeting that has been laid over the bottom gravel and sand. This prevents a disturbance of the bottom materials.
6. At this point, it is possible to plant the bottom with required plants. An aquarium without plants will be an aquarium with low oxygen content and, therefore, a poor aquarium.

   Select the plants and push the roots deep into the gravel. If the plants are not with roots, push the stem beneath the gravel and weight it down with a stone. Leave these plants in the tank for a week or so until the plants are well established. Be careful not to overcrowd the tank with plants. Some suitable plants may include those in the chart below.

Plants for aquaria		Optimum water temperatures	
Indoor cold water and tropical tanks	Characteristics	°C	°F
Vallisneria spiralis (Italian water weed)	A popular grass-like plant with a stripe down the center of the twisted leaf. The flower stems are spiralized. Propagate well in tropical tanks by runners. Requires only moderate lighting	18–20	64.5–68
Sagittaria natans (Arrowhead)	A common grass-like plant with straighter, more rigid leaves. Slow growing and propagated by runners. Requires moderate to strongish light	20	68
Myriophyllum verticillatum (Water milfoils)	A fine, brittle, spiky leaved plant requiring strong light and no overcrowding. Propagated by cuttings. Intolerant of lime salts.	15–18	60.7–64.5

Indoor cold water tanks	Characteristics	Optimum water temperature	
		°C	°F
Elodea canadensis (Canadian pond weed)	A good cold water plant to act as cover for fish. Requires good light. Propagates well by cuttings	15	60.7
Ceratophyllum (Hornwort)	A dense growing brittle plant, requiring plenty of space and strong light. Propagates by cuttings	12–18	53.5–64.5
Lemna (Duckweed)	Floating duckweeds that reproduce vegetatively. Eaten by Goldfish. It can overgrow the tank surface		

Tropical tanks	Characteristics	Optimum water temperature	
		°C	°F
Cabomba	A brilliant green plant with fan-shaped leaves which requires strong light and soft water. A brittle plant that can be propagated by cuttings	20	68
Limnophila	Resembles *Cabomba*, but has no fan-shaped leaves and grows better. A decorative plant which likes strong light. Propagate by cuttings	20	68

7. The filtering devices must now be allowed to run until the water is cleared.
   Add some appropriate scavengers, like snails, because they will keep the tank clear of green algal growths on the glass and rocks. A snail for every 2 liters of water should be enough. They reproduce rapidly and may need reducing in number later. Some common temperate snails are listed below:

Tank scavengers	
Bladder snails* (*Physa fontinalis*)	Small active snail with a patchy shell. Rapid multiplication throughout the tank. Useful scavenger
Ramshorn snails* (*Planorbis* types)	Black Ramshorn snail (*P. corneus*). White Ramshorn snail (*P. albus*). Useful in temperate and warm water
Pond snails* (*Limnea stagnalis*)	Larger snails that can be kept in cold and warm water. It can destroy many useful plants as well as waste foods and algae. The droppings are a good culture for infusoria

Caution: Snails eat fish eggs!

8. The tank may now be stocked with required animals, taking care not to add too many and not to mix incompatible species.
9. The feeding of aquaria stock must be carried out regularly, but not overdone to the extent that fouling may occur. This may kill off the stock. Feeding a little often is better than a lot infrequently.

1.5.3. SEA-WATER AQUARIA

Marine aquaria are not quite so straight-forward to set up as are the fresh-water types. The practicalities and problems can be outlined as follows:

(*a*) *Sources of sea water* are not always readily available. In some parts of the USA the nearest salt-water source may be over a thousand miles distance away. Sea water obtained from marine sources needs to be stored in the dark for a couple of weeks or so, before use. This will allow living organisms to die off. The natural sea water can be used in stock tanks or added to salt water prepared by other methods.

"Instant ocean" or synthetic sea water can be made up from proprietary salt mixtures. To this can be added small quantities of natural sea water as already mentioned.

(*b*) *Salt-water corrosion* is a major consideration when setting up a marine aquarium. The sea water will attack many metals and those metals may then come into the water and poison the inhabitants. For this reason, all equipment needs to be plastic and/or glass. This also applies to the water lift pumps that circulate the sea water.

(*c*) *Evaporation* of salt water will tend to change the density (maintain near S.G. 1.017–1.025) of the aquarium contents. To monitor this, it can be helpful to have a level marked on the tank side and then keep topping up with distilled water or natural sea water. Test the water with a hydrometer from time to time.

(*d*) *pH value* for sea water is best kept between 8.2 and 8.3, but there is a tendency for an acidity to be created by the living processes of the inhabitants. The pH is brought back to an alkaline reading by the appropriate additions of sodium bicarbonate or dilute sodium hydroxide (see aquarist manuals.)

(*e*) *Temperature* control is important for marine aquaria. For temperate waters 10°C–14°C (50–57.2°F) should be maintained. For tropical waters temperatures between 22°C and 25°C (71.6–77.0°F). Marine animals are not very tolerant of water-temperature inaccuracies and so a means of "refrigerating" the circulating water is necessary for the temperate sea-water tank.

(*f*) *Water circulation* has to be maintained in order that "used water" may be filtered of particles and toxins. This is achieved by pumping the water up from a storage reservoir into a gravity feed tank. This water is then passed through a grit bacteria filter before entering the aquaria. As mentioned earlier, a special non-corrosible cooling system has to be included in the circulatory set-up to maintain the required low temperatures. This is a closed circulatory system and the water has been described as being of value for years, provided additions of natural sea water are made from time to time to make up for losses (see Mahoney in bibliography).

(*g*) *Oxygenation* of sea water must be achieved with the aid of mechanical aerators because marine plants are poor oxygenators.

1.6. Terraria (vivaria)

A terrarium is an artificial land habitat in the laboratory. There can be different types of terraria:

> Desert terrarium.
> Woodland terrarium.
> Bog terrarium.

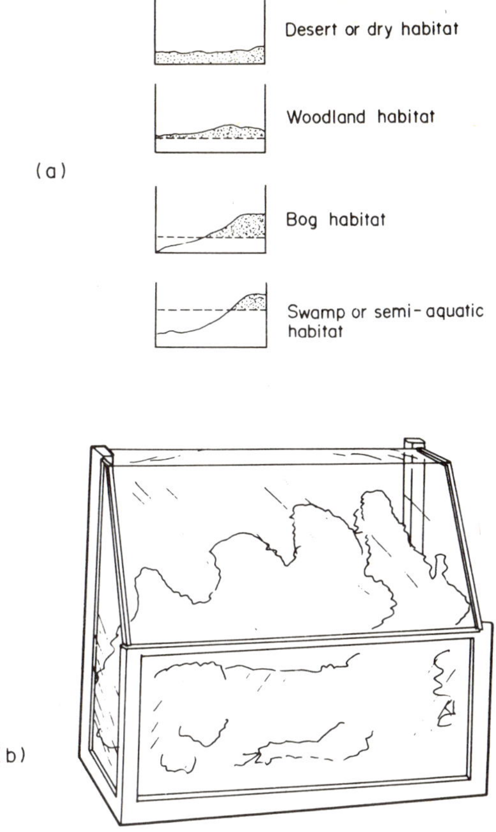

FIG. 12. Vivaria. (*a*) Different types of vivaria. (*b*) Glass vivarium with metal frame

The glass enclosed terrarium can be specially constructed or an aquarium may be used. The larger the vessel, the easier it is to provide adequate plants and other environmental features. It should be able to hold water in case a bog or pond is to be included. The open top should be covered by a sheet of glass to maintain the temperature and humidity within the tank. This cover should have a small opening to permit some air circulation. A useful vivarium can be constructed from a porcelain sink.

1.6.1. SETTING UP TERRARIA

The base material is the most important as it is the foundation in which the plants must thrive. The soil must not be too heavy and muddy as it must permit air to circulate near the plant roots, and allow water drainage.

The entire bottom of the tank should be covered and filled up to about a quarter height of the container. Heap the base material up towards the rear, sloping towards the front glass. Produce hills and hollows to add interest to the habitat. Add some rocks, stones, and logs here and there.

1.6.2. THE DESERT TERRARIUM

This habitat is necessary to maintain laboratory specimens, such as horned "toads", lizards, and small snakes. The bottom of the tank is covered with an inch or so (3 cm) of coarse sand. On top of this spread about half an inch (1.5 cm) of sand. Add a rock or two and some suitable cacti. Sprinkle the cacti roots with water before planting them. Keep them dampened somewhat.

The terrarium needs to be kept light and warm, between 68°F and 85°F (20°C and 30°C).

1.6.3. THE WOODLAND TERRARIUM

This type of habitat can be constructed to house a great variety of animals. The bottom of the tank should be covered with a foundation layer made up as follows: one part sand, three parts humus, and one part coarse gravel. This mixture should be moistened and suitable plants added to produce a natural, uncrowded habitat. Avoid having large leaved plants that may overgrow and obscure the tank contents.

The plants suitable for a laboratory terrarium may be mosses, liverworts, lichens, club mosses, and ferns. Other dwarf plants may be used to provide base cover. The plants should be sprinkled with water, the glass top put into position, so as to allow air circulation. The tank can then be left in a cool position to allow the plants to establish.

The animals that can be added to the tank for laboratory study may include the following: newts, toads, salamanders, snails, small snakes, slugs, beetles. If amphibious frogs are to be kept, then a small pond area should be included within the tank.

1.6.4. THE BOG TERRARIUM

This type of habitat is more suitable for the amphibian species like newts, frogs, toads, salamanders (for further details, see p. 197).

The foundation layer in this tank is more acid than in others. The acid bottom layer is laid out as follows. First cover the bottom with a drainage layer of gravel. Cover this with a layer of soil mixture made up of one part Sphagnum and two parts acid soil. This is then thoroughly soaked with water allowing the excess to remain in the drainage layer of gravel. Acid soil plants can be rooted. This can include Venus Flytrap, Pitcher Plant, and Sundew. Stand the tank in a cool place and cover with a glass sheet.

Those terraria that contain water can sometimes cause problems. If there is too much water, molds can grow on plants. This can sometimes be treated by sprinkling the tank with sulphur dust. Remove the cover for parts of the day in order to allow excess water to evaporate.

If there is excessive heat some of the plants may grow thin and weak.

1.7. Project program

ANIMAL HOUSE MATERIALS

This appendix to animal accommodations is concerned with those physical aspects of the animal house that may change from place to place, and with time. Technology often changes to adjust to the materials in use. Metals replaced wood and plastics have replaced metals.

The objectives of this project and practical program are as follows:

(i) To permit the student an opportunity to study the materials used in the construction of an animal house.

(ii) To study the materials used in the construction of cages and accessories.

(i) *Animal house construction materials*

When considering building an animal house, either as an extension of an existing building or as an independent structure, the major concern is cost. It is obvious that the final cost price will include a heavy bill for labor. Bearing this in mind then one has an idea of the types of materials that can be afforded. Students are encouraged to use the outline below in order to produce an account of modern construction materials used in the animal house. Information for this project can be obtained from trade catalogues and other literature.

Factors influencing the choice of building materials are:

(*a*) Site location.

(*b*) Climatic conditions in the area: temperature variations, humidity, rain, frost, sun, wind, electric storms, pollution.

(*c*) Pests in the locality: insects, rodents, reptiles.

(*d*) Types of animals being housed.

External walls can be cavity type with an outer brick layer and an inner block layer. These walls can be insulated by having the cavities filled with a proprietary injection foam/plastic material.

Internal walls can also be built of brickwork or hard-cast blocks that are strong enough to carry the roofing (less roof beams) and to carry cage racking and shelves. Wall covering for internal walls will depend upon the finance available. There is, however, a minimal covering which will serve the purposes of most, a coat of paint. This will not be practical for some animal house purposes but a paint that is impervious, washable, and not easily damaged and not too expensive is widely used. The plastic based paints are very practical.

Flooring materials have been discussed earlier as being one of the most difficult items to make decisions about because the floor is subject to such a wide range of assaults. It must be strong, not crack or throw up a lot of dust. It must withstand the frequent cleaning operations.

There are a multiplicity of industrial flooring materials ranging through asphalt, thermoplastic tiles to epoxy-resin substances that can be used to coat floor and wall in a continuous dust-free skin. This latter covering has been described as having a tendency to become slippery if too frequently polished.

Windows may be thought of as an unnecessary expense and a liability as a potential entry point for pests. If windows are to be included then perhaps aluminum-framed double glazing would be the best answer. An external insect screen would be a necessity if these windows are to be opened.

Doors and their frames are likely to receive rough treatment and so should be resistant as well as pest-proof. Metal door frames with hard wood doors are often used. The lower half of the door being protected by aluminum sheeting to withstand trolley knocks and so forth. A viewing panel of armoured glass or perspex can be a useful extra. Alternatively the door may be constructed of aluminum throughout.

Ceilings can for convenience be of the suspended type with services concealed above the ceiling panels. This is a common practice in laboratories and offices but is most inadvisable in animal rooms because the dead space above the room acts as a reservoir of dust and any micro-organisms or pests.

Using the outline just presented, students should make reference to the reading list at the end of the book and consult trade literature available at work or at college in order to produce a project study of animal house construction materials.

(ii) *Animal cage and accessories—construction materials*

The materials employed for cage construction, and the accessories, like feeding and drinking equipment, require to withstand the attentions of chewing animals and the wide range of chemicals that are in use, as well as urine and feces. In this section of the project students are advised to obtain a cross-section of the trade literature in order to produce an account of the most suitable materials for cage construction and so forth. An outline of the topic is presented below.

The characteristics of cage materials can be summarized as follows:

(*a*) Strong enough to hold heavy and active animals.
(*b*) Durable enough to give lengthy service without being too heavy for easy movement.
(*c*) Resistant to attack by acids, alkalis, and detergents or any chemicals likely to contact the equipment.
(*d*) Capable of heat sterilization without deformation.
(*e*) Hard enough to resist the gnawing teeth of animals.
(*f*) Manufacture price and repair cost should be within acceptable margins.

The materials that have been used for cage and accessory construction are listed below. Clearly all of these materials do not provide the characteristics required of the ideal.

Material	Characteristics	Uses
Wood and wood products	Many disadvantages. May be gnawed and chewed. Easily penetrated by urine and disease-causing microorganisms. Difficult to sterilize without distortion. It has a comparatively short life	Can be used as perches or larger mammal rest boards as they are fairly easy and cheap to replace if contaminated

TABLE (*contd*)

Material	Characteristics	Uses
Galvanized steel	This is mild steel that is hot dipped into molten zinc. The covering of zinc is resistant to corrosion by alkalis, but not to urine acids. The zinc oxidizes and forms a protective zinc oxide coat, but the underlying steel rusts quickly when the zinc coat is broken. Corrosion weak spots occur at joints and bolt points. This can be reduced by hot-dip-galvanizing the manufactured product. Heavy.	Cages and accessories such as trays
Aluminum and aluminum products	A soft metal that is easily damaged. It resists corrosion by acids but not by alkalis. Alloys are less disadvantageous	Used in some small mammal caging. Alloys generally used
Stainless steel	A tough, hard-wearing material that is virtually indestructible but not cheap to manufacture. Given the right price this is an ideal material when used with glass and plastics. Heavy.	Cage parts and accessories such as drinking tubes.
Polycarbonate—tough plastic	A good resistant, transparent, autoclavable material. Shatter-proof and easily cleaned. It almost has the insulating warmth of wood	Used for cage boxes
Polypropylene—pliant plastic	Translucent plastic that can withstand autoclaving. Shatter-proof. Good resistance to chemicals.	Used for cage boxes
Polystyrene—brittle plastic	Opaque plastic with a low tolerance of high temperatures. Cannot be autoclaved	Used for short-term disposable boxes
Fiber glass products	Opaque reinforced plastic that resists high temperatures and chemicals	Cage boxes
Glass	A fragile and heavy material. Resistant to heat and chemicals	Useful for vivaria and aquaria

Animal accommodations—summary

Animal housing needs to be designed in as flexible a manner as possible in order to permit a variety of species to be housed, and to take into account fluctuations in total animal populations.

The animal house will have different functional areas and the floor space needs to be divided intelligently between the different areas.

The animal house rooms will have specific functions and the traffic between these rooms will necessitate attention to corridor layout and design.

The materials and services within the animal rooms should be resistant to damage and corrosion.

The caging in animal rooms should be stored on racks or shelving that can be easily maintained.

Animal pens which house animals at floor level need to be laid out in such a manner as to be easy to maintain and easy to prevent cross-infections.

Aquaria and terraria are laboratory "imitations" of habitats for fresh water, sea water, and land-dwelling animals. These containers need to be as "natural" as possible, escape proof and easily maintained.

Project work on suitable construction materials for animal houses, cages, and accessories was outlined.

STUDY OBJECTIVES

UNIT 2:

Animal Care Routines

(*a*) Describes the range of work to be done within the different type of animal house; conventional, barrier maintained, and experimental.
(*b*) Describes the allocation of work to different grades of technician.
(*c*) Describes the allocation of time for the various work to be carried out in the animal house.
(*d*) Describes animal house record keeping as administrative records and animal records.
(*e*) Describes individual animal identification methods.
(*f*) Describes animal care routines for transportation.
(*g*) Describes some of the hazards in animal house work.
(*h*) Suggests practical work.

UNIT 2:

Animal Care Routines

LABORATORY animals, mammals particularly, are kept in accommodations that may be categorized as follows:

> Conventional.
> Barrier-maintained.
> Experimental.

The different categories of animal housing will demand different routines of husbandry.

Animal care routines are here reviewed in the following manner:

2.1. Conventional animal house duties.
2.2. Barrier-maintained animal house duties.
2.3. Experimental animal house duties.
2.4. Work allocation.
2.5. Time allocation.
2.6. Record keeping.
2.7. Routine care of non-mammal species.
2.8. Identification methods for individual animals.
2.9. Transporting animals.
2.10. Hazards in the animal house.
2.11. Practical program (bedding, instruments).

There are as many different routines as there are animal species and animal houses. The information outlined here deals mainly with mammals.

2.1. "Conventional" animal house duties

The duties performed by the animal technician in the conventional animal house may be used as a basis for discussion. These duties are common to all types of animal house, but additional standards of hygiene and expertise are appropriate to S.P.F. and experimental animal houses.

The conventional animal house will, for the purposes of this study, be regarded as one where no special procedures have to be observed in order to prevent the animals being infected with bacteria, virus, or other micro-organisms. In such an animal house there are usually no special experimental procedures or research programs in operation. This type of animal house will include most school, college, and some commercial units.

The routine duties in need of attention may be listed as below. The frequency with which these routines are put into operation will depend upon the species of animal and the purpose for which they are being kept. Before running through these routines check temperatures and humidities.

Check items	Checks and actions
Water	Replace all water-bottles with clean bottles, containing tap-filled, freshly drawn water Check the bottles and drinking spouts for blocks and leaks. Adjust (if necessary) the water flow from the automatic watering device Renew the open dish water supplies provided for those animals unable to use water-bottles.
Accidents, litters, etc.	All cages require to be checked for any "accidents" or diseased stock The presence of any new litters should be noted in terms of the date and the cage number The litter numbers may have to be checked at a later date.
Weighing litters, changing breeding pairs, cage labelling	Young growing animals may need to be weighed daily in order to monitor progress The pairing of selected animals for breeding will need to be caged The cages will have to be suitably labelled with dates and relevant information Records will need to be written up
Routine cage and bedding changes	Cages and bedding will need to be changed on a regular basis depending upon the demands of the species being housed. A daily check ought to be made of the cage (direct) bedding and of the litter (indirect) bedding for contamination by urine and feces
Food	Check the food in the cage containers for quantity as well as quality. Ensure the containers are correctly positioned. Remove any contaminated food In some laboratory animal houses, it may be the practice to place weighed quantities of food into the containers. This can be a good economic procedure, but it is time consuming Special diets may need to be prepared or vegetables washed and cut up
Experimental animals	Some animals may require particular attention, after surgery or if undergoing some kind of lengthy drug treatment. Check the instructions on the cage record cards
Environmental checks	The temperature and humidity recording devices need to be checked daily and the data noted The ventilation, heating, and lighting will need regular checks The cleanliness of walls, floors, and shelving requires daily attention and action taken in order to maintain a high level of hygiene

The duties just listed refer, generally, to the immediate, daily needs of the animals rather than to the maintenance of equipment.

The duties now considered are those that can make the life of the junior technician less than attractive. The routine cleansing operations of cages, instruments, and equipment. In some animal houses these duties are not part of the technician's work because a support cleaning staff is employed to carry out these tasks.

Check Items	Checks and Actions
Cage cleaning	All cages and accessories need to be cleaned and sterilized regularly. These items must be transported within the "dirty area" and stored in preparation for treatment The cleaning may be manual or, in large animal houses, by means of an automatic hot water or steam sterilizer

TABLE (*contd*)

Check Items	Checks and Actions
Cleaning racks and shelves	Before fresh cages are racked or shelved, the area needs to be cleaned to remove dust, fur, and other contaminants
	Particular attention is required in areas where dirt accumulates, such as in angle-iron supports, at the junction of shelving and the wall, and around the wheels of mobile rack units
	Whilst the racks and shelves are being cleansed, areas of exposed wall and floor can receive attention. They may be wet washed, "dry" cloth, or vacuum cleaned
Cleaning the animal room	Whilst the cages and racks are being cleaned, all floor debris must be removed by sweeping, mopping, or vacuum. Any dirt accumulating on window shelves, switches, pipes, or in sliding-door runners must be disposed of
	From time to time the animal room may need to be sealed off and fumigated
	Remove dust and fur trapped in ventilation outlets
	If in-room sinks are provided, check for blockages and maintain them in a hygienic condition
	The taps from which drinking water is drawn should be particularly free from contaminants, as well as the water gulleys and drains
	Check that overhead lights are not harboring dust
Cleaning animal room equipment	The equipment used to clean the animal facilities will itself require maintaining in good condition. Brooms, buckets, vacuum cleaners, and cage washers will all need regular attention. All protective clothing will need regular laundering
	Weighing machines used for monitoring the progress of litters will have to be kept free from dust and dirt if they are to continue to be accurate and useful
	Specialist rooms such as surgery, preparation rooms, stores and offices will all need the regular attention of personnel

In addition to the cleaning duties just reviewed, there are the routine tasks of renewing stock. The reception area of the animal house will be a very busy place from time to time during the week when bags of animal feed and bedding are delivered. If the animal house has been designed with personnel in mind, the input of materials should be a relatively swift and uncomplicated process. It can be the case that makeshift animal houses have difficult delivery points, and so renewing stocks of feed and bedding becomes a cumbersome task which is followed by much floor cleaning.

2.2. Barrier maintained animal houses

The degree to which the animal house is "isolated" will dictate the duties and behavior of the animal worker. The S.P.F. unit will contain specific pathogen-free animals and consequently the duties and disciplines will be more strictly regulated.

All the duties discussed for the conventional animal house apply to the barrier-maintained unit. The time taken for a given set of duties for a given set of animals will be longer in the barrier-type animal house because of the higher degree of hygiene.

An important extra duty of the barrier animal house technician is that of maintaining a high standard of personal hygiene by showering and the changing of clothes before and after periods of work within the animal room.

2.3. Experimental animal house duties

The animal house holding species undergoing experimental procedures (such as controlled diet, or having regular records made of their physiological function) requires the same servicing as any conventional animal house. In addition to the regular maintenance duties there are the routine tests or records to be made.

The extra duties that may be involved in an experimental animal house can include some of the following:

> Servicing metabolic cages.
> Dispensing radio-active substances.
> Collection of specimens.
> Recording body weights.
> Taking body temperatures.

The duties listed are time-consuming, and may necessitate the technician supervising fewer numbers of animals.

2.4. Work allocation

The best use of available personnel within an animal house is a responsibility of the senior, or administrator. It is of some importance that a career structure be seen to exist in the larger institutions whereby advancement is available to those showing interest and academic progression. It could be argued that animal husbandry is a skill and comes with experience. This is true, but knowledge behind the skill not only provides personal satisfaction, but also gives meaning to some of the more mundane tasks.

Duties may be allocated in such a way as to ensure that the more experienced, trained technician performs the less attractive tasks less frequently. The more junior personnel would thereby be allocated a rota of work that contains a heavier program of routine cleaning and maintenance. This may encourage higher standards.

Some animal houses grade their technicians in the following manner and allocate the work accordingly.

Animals aids or junior animal technicians may be newcomers to the animal house who work under supervision in the routine maintenance of animals and animal rooms. This grade of work can also be a permanent position, rather than a junior scale of animal technician. An animal unit fortunate enough to employ animal aids would be in a position to spend more time training their technicians in skills other than routine maintenance.

Animal technicians may be the more experienced personnel who work without supervision on routine maintenance duties, as well as breeding and recording work. They may be required to prepare animals for experimental work and to supervize animals within an experimental program.

Senior animal technicians may be experienced personnel with proven academic training to enable them to administer and train other technicians. They may be called upon to supervize the experimental program and allocate duties. A senior may be responsible for the preparation of operating theatres, of administering anesthetics and drugs. He or she should be familiar with the organization of the animal unit and the regulations under the law that govern its operation.

2.5. Time allocation

The time allocated to different technicians, working in different types of animal house, doing different tasks is no easy thing to generalize about. It may seem a somewhat academic exercise anyway, and somewhat remote to employees within small animal units where only two or three persons do everything.

Attempts have been made to estimate the required time for the various duties of the animal worker. The duties, such as shelf and cage cleaning, feeding, and watering have been calculated for workers in the different type of animal house. It becomes clear that the S.P.F. and experimental animal technician requires more time for some individual duties than do workers in the conventional animal house. S.P.F. technicians, for instance, spend more time in personal hygiene preparation than do other technicians.

A timed assessment of the duties of animal technicians has been the subject of a study at the Medical Research Council in Carshalton, England (see Bibliography). This study is specific to the conditions that exist at that institution, but gives one a clear insight into the time requirements for the variety of routine duties in different animal houses.

2.6. Record keeping

The coordination center of the animal unit is generally the office of the senior animal house administrator. Even the smallest of animal breeding areas will require some documentation of the stock and the coordination center may then be a card-index file box. The keeping of records can be put into two categories:

Administrative records.
Animal records.

2.6.1. ADMINISTRATIVE RECORDS

The "clerking" side of an animal house will be almost a full-time task, especially in large and busy production units. In the smaller animal house with fewer personnel, this is an insignificant time consumer, but is nevertheless important.

A diary of daily events is a valuable aid to the efficient running of an animal unit. This day-book should contain memoranda, work assignments, delivery dates, shipment dates, equipment notes such as breakdowns or repairs. A log of any experimental program may be included.

An order book with a record of all orders pending, filed chronologically. Cash records may also be kept.

An inventory book is one way of keeping a check on equipment, bedding, and food. This type of stock book will help in costing a breeding program.

Environmental records, such as temperature, humidity, and lighting may be kept if needed on a day-to-day basis as a check that the control devices are functioning correctly.

Researchers' requests are best documented and dated so that animals are available when required.

Personnel records may need to be kept if large numbers are involved. These documents can record holiday periods, work assignments, and details of study time.

2.6.2. ANIMAL RECORDS

These records will be confined to information about the animals in care—stock, breeding, or experimental.

The animal unit will always be interested in cost, and efficiency can only be assessed by keeping records. These records should at least tell us the following:

(*a*) Each week: $\begin{cases} \text{how many young are produced per female?} \\ \text{how many young are weaned per female?} \end{cases}$

(*b*) Each week: how many usable animals are wasted?

(*c*) Each week: what are the trends in production?

Breeding records, in brief, must be kept on the cages, and in more detail on filed record cards that carry the information indicated below:

Parents: Origin and genetic characteristics
                 Date of birth

Any speciality or peculiarity of the parents, such as litter sizes.

Offspring: Phenotype characters such as coat colors.
                 Date of birth.
                 Number born; still and live births.
                 Number and date of weaning.
                 Numbers dying; date of death, with reasons, if known.
                 Disposal or use in breeding program with dates.

In order to elaborate further upon a system of keeping records of animal breeding, one could do worse than make reference to a system that has been used successfully for many years. The record system referred to here is that described by Dr. Margaret E. Wallace of the Genetics Department, University of Cambridge (see Bibliography). A system of this type could be adapted for use with any livestock in long or short term breeding situations.

2.6.3. THE "CAMBRIDGE" RECORD SYSTEM

The essence of a good recording system is that it provides the information quickly and simply. The records will be needed daily and over longer periods of time. For this reason, temporary and permanent records are kept. In the interests of speed, the information attached to the cages, as temporary records, should be displayed in such a way as to be informative at a glance. This will necessitate coding of one sort or another to reduce the time spent reading written records.

The permanent records should be kept in more substantial loose-leaf book form, and located in the office.

(i) *Permanent records* are best kept in a loose-leaf form in order to allow flexibility. The make-up of this record book should be as durable as possible because it may need to be kept for many years.

The records kept in this file will refer to matings and litters. This information can be displayed as shown in Fig. 13.0.

The mating record of each pair of animals is kept on a separate page, so that, for instance, the sample page shown is a fourth mating in a sequence of studies, "M".

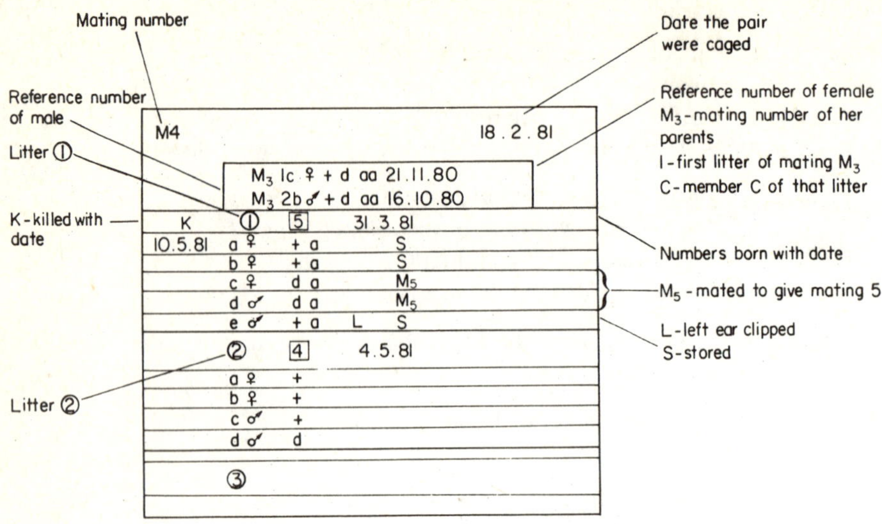

Fig. 13. A sample page in a permanent record file (adapted from Dr. M. Wallace *Learning Genetics with Mice*, Heinemann)

KEY

M4 = mating number.

18.2.81 = date male and female caged.

M3.1c = reference number of the female: M3—mating number of her parents; 1—first litter of mating M3; c—member c of that litter.

M3.2b = reference number of the male: b—member of the second litter in mating M3.

21.11.80 and 16.10.80 dates of birth of the pair in mating M4.

Line 1–5, 31.3.81 indicates litter 1 has 5 mice born 31.3.81.

L = left ear clip.

S = stored.

K = killed.

M5 = mated to form mating M5.

$^{+d}_{aa}$ = genotype.

The litter records of this mating are set out to show the sexes, date of birth, and the phenotypes. Space should be clear to the right of the phenotype symbols in which the genotype characters may be written.

The terminal fate of the animals is indicated on the extreme left (K = killed; D = found dead; G = lost).

(ii) *Temporary records* can be kept on cage cards. Such cards have the numbers 1 to 12 printed on the bottom edge. (They can be obtained at Cambridge.) The cards carry the details of the mating number, the reference numbers of the paired animals, the phenotypes, and any other characteristics.

The cage contents are indicated on the cards by means of the colored plastic paper clips. At a glance, an observer can tell what state of affairs exists from the card clipped to the cage.

The use of this coding system is explained by means of Fig. 14, and the accompanying table.

The four colored paper clips can be used with the broad or narrow side showing. When a litter is present in the cage the details are indicated by the positioning of the clips; as indicated in the table (see p. 45):

Clip color-coding

Color	Clip side visible	Position on the card	Information
Red	Broad	Right, between symbol and number	Female with male, litter expected within a week
Red	Narrow	Right, between symbol and number	Female with male. Female probably pregnant
Yellow	Broad	Right, between symbol and number	Female separated from male. Female nursing
Yellow	Narrow	Right, between symbol and number	Female separated from male. Female pregnant
Blue	Broad	Base, over relevant number	Covered number is the number of young in litter. Litter is old—to be separated, classified
Blue	Narrow	Base, over relevant number	Covered number is the number of young in litter. Litter is young—not due for separation, un-classified
Green	Broad	Left, below reference numbers, etc., of the pair	Cage should be inspected daily (i.e. poor mother—may need fostering)
Green	Narrow	Left, below reference numbers, etc., of the pair	Disease suspected in litter or adults. Culling decision needed

The lay-out and contents of these permanent records will vary from institution to institution. They should, however, provide details of some or all of the following:

Locality of animals.   Researcher's name.   Licence holder
Breeding history.   Special instructions.   and details.
Pathology.   Dates.
Experimental data.   Identification code.

Cage record cards will also vary from institution to institution, but in general they should provide the following information:

Identification.   Breeding data.   Licence holder
Strain.   Experiment.   (details).
Occupants, sexes.   Researcher's name.

2.7. Routine work with non-mammal species

Much of the work that is described as "animal technology" is generally thought to be connected with mammals. Clearly there are many animal technicians and laboratory workers whose duties are unconnected with mammals.

The following is a summary of the duties associated with the maintenance of non-mammals in vivaria and aquaria. For detailed information the reader is referred to Unit 5 and the relevant species Data Capsules at the end of the book.

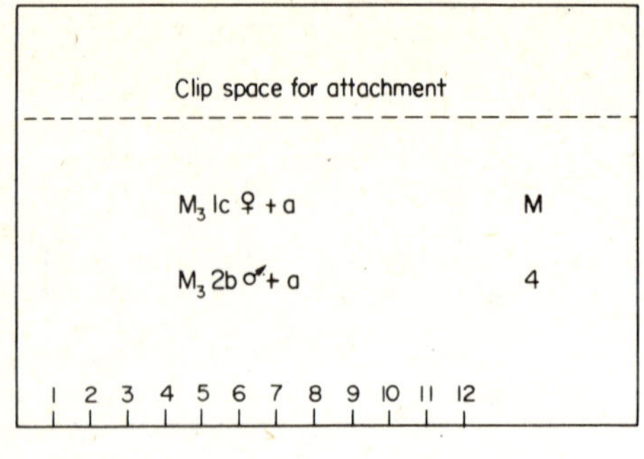

(a) Mating card

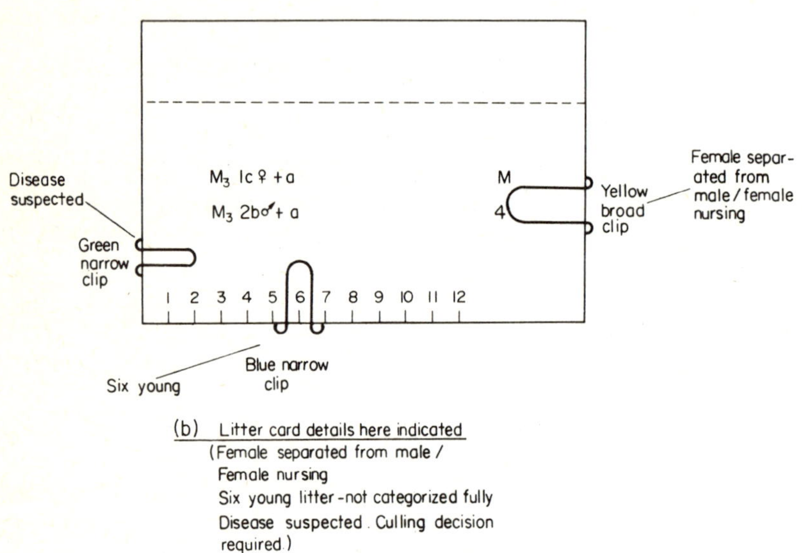

(b) Litter card details here indicated
(Female separated from male /
Female nursing
Six young litter -not categorized fully
Disease suspected . Culling decision
required)

Fig. 14. A cage card with details—see accompanying adapted "code table" (adapted from Dr. M. Wallace *Learning Genetics with Mice*, Heinemann). (*a*) Mating card. (*b*) Litter card with information (female separated from male/female nursing; six young litter—not fully categorized; disease suspected; culling decision required)

2.7.1. ROUTINE WORK WITH VIVARIA SPECIES

Terrestrial non-mammals are maintained in the laboratory in terraria (vivaria).

The animals kept range through amphibians, reptiles, arthropods to worms. The duties associated with this wide range of animals are all connected with providing the optimum temperatures, humidities, and food. Routine duties include cleaning out in order to prevent the vivarium being invaded by unwanted organisms, such as molds and mites.

The duties associated with the vivarium room are summarized below. For details of a particular species use the Glossary/Index and the Data Capsules at the end of the book.

(i) *Daily checks* on the room and vivaria temperatures and humidities. Check the contents of the vivaria, removing any dead or decaying material. Sub-culture, if appropriate.

(ii) *Weekly duties* involve cleaning out some cages and tanks to ensure that no feces or debris accumulate.

Renew water supply tubes (if appropriate). Renew culture medium according to previously planned schedule (e.g. Drosophila, beetle, or "worm" media). Empty containers and sterilize them in accordance with previously planned work schedules kept in the "work log book".

Sub-culture any stocks that are getting overcrowded.

2.7.2. ROUTINE WORK WITH AQUARIA SPECIES

The management of aquatic organisms calls for routines that are very different from those associated with the terrestrial mammals. The most obvious difference, and blessing perhaps, is that they do not require such regular cleaning out. Neither do they need a water supply that requires checking, nor a continuous supply of food. In fact too regular tank cleaning and over feeding are to be discouraged.

The duties associated with the aquarium room are outlined below.

(i) *Daily checks* of the water temperature and appropriate adjustments of the thermostat. Checks of the tank contents; removing any dead fish, plants, or scavengers. Checks of the oxygenators and any water circulating equipment.

Feeding at set times each day, and then only in sufficient quantities to be totally consumed.

(ii) *Weekly duties* involve cleaning and checking. A siphon, or a dip tube can be used to lift up debris from the bottom of the tank.

Scrape all the covering growths from the viewing walls, but leave some behind for the fish to feed on. The algal growths are generally nothing to be too worried about, unless overgrowth or decay of these algae occurs.

Cloudy tank water needs to be cleared up. This tends to occur when the tank is overcrowded. If the numbers are reduced and more light is supplied to the water, this cloudiness should disappear, if the problem has not gone too far. Some aquarists suggest that if a piece of clean coal is put into the cloudy water it acts as an effective clearing device!

Check the filters, especially the polymer wood—carbon types. The filter materials may need changing. Check all the pumps. Clean air lines and the diffusers.

Check all the plants, because healthy vegetation is necessary to provide some oxygenation and the reduction of algal growths. Cut away any dead pieces of plant or overgrowths.

If for any reason the fish need to be removed, ensure that they are taken out quietly, using a net. Do not handle the fish body. Use the netting to steady the animal. Always use the same equipment with the same tank to reduce disease spread. The equipment can be disinfected by boiling in water or soaking in a solution of potassium permanganate.

Check any breeding operations that are being set up. Ensure the correct food for fry is available. This can be prepared beforehand by the culturing techniques described elsewhere (p. 196).

For the details of rearing some aquatic species turn to Unit 5 and the Data Capsules.

2.8. Identifying individual animals

If large numbers of animals are kept and it is necessary to be able to recognize an individual animal, then some identification system needs to be operated. The methods of identification here reviewed are adapted from the animal technician manuals in the UK and USA.

The methods considered are as follows:

(a)   Cage cards.
(b)   Individual coat markings.
(c)   Stains or dyes.
(d)   Ear and toe punches or notches.
(e)   Ear studs or tags.
(f)   Tattooing.
(g)   Wing and leg bands.
(h)   Wing clips.
(i)   Collars, neck bands, and chains.
(j)   Filing.
(k)   Painting.
(l)   Shaving.
(m) Beads.
(n)   Branding.

The method must suit the animal. It must be easy to apply, harmless, permanent, and readily seen by an observer.

Before outlining suitable identification methods for named common laboratory animals, it is necessary to explain what the above methods involve.

(a) *Cage cards* should be fixed to all cages. Relevant data concerning the cage occupants must be put on these cards irrespective of the individual markings used on specific animals.

(b) *Individual coat markings* or other characteristics are only useful if one is dealing with small numbers of animals. It necessitates having a reference card system with details and sketches of the individual's characteristics.

(c) *Stains or dyes* are useful if they do no harm to the animal or interfere with the animal's well-being in any way. They are easy to apply, but without some improvized coding system this method has limited value for large numbers of animals. The stain tends to fade or disappear with time, and is virtually useless on dark coats. The stains used for this type of work are as follows:

Stain	Color
Trypan blue	Blue
Picric acid or chrysoidin (saturated)	Yellow
Fuchsin	Red
Methyl violet (gentian violet)	Violet
Brilliant green, malachite green	Green

These are made up as 3–5% solutions in 70% alcohol.

(*d*) *Ear and toe punches or notches* are only useful if a coding system is established, and understood by all those having access to the animals. This method involves punching holes through the ear tissues or clipping notches in the edge of the ear. (Not used with rabbits because of the presence of significant blood vessels in the ears.)

Amputating toes or punching toes can be used for purposes of recognition in accordance with an agreed scheme. This is a less desirable method.

The application of an ear-punching method is shown in Fig. 15.0. This is carried out using special "ticket punch" type appliances made specially for the job.

(*e*) *Ear studs or tags* are clamped onto the ear near the head where they are difficult to be tampered with. These aluminum tags can be color-coded, lettered, or numbered.

(*f*) *Tattooing* on the inner surface of the ear is sometimes used. This can be done by means of a pair of pliers or forceps manually, or by means of hand-held electro-vibrators.

The letter size and coloring needs to be changed for different animals. The dark pigmented skin is tattooed with white or yellow inks, and the lighter skins with blacks or reds.

(*g*) *Wing and leg bands* are used on birds. The wing bands are clipped around the wing in such a manner as to be unrestrictive to the bird. They can be adjusted. They are coded by color or symbol.

The leg bands are put around the legs of birds in such a manner as to be comfortable, but not loose. They are color-coded or numbered. Plastic bands may be suitable for birds, but metal rings are needed for rabbits.

(*h*) *Wing clips* are thin aluminum strips with a pointed-end that is pushed through the wing (or ear of rabbit or guinea pig). They are folded over and punch squeezed into position. They can be numbered and/or color-coded. These are sometimes described as "Ketchum tags".

(*i*) *Collars, neck bands, and chains* encircle the body and have a coded disc attached to them. These can be leather, plastic, or metal. The metal or plastic identity discs should be such that they cannot be torn off or trapped in cage meshing.

(*j*) *Filing* the shell (carapace) of turtles or tortoises to produce code notches.

(*k*) *Painting* the carapace of reptiles with a water-proof paint with identity code.

(*l*) *Shaving* patches off the coat in a coded manner can be used in short-term work.

(*m*) *Beads* are sometimes sewn into skin folds on the dorsal surface of frogs by means of nylon thread. The colored beads represent a code of identity.

(*n*) *Branding* is usually restricted to larger farm animals such as cattle and horses.

The application of these identification methods to a variety of laboratory animals is outlined in the tabulation (see p. 50).

Animal identification routines

Species	Marker	Mode of application
Fish	Natural patterns.	Recorded on card.
	Fin clip.	Dorsal or ventral fin.
	Tank isolation.	Tank record card.
Frogs	Color beads sewn into skin.	Above dorsal sac.
	Punch coded.	In web of foot.
Reptiles	Filing	Turtle shell notches.
	Paint	Turtle shell
	Punch coded	Feet
Birds:		
Day old chick	Wing band	Clipped around wing
Chicken (adult)	Wing clip	Clipped through skin
Pigeon	Leg band or ring	Fit closely, comfortably
Duck, goose and swan	Wing band or clip } Leg band or ring }	As for chicken
	Punch coded	Web of foot
Mouse	Stain	Back fur
Rat	Tattoo	Inner ear surface
	Punch and notch-coded	Ears
Guinea-pig	Stain	Back or head fur
	Ear stud coded	Close to the head
	"Ketchum tag" coded	Close to the head
	Punch and notch-coded	Ears
	Natural coat characteristics	Card records
Hamster	Punch coded	Ears
	Tattoo	Inner ear surface
Rabbit	Stain	On the back
	"Ketchum tag"	Coded tag in the ear, close to head
	Ear stud	Aluminum coded stud
	Leg ring, coded	Rear leg above the hock
	Tattoo	Inner ear surface
Ferret	Stain	On the back or head
	Tattoo	Inner ear surface
	Punch coded	Ears
Cat	Collar or chain	Coded disc around the neck
Dog	Collar	Coded disc around the neck
	Tattoo	Inner ear surface
Monkey	Tattoo	On chest, upper lip or forehead
	Chain and disc	Around the waist

2.9. Transporting animals

The breeding of laboratory animals for commercial purposes necessitates that they be transported to the "consumer". The consumer could be another breeding unit or an experimental research department. The manner in which animals are transported short or longer distances should be the subject of some concern for those with animal welfare in mind. (Transport of Animals Act 1973.)

Ear Punching System of Animal Identification
Numbers are indicated by the positioning of
the punch

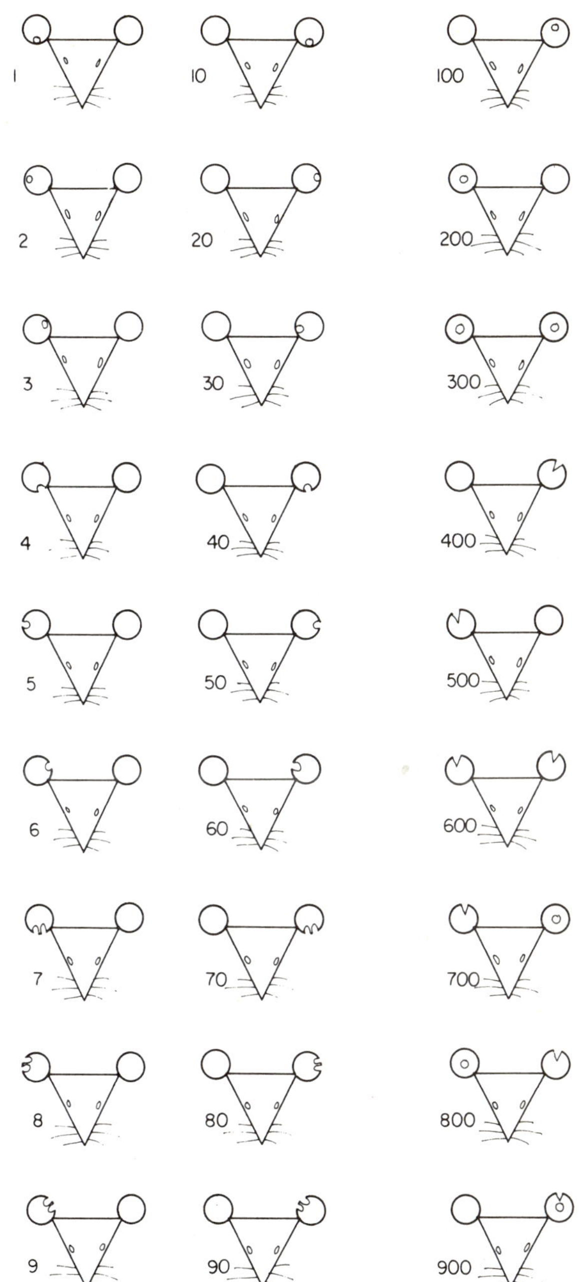

FIG. 15. Ear-punching system for individual animal identification (adapted from *IAT Manual*)

The topic of animal transportation will be considered under the following headings:

General Requirements.
Specific Requirements.
Transportation Containers.
Labels and Instructions.

2.9.1. GENERAL REQUIREMENTS

An animal in-transit requires much the same as it does in the animal house. Such requirements are, however, more difficult to provide in transit. In addition to these life-support requirements there are other considerations to be taken into account, such as the fluctuations in environmental conditions and supervision. The general requirements of an animal in transit are as follows:

(a) *Food and water* should be given to the animals just prior to their despatch. In transit they can be provided with dry pellet food, if appropriate, and succulent fruits or root crops to supply water. This applies to mammals, non-mammals generally do not require feeding in transit.

For longer trips instructions should be provided, together with appropriate watering equipment, so that personnel may water the animals at given times. A note should also be clearly displayed if the animals are *not* to be watered in transit.

(b) *Bedding and litter* needs to be provided in the containers. This can take the form of granulated peat moss or softwood sawdust scattered over the floor to a depth of 2–3 cm. This floor cover aids in the absorption of urine and provides a degree of comfort. If the animals are shy of light then hay or shredded paper should be used in order to give some refuge. Hay should in any case be used in the containers of guinea-pigs, hamsters, and rabbits.

(c) *Temperatures, humidities, and air supplies* need to be maintained at an optimum. It is not always so easy to guarantee that temperatures or humidities are kept within an acceptable range, so adequate ventilation must be ensured. The design of the containers is important in this respect as will be mentioned later. The containers must for this reason be clearly labelled "Livestock".

2.9.2. SPECIFIC REQUIREMENTS

The type, the quantity, and the quality of animal being transported will dictate specific requirements, such as the type of container, the attention required in transit, time limits for transportation, watering or no watering required, the type of bedding, and so forth. The transporter should check out national or international road, rail, and air regulations. A summary of some of the specific requirements are tabulated below.

Species	Transportation requirements
Amphibia	Need moist bedding for concealment, leaf mold, or sphagnum moss. Avoid overcrowding. Avoid packing infected animals (red leg). Use wood or wood-substitute container with plenty of metal gauze ventilation holes
	Aquatic amphibia (axolotls) should be transported in tough double plastic bags (one inside another) containing water sufficient to cover the animals

TABLE (*contd*)

Species	Transportation requirements
	Ensure the water has a good air volume above it. Transport the tied bags inside a tough box
	Check all in-transit arrangements and arrival arrangements
Reptiles	Need temperatures in excess of 15°C (60°F). Strong wood or metal boxes containing moss or hay are suitable. Metal vents essential. No food necessary (generally) although water is required for longer journeys
	Snakes can be despatched in strong sacks within strong boxes
Birds	There are regulations governing the transportation of live birds specific to the State
	For details in the UK consult "The British Transport Commission" and British Standards "Recommendations for the Carriage of Live Animals by Air". Food and water needs to be supplied for long journeys
	There are recommended crates and containers for various sizes of birds
	Avoid excesses of temperature. Ensure ventilation
Mice, rats	Small cardboard boxes with metal vents located on all sides are suitable for short journeys. Tougher materials like plastic are more suitable for longer periods to avoid urine spoilage. The packing density of the animals must be in accordance with calculated values (see density chart from *UFAW Handbook* (p. 54)
	If the animals are in transit in hot climates the packing densities need to be reduced
Guinea-Pigs	Disposable cardboard boxes with metal side vents are suitable for shorter periods of time
Hamsters	For longer periods of time wood or plastic containers are used
	Hay is needed for these animals
Rabbits	These require large strong boxes with adequate vent holes. They generally have hinged lift-up tops
Ferrets	They require strong boxes with no opening lid to prevent tampering in transit. The box also needs adequate vent-holes
Cats, dogs	Wooden boxes or wicker baskets equipped with watering equipment can be used. They should have solid floors.

From the above outline it can be seen that all the species have the following requirements in common during transportation.

(*a*) The avoidance of extremes and fluctuations of temperature.

(*b*) Adequate ventilation under all circumstances of stowage and storage.

The design of animal transportation containers should take these requirements into account.

2.9.3. TRANSPORTATION CONTAINERS

There are a number of animal transportation containers available which should ideally conform to the following specifications:

(i) They should be *crush proof* and not easily destroyed by the animals (reinforced cardboard, wood, metal, plastics).

(ii) They should have *insulating* properties to keep the animals warm in cooling environments. Wood and card is a warmer material.

(iii) They should have an *internal lining* to prevent urine or water spoilage. This can be a plastic or metal grid liner.

(iv) They should be *reusable containers* constructed from metal or plastic because they can be sterilized before reuse but they are more expensive. The disposable cardboard containers or the plastic-card type container are cheaper but not as strong.

(v) *The shape* of the container can be drum-shaped or rectangular. The disadvantage of the latter is that some animals tend to heap in the corners, this is overcome by the drum shape. The outer walls can be sloped to give a top wider than the base so that when they are stored side by side some air space remains between the containers.

(vi) *Air vents* can be located in the lid area provided there are spacers to permit air circulation if the containers are stored one on top of another.

The best situation for air vents is towards the top of side walls so that heaping animals or bedding cannot block the vents. This position also reduces draughts in the base of the container. There are specialist transport containers with filter vents that are used to carry animals that have been reared under barrier conditions.

2.9.4. LABELS AND INSTRUCTIONS

When animals are to be shipped their transportation containers need to be labelled in order to convey the following information:

(i) Name, full address, and telephone number of consignee (forwarding address).

(ii) Name, full address, and telephone number of consignor (sender's address).

(iii) Date and time of containerization.

(iv) "Livestock" label in large red letters.

(v) Descriptions of the contents—species, sexes, numbers, ages. For international despatch pictorial instructions are necessary.

(vi) Instructions for in-transit attention to include: watering information or *no* watering; do *not* open or do *not* feed; avoid extremes of temperature or draughts; *not* to be left out in the open.

(vii) Customs and veterinary information.

(viii) Purchaser's information, order numbers, insurance, etc.

Individual institutions will have their own system of labelling for national and international transportation, but clarity and ease of understanding is a priority whatever method is employed.

Density chart for animals in transport

Species	Animal weights	Maximum numbers per compartment	Space per animal per cm^2 (per in.^{2})	Height of the box cm (in.)
Mice	15–20 g	25	20 (3)	13 (5)
	20–35 g	25	26 (4)	13 (5)
Rats	35–50 g	25	40 (6)	13 (5)
	50–150 g	25	52 (8)	13 (5)
	Adult	12	100 (16)	13 (5)
Guinea-pigs	170–280 g	12	90 (14)	15 (6)
	280–420 g	12	160 (25)	15 (6)
	Over 420 g	12	230 (36)	15 (6)
Rabbits	Under 2.5 kg	4	770 (120)	20 (8)
	2.5–5.0 kg	2	970–1160 (150–180)	25 (10)
	Over 5.0 kg	1	1400 (220)	30 (12)

(Adapted from *UFAW Handbook*)

2.10. Hazards in the animal house

The risks associated with animal house work can be grouped as follows:
 (i) Problems of disease from the animals.
 (ii) Problems connected with the handling of animals.
 (iii) Problems of the work environment.

Animal house work is no more hazardous than work in any other scientific environment provided common sense and knowledge is applied. The obvious hazards would seem to be those associated with disease, but in practice it might be fair to say that other aspects of the work can be more hazardous. The hazardous areas of animal house work are summarized below.

(i) There are *problems of disease* that may be picked up from animals (the zoonoses) if the elements of hygienic practice are not carried out. In order to avoid duplication of information it is sufficient to mention that the zoonoses are discussed more fully elsewhere (p. 107).

The procedures of antisepsis and sterilization will prevent a large proportion of pathogens being available to infect animal workers (p. 86).

(ii) There are *problems connected with handling* the animals if the stock has not been handled regularly and become frightened and aggressive when approached. Cuts, scratches, and abrasions caused by animals must be treated seriously, particularly if inflicted by species known to harbour pathogens such as virus B or tuberculosis. Medical coverage should be available to workers, including appropriate medication like anti-tetanus treatment.

Recommended handling methods are considered elsewhere (p. 229).

(iii) There are *problems of the work environment* even in the best-designed establishments. In the animal house situation there are specific hazards that may be all too often discovered *after* injury.

The following is not a complete list of all hazards associated with the work environment, nor of the preventative measures, but the contents are worthy of attention. Animal house workers should be aware of the Health and Safety at Work Act 1974 as well as the Howie Report and Dangerous Pathogens Recommendations. In the UK Her Majesty's Stationery Office (H.M.S.O.), 49 High Holborn, London WC1V 6HB supplies useful publications in this context (i.e. *Safety in Science Laboratories*).

Hazard	*Preventative actions*
Damage to clothing	Wear comfortable overalls without lengthy projections that can be trapped. Wear industrial shoes or water-proof footwear
Burns and scalds	Wear hair covering. Wear gloves when handling hot items near autoclaves or incinerators
	Wear gloves and goggles when using chemicals
	Cover the arms near hot steam pipes
Dust inhalation	Wear face masks to filter out dust
Cut wrists and fingers	Care when lifting cages, bins, etc., that may have jagged metal edges. Use gloves. Care when using sharp knives for cutting vegetables or scraping trays
	Caution when putting the arms into cages as the edges may be sharp
	Learn how to push glass tubes into rubber bungs without having the broken end plunged into the palm of the hand!
	Be careful putting your hands into sinks or waste bins; broken glass may be in there. Always put broken glass in a special container

TABLE (*contd*)

Hazard	Preventative actions
Gas inhalation or explosion	Ensure that no gas leakages are present and *never* use flame if in doubt. Have gas cylinders stored safely and away from heat. Care when using anesthetics and *never* smoke
Electrical dangers	Check that flash-proof switches, etc., are correctly insulated in rooms where inflammables are used. Check all electrical insulation on vacuum cleaners and so forth to prevent shocks
Storage dangers	Bedding and other combustibles must be stored away from heat or fire danger. Chemicals should be stored safely for easy access, not above the head nor on the floor where they can be kicked
Behavioral dangers	Never lift heavy objects in such a manner as to strain the back or abdomen. Learn how to lift
	Do not overreach to unload cage racks
	Care on ladders
	Caution with trolleys on slopes and do not overload them
	Be careful on slippery floors especially in the wash area

To this summary can be added the hazards associated with radio-active materials, X-rays, and surgical instruments. A first-aid cabinet should be in an obvious place and well stocked.

2.11. Practical program

ANIMAL BEDDING AND ANIMAL ROOM INSTRUMENTS

This program will involve the student in a practical study of those items that are regularly encountered as a matter of routine in the animal house. When the study is completed a written record of the exercises should be produced and kept for future reference.

The objectives of this program are as follows:

(i) To enable the student to make a closer study of bedding and nesting materials and to discuss advantageous and disadvantageous characteristics.

(ii) To give the student an opportunity to examine, use and maintain instruments employed in the animal house; including the following:

(*a*) Thermometers.
(*b*) Hygrometers.
(*c*) Weighing machines.
(*d*) Microscopes.

(i) *Bedding and nesting materials*

Bedding and nesting materials provided for the animal house will largely depend upon availability. Materials used in the UK will differ from those used in the US; for instance, because of the differences in plant products available. Whatever bedding is chosen it should conform to some or all of the following criteria.

(*a*) *Easily available.* The materials should not be chosen if only seasonally available as a change of bedding material in a colony will represent a new element in the experimental study of animals. It should be easily transportable and easy to store.

(*b*) *Inexpensive.*

(*c*) *Non nutritive.* If chewed and ingested the bedding should not represent an attractive source of any nutrient.

(*d*) *Absorbent.* It should be capable of absorbing a fair amount of moisture such as may be provided by urine and feces.

(*e*) *Harmless to animals.* It should not contain any toxic or staining substances. It should be free from any potential pathogenic organisms.

(*f*) *Comfortable.* If the materials are to be used as direct bedding it should not contain any injurious materials. It should be a good heat insulator.

(*g*) *Easily disposable.* Dirty bedding should be easily removed from the animals and incinerated in order that it does not become a health hazard.

Bedding is often described as *direct* and *indirect*. Those materials that come into contact with the animal are described as direct bedding. The materials used in trays beneath grill-floor cages are described as indirect bedding. Both types should be sterilized before use.

A range of bedding and nesting materials that could be examined by students are tabulated below with some information relating to their characteristics.

Bedding	Direct (D), Indirect (I)	Absorbency	Disposable by incinerator	Notes
Sawdust	D and I	Very good	Yes	Must be sterilized after delivery from machinery at a reliable supplier. Stored sawdust may be contaminated. Caution—some woods may contain resinous oil that can be noxious. Use freshly cut white soft wood. Avoid dusty material
Soft wood shavings	D and I	Good	Yes	Often used for rodents and other small mammals for bedding and nesting. Supplies may vary in composition
Hard wood chips	D and I	Excellent	Yes	Useful in cat soil pans
Ground corn cob	D and I	Excellent	Yes	Various particle sizes available
Wood wool	D	Fair	Yes	Useful in the finer grade for rodents.
Peat moss	I	Excellent	Yes	Very good at reducing animal house odours. Expensive
Cotton waste	D	Excellent	Yes	Can be a hazard with young animals becoming entangled
Chopped straw	D and I	Fair	Yes	Useful on the floors of pens
Paper products	D and I	Excellent	Yes	Useful as nesting for small rodents. No dust given off

Whilst students are examining various forms of bedding they should record a recommended sterilizing method as well as any other advantages or disadvantages.

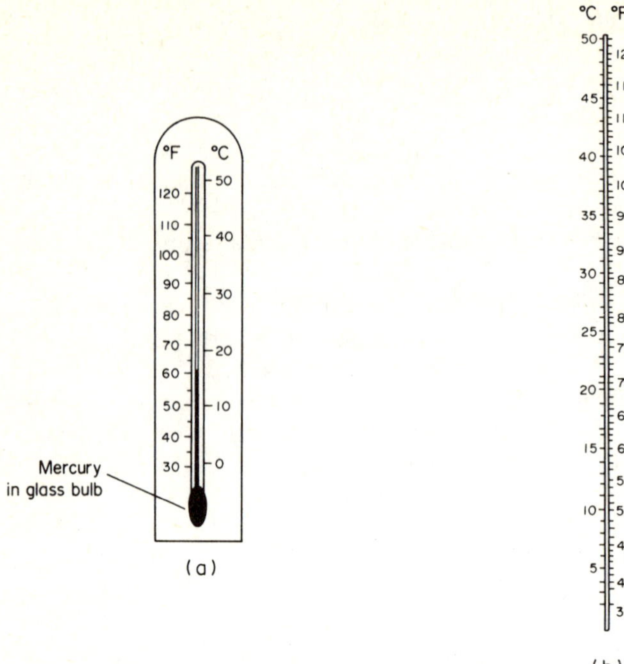

FIG. 16. (*a*) Room thermometer (Centigrade/Fahrenheit). (*b*) Conversion scale for temperatures

(ii) *Animal house instruments*

This section of the program will necessitate the availability of certain pieces of animal room equipment. Students are encouraged to make a close study of the instruments now mentioned.

A room thermometer. The air temperature of an animal room needs to be maintained at an advised value. Recommended air temperatures for different animal species are recorded elsewhere. A measurement of the air temperature at any given moment is registered on a *mercury thermometer* such as that shown in Fig. 16. The scale of temperature is either Celsius (centigrade) or Fahrenheit. It is general in scientific work to use the Celsius scale.

A comparison of the two scales and a method of converting one to another is shown below.

A fairly simple way to convert the temperature scales is as follows:

Fahrenheit to Centigrade
Subtract 32 ($°C = °F - 32 \times \frac{5}{9}$). For example: $122\,°F - 32 = 90$
Multiply by 5 $90 \times 5 = 450$
Divide by 9 450

$$\frac{450}{9} = 50\,°C$$

Centigrade to Fahrenheit
Multiply by 9 ($°F = \frac{9}{5}°C + 32$). For example: 50 °C × 9 = 450
Divide by 5   450   = 90
                                                         ———
                                                          5
Add 32   90   + 32 = 122 °F

Students are advised to practice other conversions. The mercury thermometer functions by the fluid silvery metal expanding when heated and moving up the capillary stem. Upon cooling the metal contracts and moves down the stem.

A Maximum–Minimum Thermometer. The air temperatures in the animal house may fluctuate throughout 24 hours because of draughts or warm pockets of air. If this is the case, the information must be known to the technician. A thermometer that records the highest and lowest temperatures over a period of time is the max.–min. thermometer. The ordinary wall-mounted type as shown in Fig. 17a does not say when, nor for how long, the temperatures remained at a given value. For this purpose a *thermograph* type is employed. This type produces a continuous pen recording on a rotating drum showing the temperature at a given time.

Students are advised to examine these types of temperature recorders closely.

A clinical thermometer. **Recording the body temperature of an animal requires a special type of "bullnose" thermometer.** This thermometer has a column of expanding mercury as do most others, but in this case the mercury cannot fall back into the bulb unless shaken sharply. There is a constriction in the capillary tube to prevent the mercury falling and so the body temperature can be read even when the thermometer is removed from the animal's rectum or wherever.

Before a reading is taken with the clinical thermometer it must first be shaken to push the mercury back down into the bulb. The instrument should be disinfected and lubricated with a light grease or soap before use. It should be washed and cleaned with an antiseptic after use. This type of thermometer must never be subject to high temperatures such as boiling water otherwise it will burst.

Students may check the rectal temperature readings of various animals by turning to the appropriate section of the book. An instructor will need to show students the best way in which to insert the thermometer for rectal recordings.

A wet and dry bulb thermometer (wet-dry bulb hygrometer). This type of thermometer arrangement enables the user to determine the humidity of the atmosphere—that is, the amount of dampness in the air. The best way in which this can be done is to expose the two thermometers to moving air. This can be done by whirling the instruments in the air mounted on a structure that resembles a football rattle. This is called a *whirling hygrometer.*

The wet bulb will show a lower temperature reading than the dry bulb because of the evaporation of water from the wet one. Knowing these two different temperatures it is possible to calculate the *relative humidity.* This can be simply done by using the chart or slide rule that is supplied with the instrument. Full instructions are supplied with these charts so it is inappropriate to go into detail here.

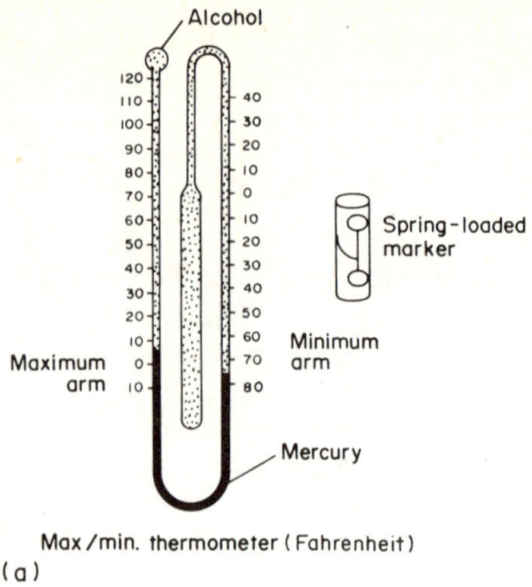

Max/min. thermometer (Fahrenheit)
(a)

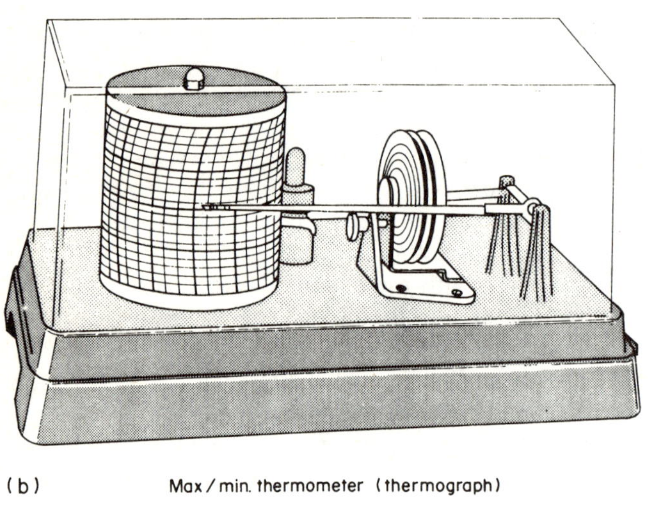

(b)   Max/min. thermometer (thermograph)

FIG. 17. Max/min thermometers. (*a*) Wall type. (*b*) Thermograph type

Students are recommended to determine the relative humidity in different places and at different times.

Weighing machines. In some laboratories a regular routine is the daily weighing of the animals and their diets. It is in the interests of the technician that the weighing apparatus be well maintained and be easy to use. Students are encouraged to use as many types of weighing machines as are commonly found in the animal house.

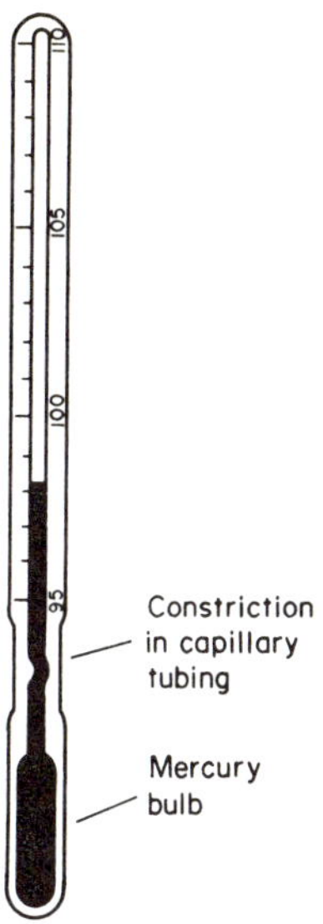

FIG. 18. A clinical thermometer (human type)

Each type of balance needs to be cleaned regularly. It must not be allowed to corrode, nor must any of the weights. It must be levelled before it is used.

A useful and quick method of weighing small animals can be carried out as follows. It involves having makeshift weighing cylinders or having custom made weighing cylinders of the sort here described.

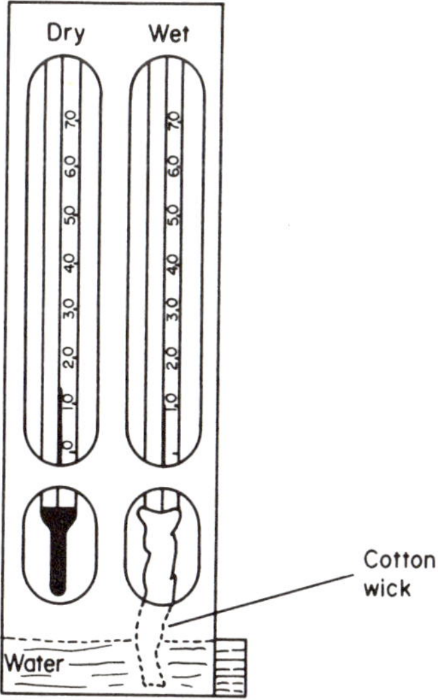

FIG. 19. Wet and Dry Bulb Thermometer (hygrometer)

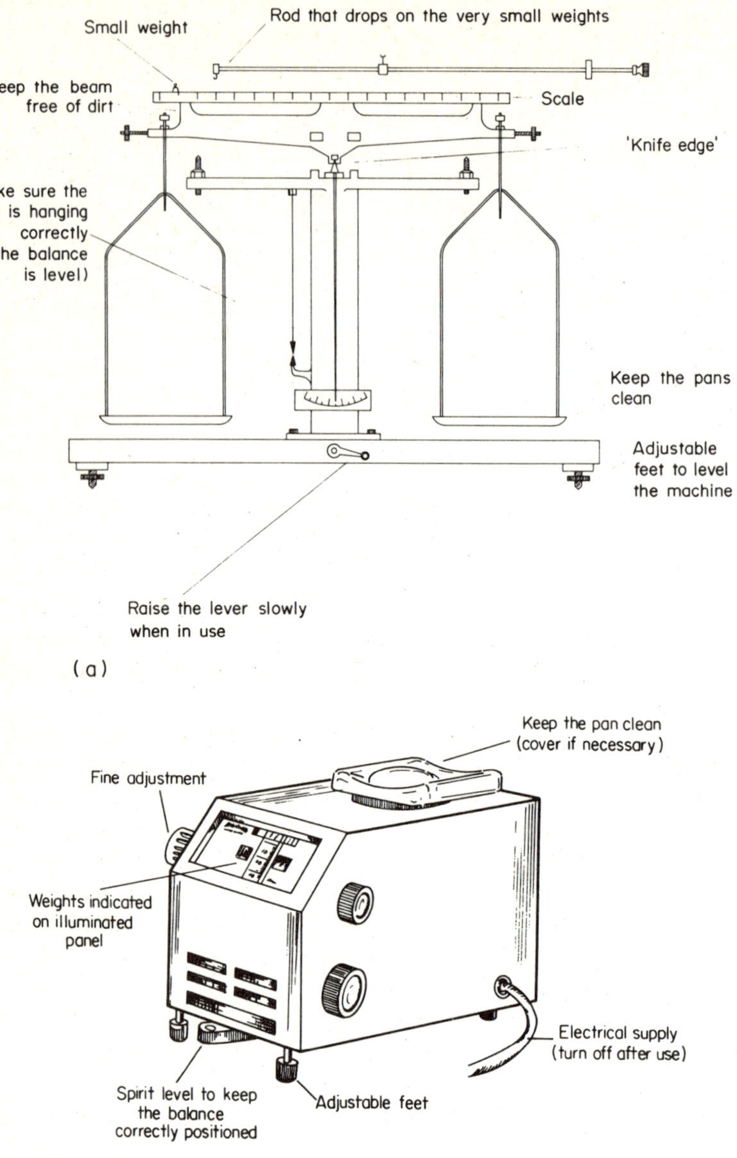

Fig. 20. Weighing machines (drawn by Ross Mackay, Oxford). (a) Chemical beam balance. (b) Top pan automatic balance

The weighed cylinder is laid on the table and the animal permitted to crawl into it. The cylinder is then put into the vertical position standing in a push-fit base. The animal can be weighed in this manner by reading off the increase in weight of the cylinder due to the animal (see Bibliography, p. 301). This method may not suit the needs of many work situations. Students are encouraged to compare the accuracies of rat weights by different weighing techniques.

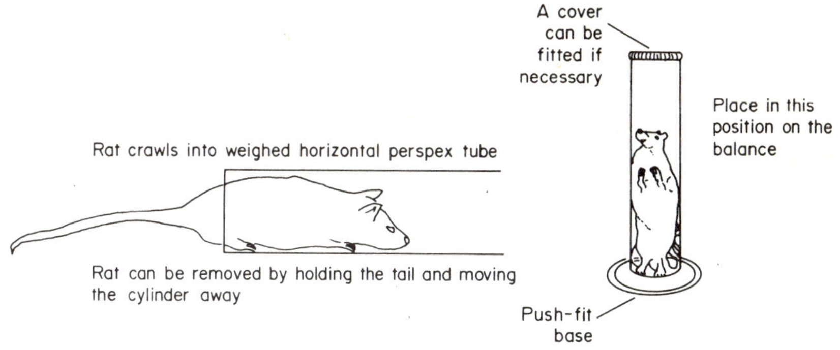

FIG. 21. Weighing a rat (adapted with permission from *IAT Journal*, vol. 27, no. 1, May 1976)

The microscope. Some animal rooms will have reason to keep and use microscopes. If these microscopes are to be used for a detailed examination of prepared slides of microorganisms then they should be maintained in the best possible order and stored in a protective place.

A lengthy study of the microscope is not really appropriate here and so the information will be confined to stating the parts of the microscope, how it is used and maintained in working order.

The eyepiece is fitted into the body tube. The magnification value is usually written on the eyepiece; it can be $\times 4$, $\times 8$, $\times 10$, or $\times 15$. These eyepieces may get dusty so they need to be removed and polished with lens tissues before use.

The objective lenses are fitted on a rotating device so that the lenses can be changed around easily. These lenses are the most expensive part of the microscope and their magnifications are shown on the side as $\times 10$, $\times 20$, $\times 40$, $\times 90$. To work out the approximate total magnification of the microscope in use it is only a matter of multiplying the eyepiece value by the objective value. The objective lenses must always be kept clean with lens tissues. If need be wipe with a little alcohol or ether to remove greasy spots.

The condenser lenses are located beneath the platform or stage and their job is to focus light reflected from the mirror up through the specimen.

Setting up the microscope involves a careful sequence of events otherwise the frustrating experience of looking at the name on an electric lamp bulb may result. Carry out the following actions and then it should be possible to perform low- and high-power work.

Low-power microscope work.

1. Place the specimen on the stage.
2. Adjust the mirror so that the light strikes the flat side and is reflected up into the stage. (Open any diaphragms, remove any filters first.)
3. Rotate the objective lenses so that the lowest magnifying one is in the operating position.

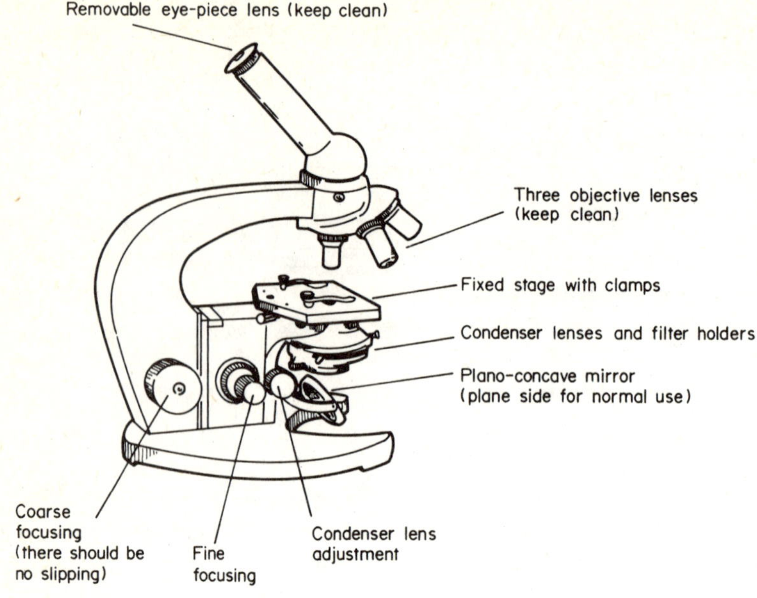

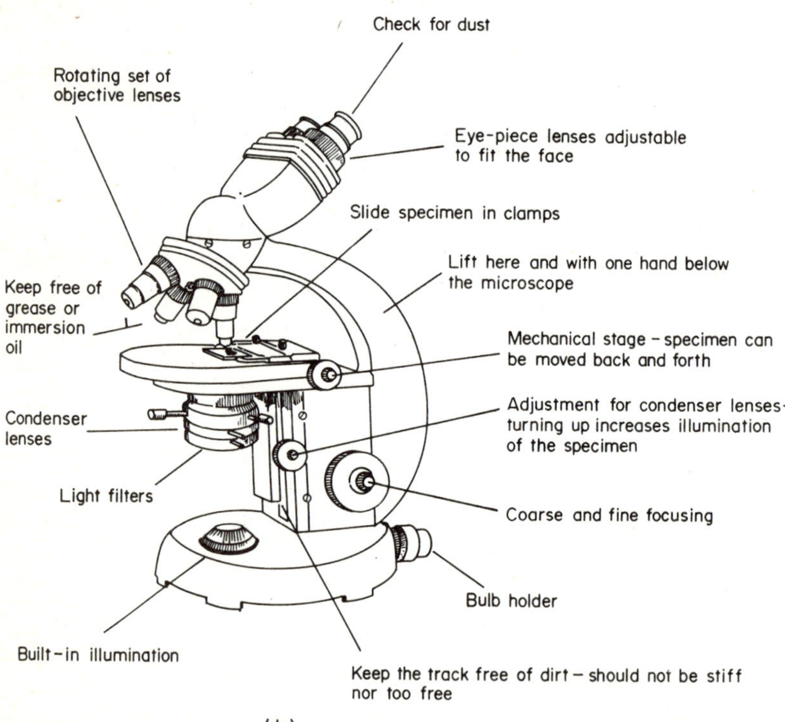

Fig. 22. Microscopes (drawn by Ross Mackay, Oxford). (*a*) Monocular student microscope. (*b*) Binocular student or research microscope

4. Bring the lenses into focus. Turn the focus knob (the bigger one) until the lenses are a finger's width away from the specimen. Look down the eyepiece and turn the focus knob to bring the lenses upwards until the specimen comes into clear focus.
DO NOT LOOK DOWN AND TURN DOWN

5. Adjust the condenser and mirror to improve the illumination.

High-power microscope work. In order to examine very small objects like bacteria it is necessary to use a high-power lens, such as $\times 90$. This is sometimes called the oil immersion lens because it is only used by dipping it into cedar wood oil and looking through that in order to prevent light scattering.

1. Set up the microscope as if to view a specimen under low power.
2. Turn the $\times 90$ lens into the operating position.
3. Put a drop of immersion oil on top of the cover glass on the slide. Lower the lens into this oil gently whilst viewing from the outside.
4. Look down the microscope and operate the smaller focus knob (fine adjustment) and bring the specimen into clear view. Bright illumination is a great help when magnifying to such a high power.

Students are recommended to try out the microscope techniques using specimens of parasitic pests as described in a later unit (p. 68).

Animal care routines—summary

Animal houses can be categorized as conventional, barrier maintained, and experimental.

Animal care routines are the same for all types of animal house with extras for those barrier maintained and experimental.

The routines include time-consuming maintenance and cleaning as well as checking the condition of animals in experimental programs.

Barrier maintained animal houses may produce S.P.F. animals. In this type of animal unit, the routines are complicated by the high standard of hygiene that must be maintained.

Experimental animal houses involve the technician in collecting samples and recording physiological data.

Work allocation will depend upon the type of animal house in the grading system of animal care personnel.

Time allocation for animal house duties will depend upon the type of unit concerned, and might well be an academic exercise rather than a work-day reality.

Record keeping is an essential aspect of the duties of administrative personnel for financial and scientific reasons.

Various techniques for identifying individual animals in a large group were described. These methods included ear-clipping, staining, tattooing, and collars. Transporting animals in standardized containers was described to ensure minimum stress to the animals in transit.

Hazards in the animal house were described as ranging from the animals to physical and chemical agents.

Practical work connected with bedding and nesting materials as well as animal house instruments was described.

STUDY OBJECTIVES

UNIT 3:

Animal Health and Hygiene

(*a*) Describes organisms that are parasites and/or pests.
(*b*) Outlines methods of eradicating and/or controlling parasites and pests.
(*c*) Describes routine cleaning and sterilizing procedures.
(*d*) Describes selected animal diseases and the signs of ill health.
(*e*) Outlines the zoonoses.
(*f*) Describes the animal body defenses and the immune response.
(*g*) Describes the specific pathogen free animal and barrier maintained animals.
(*h*) Describes gnotobiotic animals and isolators.
(*i*) Suggests practical work.

UNIT 3:

Animal Health and Hygiene

IN order to maintain an animal in a healthy condition it is necessary to know what
conditions are responsible for ill health. This unit is concerned with the causal agents
of disease and the way in which they can be eliminated or controlled. The study is
organized as follows:

3.1. Parasites and pests.
3.2. Eradication and control of pests.
3.3. Cleaning and sterilizing.
3.4. Animal diseases.
3.5. Signs of ill health.
3.6. Zoonoses.
3.7. Animal body defenses.
3.8. Specific pathogen free animals.
3.9. Gnotobiotics.
3.10. Categories of laboratory reared mammals.
3.11. Practical program (aseptic techniques).

3.1. Parasites and pests

Parasitism is defined as a relationship between organisms in which one organism
lives in or on another organism, of a different species, and at the expense of the other
organism.

The parasite feeds on the cells, tissues, or fluids of another organism, the host, which
is commonly harmed in the process. The parasite is dependent upon the host and lives
in intimate physical and chemical contact with the host. All major groups of plants
and animals are susceptible to attack by parasites. For convenience, parasites are
often classified as ectoparasites and endoparasites.

Ectoparasites (external parasites) live on the surface of the body of their hosts, suck
blood or feed upon hair, feathers, skin, or secretions of the skin.

Endoparasites (internal parasites) live inside the body, occupy the digestive tract or
other cavities of the body, or live in various organs, blood, tissues, or even within cells.

No easy distinction can be made between these two types of parasites since inhabitants of the mouth and the nasal cavities, and such mites and worms that burrow just
beneath the skin, may be placed in either category. Parasitic organisms may be plants
or animals.

68

Organisms, parasitic and non-parasitic, that may be pests in the animal rooms are outlined in tabular form below. An animal or plant becomes a pest if its presence has detrimental effects upon the host or the food of the host.

Parasitic pests	General notes	Examples of disease or disorder
Viruses	A sub-microscopic parasite living within cells	β (beta) virus disease in monkeys; distemper, hard pad, and jaundice in dogs
Bacteria	A microscopic parasite living within tissue fluids	Intestinal disorders in rats; snuffles in rabbits
Fungi	A parasitic plant living within the skin or outgrowths of the skin or within internal organs	Ringworm in mice
Protozoans	Single celled animal parasites living in the tissues of the intestine or the internal organs	Intestinal coccidiosis in rabbits
Platyhelminthes	Parasitic flatworms living within organs such as the liver or intestine	Tapeworms in dogs and cats; bladderworms in rabbits
Nematodes	Parasitic roundworms living within the intestines and other organs	Toxocara worms in dogs and cats; hookworms in monkeys
Arthropoda	Arachnids: eight-legged arthropods	
	Acarina: ectoparasitic mites feed on skin tissue; ectoparasitic ticks feed on blood and transmit disease	Mange and ear "canker" Tick bite fevers
	Insects: six-legged arthropods	
	Hemiptera: bugs, insects that suck blood and may transmit disease	Anemia; infectious jaundice
	Anoplura: blood-sucking lice, they can transmit disease	Skin irritation; transmit typhus in rats
	Mallophaga: chewing lice	Skin irritation and allergy reactions
	Siphonaptera: blood-sucking insects, the fleas	Transmit typhus in rats; transmit dog tapeworm eggs; skin inflammation
	Diptera: the true flies that may suck blood or deposit their eggs in living flesh	Mosquitoes transmit animal malaria; tsetse flies transmit sleeping sickness; bot and warble fly larvae burrow in the skin of mammals

Non-parasitic pests	General notes	Infested product
Arachnids (eight-legged arthropods)	Acarina: mites	Cereals—wheat, flour
Insects (six-legged arthropods)	Orthoptera: cockroaches, crickets, locusts	Foodstuffs in warm moist areas
	Coleoptera: beetles, weevils	Cereals, rice, bread, meat
	Lepidoptera: moths	Cereals, beans, flour
Mammals	Rodents: wild mice and rats	Foodstuffs in store

From the foregoing it becomes clear that a major task of the animal technologist is to prevent infestation of the animals and their foodstuffs by parasitic and non-parasitic pests. In order to take appropriate actions against infestation it is necessary to have some more detailed knowledge of the pests that one hopes to prevent or eradicate. To this end we now examine a few pests more fully.

Viruses

This large and variable group of infectious agents are parasites that inhabit the interior of cells. They are so small that they can pass through the pores of filters which do not permit the passage of bacteria. The largest virus is less than a quarter the size of a typhoid bacterium.

They cause disease in insects, fish, microorganisms, plants, humans, and other animals. Many viruses appear to do no particular harm to their hosts. Those viruses that do invade the cells of man and animals to produce particularly unpleasant disease range from pox to cancer.

Virus diseases		
Man	Animals	Comparative sizes
Small pox	Hog cholera	Yeast cell, 5000 nm
Chicken pox	Foot and Mouth disease	Typhoid bacterium, 1000 nm
Herpes simplex (cold sores)	Herpes B (in monkeys)	Small pox, 200 × 300 nm
Rabies	Rabies	Herpes simplex, 110 nm
Yellow fever	Canine distemper	
Mumps	Myxomatosis	Influenza, 80–120 nm
Rubella (German measles)	Sheep pox	Poliomyelitis 28 nm
Rubeola (measles)		
Influenza	Swine influenza	Foot and mouth, 10 nm
Infectious hepatitis (jaundice)	Infectious hepatitis	

Previous to the development of the electron microscope it was not possible to see viruses. Specimens can now be magnified as much as 100,000 times and photographed.

Bacteria

Bacterial cells show considerable variation in size, shape, structure, and arrangement. Individual bacterial cells may have one of three forms:

elipsoidal or spherical, called *cocci* (coccus-singular);
cylindrical or rod like, called *bacilli* (bacillus-singular);
spiral or helical, called *spirilla* (spirillum-singular)

The coccal cells may be arranged in groups, if they are so arranged then they are given the following names:

Coccus	Disease example
Pairs of cocci: diplococci	Pneumonia
Chains of cocci: streptococci	Tonsillitis, scarlet fever
Bunches of cocci: staphylococci	Gastroenteritis, boils

The bacilli do not arrange themselves in the patterns as shown by the cocci.

Bacillus	Disease example
Corynebacterium diphtheriae	Diphtheria
Bacillus *piliformis*	Tyzzer's disease in mice and rats
Clostridium *welchi*	Food poisoning
Clostridium tetani	Tetanus
Salmonella typhimurium	Animal typhoid
Mycobacterium tuberculosis	Tuberculosis

The spirilla are generally unattached individual structures. They vary in the lengths and tightness of their spiral form. The short, incomplete spirals are known as comma-bacteria or *vibrios*.

Spirillum	Disease example
Treponema cuniculi	Syphilis (rabbits)
Leptospira species	Leptospirosis

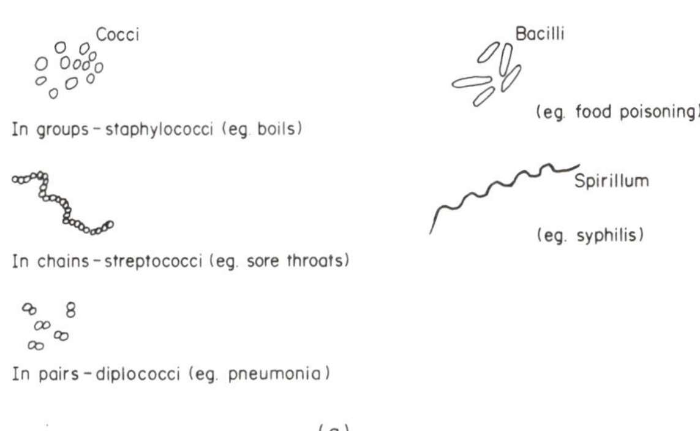

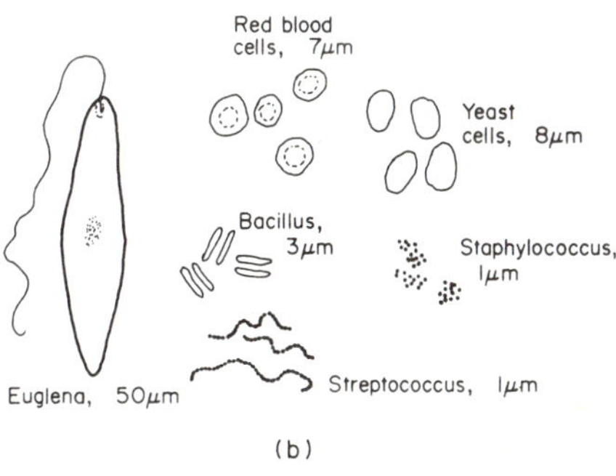

Fig. 23. Bacteria. (*a*) Types of bacteria. (*b*) Sizes of bacteria compared with other cells (adapted from Humphries's *Bacteriology*, John Murray)

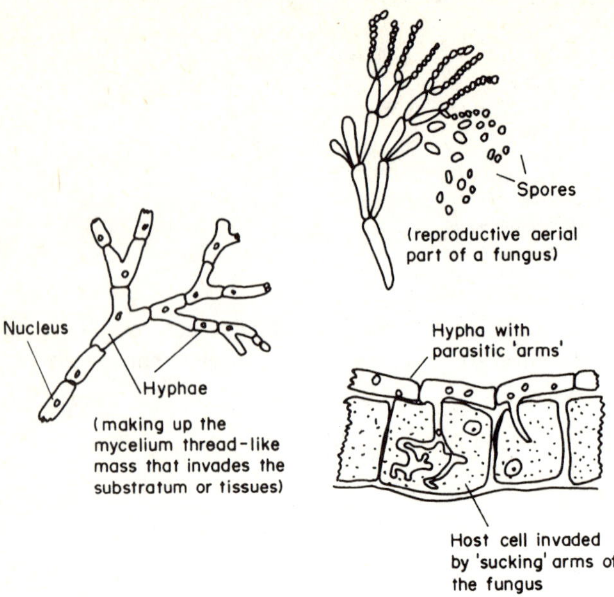

FIG. 24. Fungus—showing the parts of a mold-like fungus such as may infest foodstuffs

Fungi

The molds are a wide group of microorganisms familiar as growths on cheese and bread, or in the garden as toadstools, or on fruits as yeasts. The fungi can be parasites obtaining their food from a living host (fungus in "athletes foot") or *saprophytes* obtaining their food from non-living organic matter (mildews). The fungal growth is usually thread-like (*hyphae*) and branched. A mass of hyphal threads is called a *mycelium.*

Fungus	Disease example
Trichophyton mentagrophytes	Ringworm
Absidia ramosa	Moldy hay disease

Protozoa

These single-celled animals are microscopic in size. The wide variety of protozoans can be organized for study purposes into groups based upon the way in which they move.

Rhizopoda, move by "sliding" actions, e.g. *Ameba* species (dysentery).
Ciliata, move by a beating action of whisker-like cilia, e.g. *Balantidium.*
Flagellate, move by a beating action of a whip-like flagellum, e.g. *Trypanosoma* species (sleeping sickness).
Sporozoa, no movement, e.g. *Plasmodium* species (malaria).

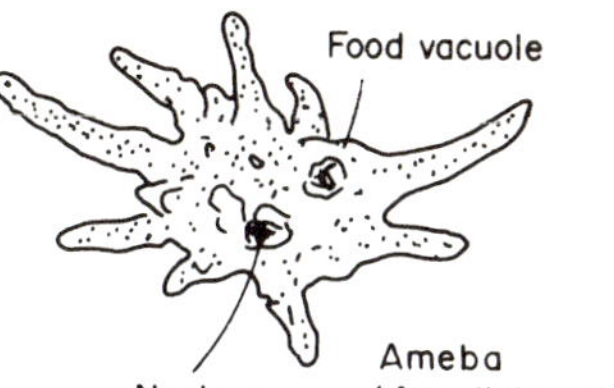

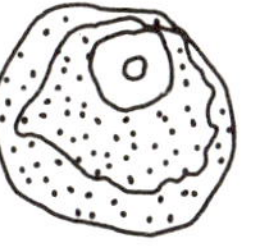

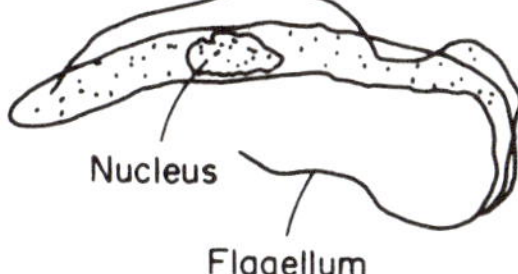

FIG. 25. Protozoa

Protozoan	Disease example
Entameba histolytica	Amebic dysentery
Trypanosoma gambiense	Sleeping sickness (transmitted by tsetse fly)
Trypanosoma cruzi	Chaga's disease (transmitted by a "bug")
Plasmodium vivax	Malaria (transmitted by anopheles mosquito)

Platyhelminthes

The parasitic flatworms fall into two main groups:

Trematoda—the flukes
Cestoda—the tapeworms

Platyhelminth	Disease example
Schistosoma japonicum	Blood flukes in man, dogs, cats (transmitted by way of water snail)
Fasciola hepatica	Liver fluke (transmitted by way of snail)
Dipylidium caninum (tapeworm)	Adult worms in dogs and cats (transmitted by flea)
Taenia taenia formis (tapeworm)	Cysts in the liver of rats and mice. Adults in cat
Echinococcus granulosus (tapeworm)	Adult worm in dog

Nematoda

These parasitic roundworms are smooth, non-segmented, and circular in cross-section. These are the "worms" generally referred to when one speaks of "worming" an

Platyhelminthes

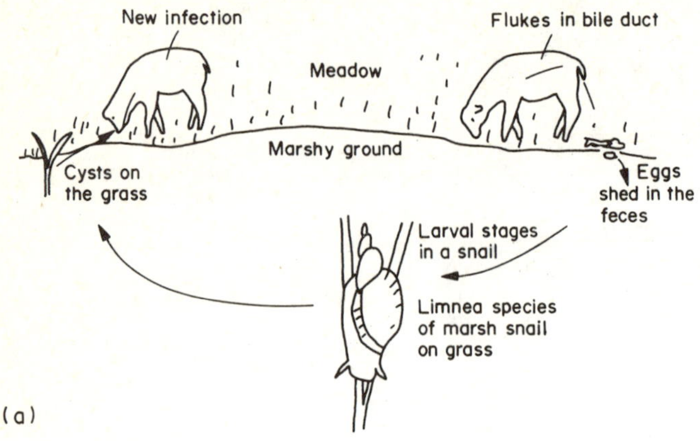

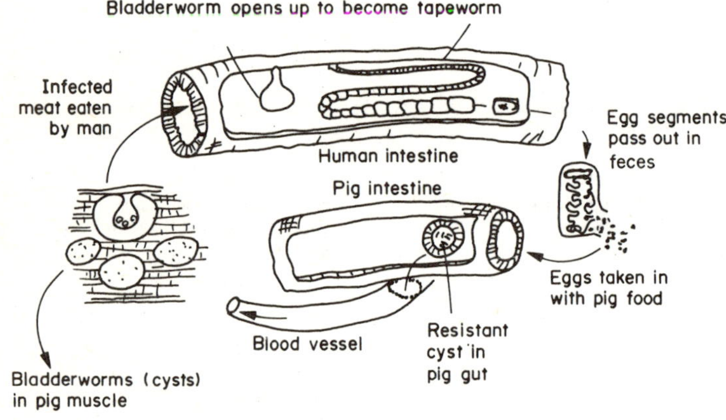

FIG. 26. "Flatworms" (*Platyhelminthes*). (*a*) Life history of the liver fluke (*Fasciola hepatica*). (*b*) Life history of pig tapeworm (*Taenia solium*)

animal. If an animal is infested with nematodes there is not always an accompanying group of symptoms that suggest a disease. In exaggerated cases of infestation the animal may show indicators of malnutrition due to the worms consuming a large portion of the animal's intestinal contents.

Nematode	Notes
Trichuris species	Whipworms in large intestine of dog, rodents, monkey, man
Toxocara canis	Large roundworm in dog (10–18 cm)
Toxocara cati	Large roundworm in cat (5–10 cm)
Uncinaria stenocephala	Hookworm in dog (0.5–1.0 cm)
Haemonchus contortus	Stomachworm or wireworm found in sheep, goats, cattle
Aspicularis tetraptera	Pinworm in mouse and rat

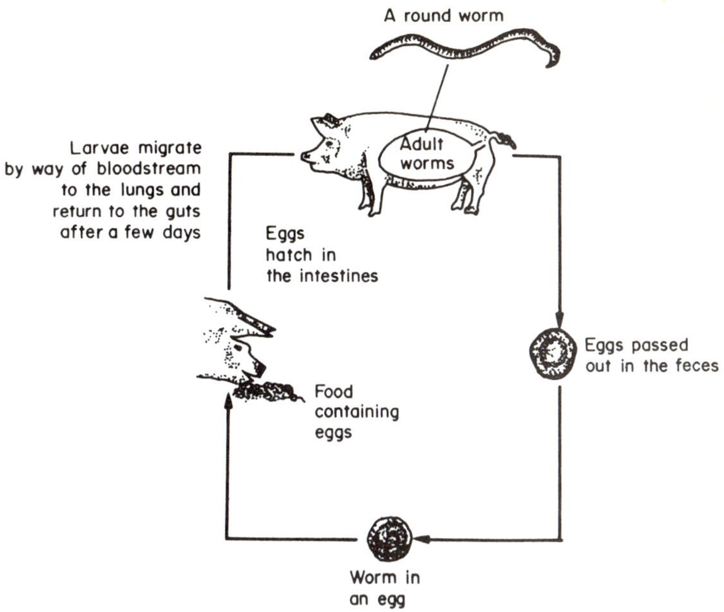

FIG. 27. "Roundworms" (Nematoda). Life history of pig nematode (*Ascaris* species)

Arthropoda

Animals that have an outer, hard exoskeleton and jointed limbs are usually arthropods. This is an extremely large group of animals. There are probably more arthropods than any other animals in the world.

The group includes crabs, spiders, scorpions, flies, centipedes, and many others. Here we study only those arthropods that may be pests in the laboratory animal area.

Arachnida are jointed limbed animals that have eight legs. Spiders, scorpions, ticks, and mites are examples. Only one group of the arachnids shows parasitic species.

Acarina

This is the only group of arachnids that has parasitic members. The ticks and mites are the examples to be examined in fuller detail.

Acarina (mites)	Notes
Sarcoptes species	Itch mites burrow in the skin "mange"
Notoëdres species	Skin burrowing mite causing "mange"
Psoropteres species	Skin surface feeding mite "ear canker"
Demodex species	Hair or face mite lives in hair follicles and sweat glands; worm-like in appearance
Otodectes species	Ear mites in dogs and cats
Dermanyssus gallinae	Blood sucking "red mite" of domestic fowls and pigeons
Cheyletiella species	A mite in the fur of rabbits, dogs, and cats. Can cause dermatitis in cat handlers.

Mites

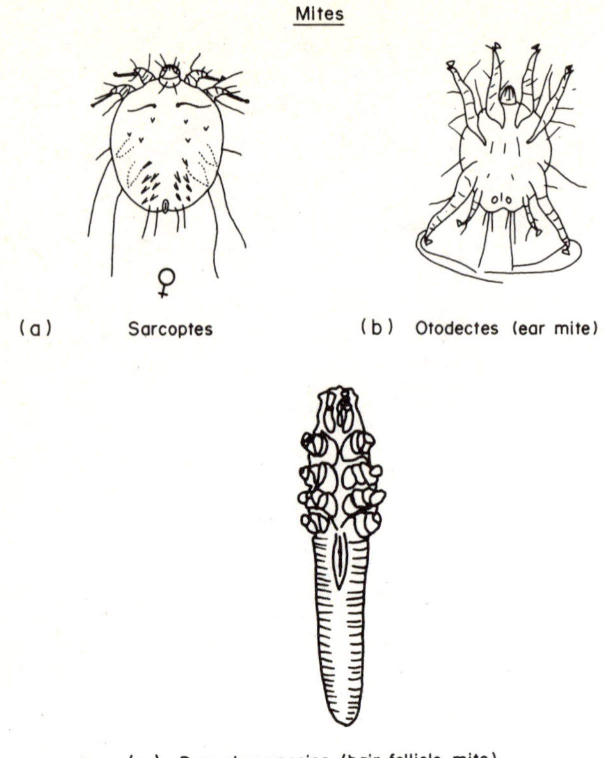

FIG. 28. Mites. (*a*) Itch mite (*Sarcoptes*) burrowing female, dorsal surface. (*b*) Ear mite (*Otodectes*) ventral surface. (*c*) Hair follicle mite (*Demodex*) ventral surface

The previously mentioned acarina were the mites, chewing, and blood sucking. The mites move over the skin feeding on hair, skin scales, and other debris. Some of them burrow beneath the skin, others suck blood. It is these activities that cause skin irritation with the accompanying symptoms of "mange" and "canker". Some important mite pests are those that feed off grain or flour in the store rooms.

Not all mites are parasitic as are the next group of acarina, the ticks. These have at least part of their life history as a parasite. They are larger and easier to recognize than mites. They are perhaps second to the flies as carriers of disease affecting animals and man. Ticks are grouped as follows:

"*soft*" *ticks* have no hard shield over their backs.
"*hard*" *ticks* have a hard shield over their backs.

Both these types have tough skin-piercing mouth pieces which can be so well embedded that they are left in the skin if the tick is pulled off.

Since many ticks live in association with wild birds these must be excluded from laboratory animal areas.

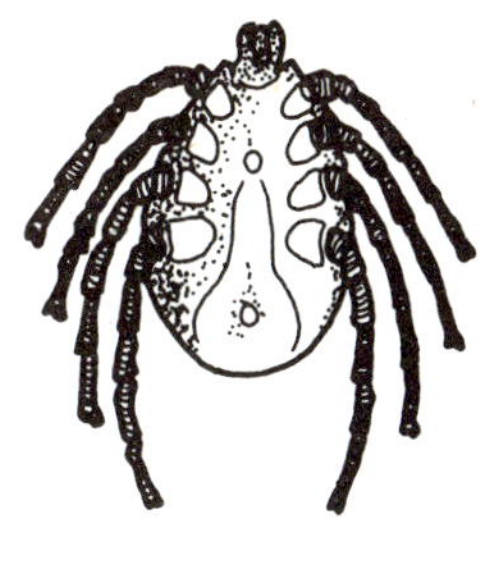

Fig. 29. Tick (*Ixodes*) ventral surface

Acarina (ticks)	Notes
Soft ticks	Inhabit the bedding and feed off the inhabitants from time to time. They generally stay in the nest.
Ornithodoros species	A soft tick that lives in close association with rodents and many domestic animals. Transmit many tick bite diseases in the USA
Argas species	Soft tick of poultry and wild birds transmitting fowl fever in warm countries
Hard ticks	These types generally remain attached to the host off whose blood they feed
Ixodes ricinus	The European castor bean-tick attacks birds and mammals. They can transmit a cattle bacterium or sheep virus
Dermacentor species	These types are responsible for transmitting a wide range of diseases from brucellosis to "tick paralysis"

Insects are jointed limbed animals that have six legs and the body divided into three distinct areas—the head, thorax, and abdomen. This is an extremely large group covering the flies, earwigs, bees, butterflies, and ants. Some insects show modifications of their body to allow them to live a parasitic or fluid-sucking existence.

Hemiptera (the true bugs) are insects with modified mouth parts that allow them to suck up fluids. They can obtain the food nutrients by sucking plant juices or animal blood. The most important of the blood suckers is known as "the bed bug".

Hemiptera (bugs)	Notes
Cimex lectularius	Common bed bug of the temperate areas. This one attacks man.
Cimex columbarius	Found in pigeon nests
Cimex pilosellus	Found in association with bats

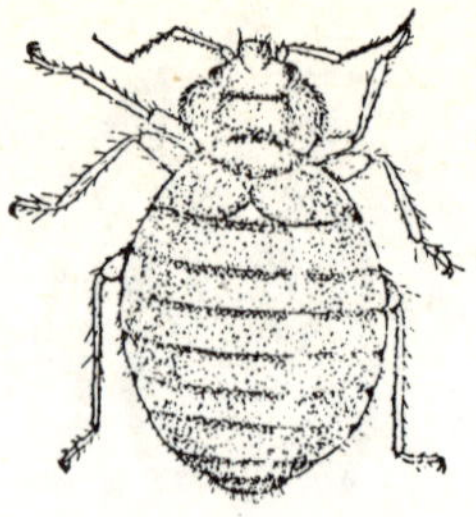

Fig. 30. Bedbug (*Cimex*)

Despite the blood-feeding habits of bed bugs it has been difficult to link the insect with the spread of any major disease in animals or man. (*Leptospirosis*, however, has been shown to pass from guinea-pig to guinea-pig by way of the bug bite.)

Anoplura (blood-sucking lice)

These very small insects have piercing mouth parts that can be withdrawn under the body when they are not in use. A large population of such lice can suck up a considerable amount of blood as well as cause allergic reactions and spread disease.

These sucking lice can be distinguished from the chewing lice by having a head that is narrower than the thorax.

The eggs of lice are called "nits" which are oval, whitish objects with a little lid at the larger end. These eggs are found cemented to hairs on mammals.

Anoplura (sucking lice)	Notes
Pediculus humanus corporis	Body louse of man
Haematopinus species	Attack pigs, horses, cattle
Hoplopleura species	Rat louse in USA
Linognathus species	Attack sheep, dogs, foxes

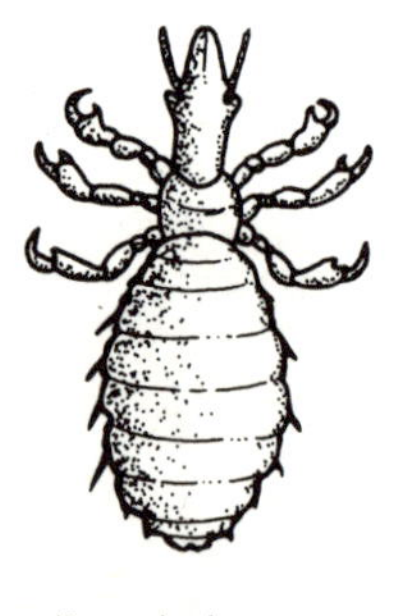

Fig. 31. Lice—biting and sucking. (*a*) Biting louse (*Trichodectes*). (*b*) Sucking louse (*Haematopinus*)

Mallophage (chewing lice)

These lice have a head that is as broad as, or broader than, the thorax of the insect. They have cutting mouth parts rather than the piercing, sucking parts that the previous types had. They live for generations on the same host feeding on feathers or skin scales. They tend to be found on the head or back areas where they cannot be picked off by the host.

Mallophaga (chewing lice)	Notes
Columbicola columbae	A pigeon louse
Trinoton querquedulae	A duck louse

Lice spread from host to host by direct body contact between infested animals.

Siphonaptera (fleas)

These are small blood-sucking insects that are compressed from side to side in order to enable them to slide between hairs and feathers. They are not only annoying pests but also act as carriers of two important human diseases, bubonic plague and typhus.

The fleas are not strictly host parasites nor nest parasites. They occur in both. They frequently occur behind the ears or at the base of the tail in cats and dogs. Rodent and bird fleas are more frequently found in the nests than on the bodies of the hosts. Animals that move around a great deal and therefore have no permanent lair tend to have less fleas. This applies to some monkeys and deer.

Siphonaptera (fleas)	Notes
Pulex irritans	Human flea
Xenopsylla species	Rat and mouse fleas
Ctenocephalides species	Dog and cat fleas
Leptopsylla species	Rat and mouse fleas

The fleas are less particular about their host than the lice. They will take blood from wherever they find it. For this reason they are important as transmitters of disease and act as intermediate hosts for tapeworm in cats and dogs (*Dipylidium*)

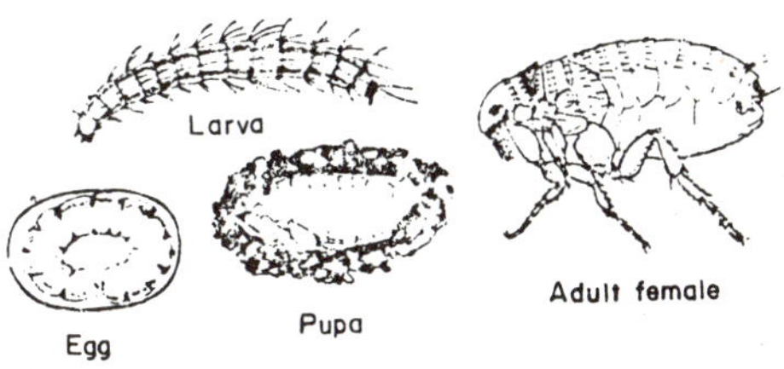

FIG. 32. Flea (*Ctenocephalides felis*—cat flea)

Diptera (true flies)

These are six legged, two winged arthropods. Some of them have a stage of their life history which feeds off living tissues. The larvae (maggots) may suck blood from young birds in the nest, others may invade wounded flesh or penetrate into ear and nose cavities. The adults may suck blood from time to time, rather than remain permanently attached to the host. They frequently cause intense irritation.

Diptera (flies)	Notes
Phlebotomus species	Sand flies of warm climates. Blood-feeding adults transmit several fevers in man and animals.
Culicoides species	Blood-sucking midges that attack both man and animals causing irritation and spreading several diseases.
Simulium species	Blood-sucking black flies that cause intense irritation and can kill large numbers of animals. They spread parasitic nematode worms.
Tabanus species	Blood-sucking horse-flies, sometimes known as deer flies. They inflict painful wounds and can spread animal blood disease (*trypanosomiasis*).
Stomoxys species	Stable flies suck blood, causing irritation and transmitting disease in domestic animals.
Glossina species	Tsetse flies suck blood and transmit "sleeping sickness" fevers in man and animal.
Anopheles species	The mosquito involved in the spread of malaria and yellow fever in man and associated diseases in animals.
Callitroga species	A screw-worm fly lays its eggs in fresh skin wounds, or in body cavities. The maggots feed on the living flesh of cattle, goats, sheep, and hogs.
Hypoderma species	Bot-flies lay their eggs in the skin of cattle. The maggots live beneath the animals' flesh.

Amongst the species referred to above there are several non-European examples. It is in the warmer climates that flies are a greater pest in the laboratory animal area.

3.2. Eradication and control of pests

The control of pests in the animal area is a big topic ranging from partial elimination of the pest to complete eradication. Control can be approached in two ways:

(i) Barriers to the pest's entry resulting from elaborate hygienic practice and thoughtful animal house design.
(ii) Attacks on any pests that may have gained entry by physical and chemical means.

FIG. 33. Fly (*Tabanus*—horse fly) female bloodsuckers

(i) *Barriers to pests.* Disease control by erecting barriers to pests is an important duty of the animal technician. With our knowledge of the pest as an agent of disease-spread it is necessary to provide conditions that actively discourage the entry of pests, and their livelihood, should they gain entry.

Some of the following factors encourage infestation by pests, and disease-spread. The reversal of these conditions should therefore create some barriers to disease-causing pests.

(*a*) Animal rooms with badly fitting doors.

(*b*) The presence of services, like gas, electricity, and water entering by way of poorly sealed pipeways.

(*c*) Badly planned drainage areas permitting the entry of pests from the outside.

(*d*) Air-circulating apparatus lacking effective filters.

(*e*) Animal stocks introduced into the animal house from suspect sources and therefore possible carriers of pests.

(*f*) Bedding and nesting brought into the animal house without first being sterilized.

(*g*) Food and water introduced to animal areas without first being sterilized.

(*h*) Poor personal hygiene on the part of the technician who may carry pests on his body or clothing or on some equipment brought from the outside.

If strict attention is paid to the points just mentioned, then the incidence of disease-causing pests should be drastically reduced. If pests do gain entry to the animal areas, poor nutrition, overcrowding, poor ventilation, and bad-handling will encourage the increase of their numbers.

A final solution to the pest infestation problem is culling, followed by appropriate sterilization of all surfaces and equipment, and the eventual introduction of new stocks of animals.

(ii) *Attacks on pests.* If pests are known to be present in the animal quarters, as they are most likely to be, at least to some degree, then they can be attacked by some means or another. This attack is only necessary if the pests are interfering with the animal's health, or the needs of the animal custodian. The type of attack can be physical or chemical.

Physical means of controlling pests may involve techniques referred to under the title of sterilization (p. 89). These include the following:

(*a*) Gamma radiation.

(*b*) Ultra-violet radiation.

(*c*) Hot air.

(*d*) Steam.

(*e*) Flaming.

Chemical methods of controlling pests may involve the use of toxic substances. These chemical agents of attack include the following:

(*a*) Formaldehyde (methanol)

(*b*) Ethylene oxide gas.

(*c*) Phenolic compounds in solution.

Once the pest has become resident upon or within the host animal then the method of chemical attack comes within the realm of medication. This field of study is outside that of the animal technician and falls within the scope of work for the animal nurse and the veterinary surgeon. For this reason, and because the proprietory names of medicines quickly become dated, only brief reference will be made here to the medication of animals.

Medicines or drugs are administered to animals by various routes:

External
{ oral (by way of mouth or nose).
  rectal (by way of anus).
  topical (by way of the skin).

By injection
(i.e. parenteral)
{ subcutaneous (beneath the skin).
  intramuscular (into the muscles).
  intraperitoneal (into the abdominal cavity).
  intravenous (into a vein).

A more detailed study of the administration of medications and drugs is presented later (p. 242).

Disease or disorder	Preventative action
Virus disorders	Vaccination to give immunity
Bacterial disorders	Sulphonamides—limited use (against streptococci)
	Antibiotics:
	Penicillin G (against streptococci and staphylococci); substitutes for penicillin G may be needed as many bacteria develop a resistance
	Streptomycin by injection (not gut-absorbed)
	Neomycin by injection
Fungal disorders	Griseofulvin—taken orally against ringworm
Protozoan disorders	Sulphonamides are sometimes given in the food against coccidiosis
Helminth disorders:	Antihelminth drugs:
Roundworms (nematodes)	Piperazine compounds for dogs and cats (e.g. Citrazine and Coopane)
Tapeworms (cestodes)	Bunamidine compounds for dogs and cats (e.g. Scoloban)
Roundworms, hookworms, pinworms, whipworms and tapeworms	Wide spectrum antihelminthics: Mebendazole tablets for cats and dogs
Mites—ear canker, mange in dogs and cats	Benzyl benzoate cream; gamma Benzene hexachloride (BHC) in ointment, spray, dusting powder, ear drops. Care should be
Ticks Lice Fleas	taken if used on the coat, because of licking

3.2.1. ANIMAL CARE PERSONNEL AND ANIMAL HEALTH

The degree to which personnel must conform to strict hygiene measures depends upon the purpose for which the animals are being kept. The animals may be kept for dissection, demonstration, or experimental purposes. In the case of experimental animals it is very important that they enter the experimental procedure with as little infection or infestation as possible. The presence of pests can influence the outcome of an experiment.

All animal care personnel should be cautious in case they are carriers of disease. The disease hazard can come from a technician's clothes or from his own body. A sore throat or a septic thumb are both sources of bacteria that could be a danger to animal stocks.

In animal areas such as the specific pathogen-free units personnel will be expected to shower and put on fresh clean clothes when reporting for work. Such persons should not keep domestic pets for fear of carrying potential pathogens to the laboratory animals from home.

In the larger units, personnel will undergo regular health screening in order to ensure no health problems that the technician may have become problems for the animals, and vice-versa.

In the context of personnel hygiene it is important to remember that eating and drinking in animal areas should be strictly forbidden. Similarly smoking should never be permitted in the vicinity of animals.

3.2.2. FOOD PESTS

Any plant or animal that causes food to deteriorate and become useless is to be regarded as a pest.

Molds are plant pests that grow on foodstuffs in warm, damp conditions. Insects are common animal food pests, as well as mice and rats. Below there are some notes on the more common arthropod pests.

Class	Order	Examples	Notes
Insecta	Orthoptera	*Cockroaches*	
		Periplaneta americana	*American cockroach*—large, dark brown (200-day life-cycle)
		Blatta orientalis	*Oriental cockroach*—large, dark brown with wingless female (300-day life-cycle)
		Blatella germanica	*German cockroach*—small, yellowish with quick life-cycle of 50 days
			These animals are dependent upon high temperatures and high humidity and in northern areas live indoors only. They eat food and contaminate it with their feces. They produce a foul odor. For details of laboratory culture techniques see. p. 211
	Coleoptera	*Beetles and Weevils*	
		Sitophilus granarius (USA)	*Granary weevil*
		Sitophilus oryza (USA)	*Rice weevil*
		Calandra granaria (UK)	Both these insects have larvae (grubs) that burrow and feed within the grain kernels of the
		Calandra oryzae (UK)	following: wheat, oats, barley, maize, rice.
			The weevils are "beetle like" in that they have hard wing covers and elbowed antennae. They are recognizable by the penetrating beak-like structure on the head.
		Dermestes (ardarius)	*Flesh-eating beetles* or larder beetle. The larval or maggot stage of the life history penetrate and feed on dried animal products.
	Lepidoptera	*Moths and Butterflies*	
		Ephestia elutella	
		Sitotroga cerealella	*Grain moth*
			These moths have larval or caterpillar stages that feed on the germ area of grains such as wheat, ground-nut, and linseed.

Class	Order	Examples	Notes
		Ephestia kuhniella	*Flour moth* The larval stage of this moth feeds on flour leaving behind a silken thread network, as do the previous type. The contaminated meal will have a recognizable sour odor.
Arachnida	Acarina	*Mites* *Tyroglyphus farinae*	*Flour mite* The larval stage of this eight-legged adult mite bores its way into grain leaving behind a powdery dust. The flour is contaminated by excreta and caste-off skins. The higher the temperature and humidity the better it is for the mites

3.2.3. CONTROL AND ERADICATION OF FOOD PESTS

The pests that invade animal quarters and spread disease are to some extent more easily controlled or eradicated, because sterilization processes that would ruin foodstuffs can be employed with cages and their accessories. The use of the steam autoclave cannot easily be used to sterilize diet without moisture penetrating the food. Nor can high temperatures be employed without ruining the food.

The most economic proposition is to store food correctly in cool, covered containers and to dispose of any materials showing signs of contamination. It could be cheaper to dispose of stock rather than to attempt decontamination. For this reason, holding smaller stocks of food is more advisable rather than running the risk of losing larger quantities.

Some action may be taken in suspect storage areas. The use of certain proprietary chemical agents is very helpful.

Treatment of storage areas	Notes
Dusting powders containing gamma—BHC	Useful against insects
Fumigation with ethylene oxide	Destroys insects
Insecticide sprays	Reduce insect populations

In older, badly designed buildings these infestations can be very difficult to eradicate and the attempts to do so are particularly disruptive to the workday.

3.3. Cleaning and sterilizing

Perhaps the most important, even though least interesting, aspect of animal technology is cleaning and sterilizing. Without these two operations disease would reduce the animal stocks to an expensive and useless population that would have to be destroyed and carefully disposed of.

In order to make sure we all understand what we mean by the terms used here, it is as well to make some kind of definitions.

Cleaning will be used to refer to those physical operations such as sweeping, dusting, or hosing-down that remove offending or offensive material from contact with the animals and their immediate surroundings.

Disinfecting will mean the application of a chemical or physical agent to prevent or inhibit the growth of a named disease-causing organism. Such a disinfectant will often have an *antiseptic* action—that is, it will stop wounds becoming *septic* or overgrown with bacteria and fungi. An antiseptic can be thought of as a disinfectant that is safe to be applied to the living tissues.

Sterilizing can be thought of as applying some chemical or physical agent in order to kill all living organisms. Sterilization will destroy all organisms, *pathogenic* (disease-causing) and *non-pathogenic*. If confusion persists concerning the use of words like *germicide, bacteriostat, antimicrobial* turn to the Glossary at the end.

3.3.1. CLEANING

Cleaning is a time-consuming operation and must be a regular and well-organized daily routine. If a routine is not established there is always the possibility that something may be omitted. That one dirty item could become the source of a later disease outbreak.

If a priority order of cleaning jobs is to be constructed then clearly those items that come into contact with the animals are high on the list for early attention, such as the cages, the bedding, the food, and the water apparatus. An animal room that shines and is odor-free is not necessarily hygienic. Attention to detail is important.

Let us run through the cleaning jobs using the animals as the focal-point and move outwards into the animal room. Much of what is mentioned here is also included in the section on "animal care routines".

Bedding and other direct contact materials such as wood-wool or sawdust will become contaminated with urine, feces, water, and food debris. This must be cleared out of the cages at least once a week, but more frequently depending upon the animal numbers. Cleaning out bedding is not dependent upon its being smelly or dirty. Infected bedding can look clean. Some bedding can be removed by vacuum machines which save the technicians having to handle the bedding. Used bedding is best burnt.

Litter and indirect contact materials such as peatmoss or sawdust will absorb urine and water spillage. It can be cleaned out of cages and litter trays by means of scrapers. It is collected in bags or waste bins. The frequency with which litter is changed will depend upon cage numbers. It may mean in excess of three times per week for some species.

The vacuum cleaner with suitable "animal screen", to prevent sucking up animals, is also a valuable aid to cleaning up litter materials. Labor-saving devices like the installed vacuum cleaners are clearly superior to the scraping-out task. The debris sucked up is passed into sealed containers outside the animal room. These containers are removed and incinerated when full.

The use of vacuum cleaners for these cage-cleaning jobs reduces the amount of apparatus that needs sterilizing after cleaning out.

Cages will need to be cleaned out regularly. A popular routine is to clean out on Mondays, Wednesdays, and Fridays. From time to time cages need to be changed and replaced by sterilized cages and fittings. The change can be made monthly or more frequently if the animals produce large amounts of excreta.

When cages are replaced by clean ones it is a good idea to have a system of labelling in order to indicate the date for the next change. If animals are kept on metal-grid floors these can be changed every day especially if they become clogged with sticky feces as in the case of ferrets and sometimes rabbits. Cage cleaning may be carried out by means of a vacuum machine to remove bedding and litter as just described. The metal parts can be washed down with a safe disinfectant and rinsed with freshly drawn water. Some cages may need cleaning with wire brushes or scrapers to remove encrusted matter. Hosing-down cages in a tiled wash-bay area is another way to clean out. High pressure steam hoses are useful metal cage cleaners. Care must be taken not to disperse the cage dirt.

Remember these cleaning methods do not render the cages sterile and thereby free of disease-causing organisms. The cleaning operation has only reduced the risk of disease by removing fecal matter and so forth that may harbor microorganisms.

Cage washers can be used to sanitize not only cages but also a large variety of accessories such as water bottles, drinking tubes, and feeding hoppers. The machines are often custom-built and automatically operated.

These washing machines do not sterilize the materials put into them. Sterilization can only be achieved by one of the methods described later.

The variety of automatic washing machines can be seen by inspecting the commercial literature.

Briefly, cage washers fall into two categories: the "tunnel-type" and the "cabinet-type". In both cases sprays of high pressure steam or hot water are blasted from many angles at the items to be cleaned. In the tunnel-type the materials are generally taken through the machine on a conveyor belt and the final stage involves a scorching hot-air blast, 100°C (312°F), to dry the equipment.

Cage-racks, walls, and floors in the vicinity of the animal cages need to be cleaned by the most convenient yet efficient method, when the cages have been removed for cleaning. Washing down using a low toxicity disinfectant will remove dust and dirt trapped by wall-fitted structures. Wet cloth cleaning (treated with 1% Tego solution) has the advantage of reducing the atmosphere dust created whilst cleaning.

The floors can be swept to remove sawdust or any loose debris; care being taken not to create dust clouds. Mopping down the floor can be useful to spread disinfecting agents into corners and edges of the animal room. The mop must be clean. Hosing down the floors is a quick way to dislodge and remove debris, but if care is not taken offensive material will merely be distributed over a wider area by the splashing. It also raises the animal room humidity and may even wet animals if they are left in the room during this process.

3.3.2. DISINFECTING

After the physical operations of cleaning, outlined previously, now comes the application of an agent, usually a chemical, that kills the growing forms but not necessarily the resistant spores of the disease-producing microorganisms.

Disinfection is the process of destroying infectious agents, usually by means of a chemical.

The result of disinfection is a sanitary condition where hopefully 99.9% of the growing bacteria have been killed. A disinfectant producing this condition is known in the USA as *sanitizer*.

Fig. 34(a)

Fig. 34. Cage washers. (a) Tunnel type (photographs supplied by John Burge (Equipment) Ltd.)

Fig. 34(b). Cabinet type

There are a great number of disinfecting agents to choose from. Before selecting our disinfectant, we should bear in mind the following:

(*a*) *Nature of the material to be treated.* Some disinfectants must not be allowed skin contact.

(*b*) *Types of microorganisms.* Chemical agents are not all equally effective against bacteria, fungi, viruses, and others.

(*c*) *Environmental conditions.* Such factors as temperature, pH, time, concentration, and the presence of organic material may all have a bearing on the rate and efficiency of microbe destruction.

In making our choice of an antimicrobial agent, it is worth making a note of the characteristics of an ideal disinfectant:

(*a*) *Antimicrobial activity.* The chemical should have the capacity to kill a broad spectrum of microbes at low concentrations.

(*b*) *Solubility.* It should be soluble in water or other solvents to the extent necessary for effective use.

(*c*) *Stability.* The disinfectant should not change its germicidal action upon standing.

(*d*) *Non-toxicity to man and other animals.* Ideally the chemical agent should be lethal to microorganisms, but harmless to man and other animals.

(*e*) *Non-combination with extraneous organic material.* Many disinfectants have an affinity for proteins and other organic material. If the disinfectant is "taken up" by other organic material little will be available for action against the microbial cells.

(*f*) *Toxicity to microorganisms at body temperatures or room temperatures.*

(*g*) *Penetrating ability.* The agent should ideally be able to penetrate beneath surfaces.

(*h*) *Non-corroding and non-staining.* The antimicrobial agent should not rust or otherwise disfigure metals nor stain or damage fabrics.

(*i*) *Deodorizing ability.* Ideally the disinfectant should be either odorless or have a pleasant smell and deodorize.

(*j*) *Detergent capacity.* A disinfectant that cleans whilst it disinfects achieves two objectives.

(*k*) *Availability.* The chemical should be available at a reasonable price in the dilutions required for use.

3.3.3. DISINFECTANTS

Some antimicrobial agents available for use in animal housework are outlined below.

(*a*) *Phenol and phenolic compounds.* These compounds probably act on bacteria by denaturing the proteins and damaging the cell membranes.

The compounds are *bactericidal* at low concentrations becoming *bacteriostatic* at higher concentrations.

The phenols have their activity reduced by alkaline conditions, the presence of soap and lowered temperatures. Organic matter like feces reduces the activity also. Crystalline phenol is made to a 2–5% solution in water and used as a general disinfectant.

Cresols are several times more germicidal than phenol, but are less soluble in water. They are used as emulsions in liquid soaps and alkalies.

Lysol is a coal tar derivative with a high toxicity and is irritant to the skin. It is used as a 3–5% solution for general work. Mixed cresols make up 50% of the commercial lysols dissolved in 22% linseed oil soap.

(*b*) *Synthetic detergents.* Chemicals, such as soaps, are called "*wetting agents*". By "wetting" is meant the ability of a chemical to lower the surface tension of water thereby enabling the water to penetrate more easily into minute spaces in fabrics and the like. This "wetting" action is characteristic of detergents which are valuable in the mechanical removal of oils and dirt, with the accompanying microorganisms from the skin, clothing, or other materials. "Tego" is an extremely successful disinfectant worthy of further study as it closely approaches "the ideal" (see Bibliography p. 301). There is a large range of synthetic detergents with germicidal agents incorporated within them.

Quaternary ammonium compounds are detergent-like substances with a high bactericidal power. They are non-toxic at lower concentrations but at higher concentrations they are liable to cause skin irritations as well as other side effects.

This group of disinfecting agents are available under a whole range of trade names—for example, Cetrimide (UK); Cetavlon (USA); Ceepryn (USA); Zephiran (USA).

The quaternaries (quats) are suitable disinfectants against most bacteria (but not so effective against spores), and they have a measure of activity against fungal and protozoan organisms. Viruses appear to be more resistant to the effects of this disinfectant. The success of these compounds as sanitizers and disinfectants may be related to the following properties:

Combined detergent and germicidal action.
Relatively low toxicity. Colorless. Non-staining.
Solubility. Non-corrosiveness. Able to be stored without losing their strength.

The "quats" are thought to kill pathogenic cells by a combination of methods—denaturing the proteins, rupturing the membranes, inhibiting the cell enzymes.

(*c*) *Halogens.* This group of disinfectants include the iodine and chlorine preparations.

The iodine germicides are active against all bacteria, and their spores by combining (denaturing) with the cell proteins. They are also fungicidal.

In use iodine crystals are dissolved in alcohol and potassium iodide solution to produce a tincture of iodine.

Chlorine as a gas or as a compound is widely used as a disinfectant. The gas is difficult to handle but is used to purify municipal water supplies. The compounds of chlorine are more easily handled and under the correct conditions of use are effective disinfectants.

Hypochlorites, such as chlorinated lime (calcium hypochlorite), are popular compounds used widely for domestic and industrial purposes. They are available as powders or liquids. These compounds provide available free chlorine gas which is the active disinfecting agent when it dissolves in water.

These compounds are unable to be stored for any length of time under animal house conditions.

Chloramines are chlorine compounds used as disinfectants. They have advantages over the hypochlorites because they are more stable in terms of their chlorine release.

A disadvantage of these chlorine compounds is that the efficiency of chlorine as a disinfectant is reduced by the presence of organic material such as feces. It is therefore important to clean cages and accessories very well before applying chlorine-type disinfectants for the reasons just mentioned.

(*d*) *Alcohols.* The alcohols are antimicrobial because they denature the cell proteins of bacteria. They dissolve fats and, therefore, damage cell membranes which have a fat content. Alcohol is also a dehydrating agent.

Ethyl alcohol in concentrations between 50 and 70% is effective against non-spore-forming bacterial cells. It is useful in reducing skin bacteria and for disinfecting glass.

Methyl alcohol is less bactericidal than ethyl alcohol and is very poisonous. It is not used much for disinfection.

3.3.4. STERILIZING

This is the process of killing *all* living cells. This is an all or nothing process. The sterilization process produces a sterile condition. If any living organisms remain that are capable of reproducing (spores) then the condition cannot be considered sterile.

Sterilization techniques may employ chemicals, wet heat, dry heat, or radiations. The technique chosen will depend upon the items to be sterilized. Plastic items, for instance, will melt in a hot-air oven. In the table below some sterilization methods are summarized.

Sterilizing agent	Possible uses
(*a*) Phenolic solutions	Cleaning hardware and floors
(*b*) Formaldehyde vapor (methanol)	Sterilizing heat sensitive items, fumigation of rooms
(*c*) Ethylene oxide gas	Sterilization of items which are heat and moisture sensitive
(*d*) Red heat (flaming)	Sterilizing metal items, incineration of bedding and carcasses
(*e*) Hot-air oven	Sterilizing glassware
(*f*) Steaming under pressure (autoclaving)	Sterilizing glassware, metal items such as cage materials
(*g*) Filtration of liquids or gases	Water filtration through porcelain; air filtration through glass wool
(*h*) Radiations—gamma, ultraviolet, infra-red	Sterilization of diets, and equipment

(*a*) *Phenolic solutions* are more effective in higher concentrations and when used under specific conditions like higher temperatures, and in the absence of organic materials such as feces. It is not always possible to provide the correct conditions for the use of phenols in a sterilizing capacity. It is difficult to ensure that the animal house floors and walls are warm and free from organic debris. High concentrations of phenol solutions are dangerous to man and animals.

(*b*) *Formaldehyde vapor* is generated by heating a commercial formalin solution. This is a 37–40% solution of formaldehyde gas in water. The gas is only stable at high concentrations and at elevated temperatures; at room temperatures it changes into a colorless solid called paraformaldehyde.

Sterilizing an enclosed area by vaporizing a formalin solution (fumigation) is best done when the room temperature is about 22°C (71.6°F) and the relative humidity is about 60–80%. Spray the walls with water to give a film of liquid in which the gas can dissolve.

The use of formaldehyde for sterilizing purposes has the disadvantages of being corrosive and limited in its ability to penetrate surfaces. It is very toxic. Masks and protective clothing should be worn. Some "neutralization" may be necessary afterwards to remove the fumigant.

(*c*) *Ethylene oxide* is a liquid at temperatures below 10.8°C (51.4°F), above this temperature it rapidly vaporizes to become a highly flammable gas. Because of its fire-hazard properties it is sold as a mixture with carbon dioxide which makes it non-flammable.

The gas is a powerful sterilizing agent used for treating materials that are sensitive to heat or moisture. The gas is effective against the usually resistant bacterial spores because it has a powerful penetrating ability. It can penetrate large packages and even some plastics.

The use of ethylene oxide requires strict attention to safety details because the gas is explosive in the presence of air. There are commercially available chambers, modified autoclaves, with controls that can maintain the desired concentration of ethylene oxide and the proper temperature and humidity.

The disadvantages of this gas as a sterilizing agent could be thought to be its explosive potential and the length of time required for the gas to be effective. Exposure time to the gas within the chamber can be as long as 12–24 hours. This depends upon concentrations of the gas.

(*d*) *Red-heat* refers to the burning of materials thereby destroying microorganisms. This technique is employed in order to sterilize needles and wires in the laboratory. Metal cage items can be flame sterilized whilst organic matter is burnt off.

Incinerators used for the disposal of carcasses, bedding, or any other infected material will destroy microorganisms.

(*e*) *Hot-air ovens* are useful because the heat supplied will penetrate to all parts of the item being sterilized, unlike the steam or gas methods where the surfaces contacted are only effectively treated. The dry heat sterilization method is useful for materials that are liable to damage or denature in moist heat; like food and bedding. Hot-air ovens may be supplied with air-circulating fans or without. The still air oven has a disadvantage in that temperatures throughout the oven may not be uniform. There can be pockets of cooler air.

In practice the dry heat oven can be used to sterilize glassware at temperatures of 160°C (320°F) over a period of 2 hours. High vacuum ovens can reach temperatures in excess of 250°C (482°F) and exposure times for sterilization can be reduced to 15–20 minutes.

(*f*) *Steaming at atmospheric pressure* with a maximum temperature of 100°C (212°F) for about 30 minutes is not adequate to kill all pathogens, especially the spore formers.

Steam under pressure (*autoclaving*) is a more effective method of destroying microorganisms. Compare the effects of autoclaving with dry heat sterilization:

Bacterial spores of botulism, food poisoning (Clostridium botulinum), are killed in 4–20 minutes of moist heat at 120°C (248°F) whereas 2 hours is required at the same temperature in dry heat.

The autoclave is designed to use steam under regulated pressure. This apparatus enables temperatures to rise above those obtainable with boiling water, 100°C (212°F), at normal pressures. It is this high temperature that kills the bacterial spores, not the pressure.

The autoclave is generally operated at approximately 15 lb per square inch (on the gauge) at 121°C (249.8°F). At this temperature for a period of 30 minutes all micro-organisms, and their normally resistant spores, are killed.

Some materials cannot be sterilized under these conditions. The time of operation to achieve sterility depends upon the type of material being sterilized, the type of container, and the quantities.

Some materials cannot be sterilized in the autoclave because the substances may be destroyed by the high temperatures (plastics) or the substances may contain fats or oils which are immiscible with water and, therefore, the steam will not reach the microorganisms. For these materials another sterilization method must be used.

Operating the autoclave

   (i) Put a quantity of water into the bottom of the pressurized metal chamber. The amount of water depends upon the size of the apparatus, this is usually indicated in the instruction manual.

  (ii) Place the articles to be sterilized into the chamber. If bottles with stoppers are to be treated ensure that the tops are not screwed down tightly otherwise they may be difficult to remove afterwards.

 (iii) Screw down the wing nuts on the lid in opposing pairs in order to ensure an even closure.

 (iv) Open the steam cock and commence the heating.

  (v) Close any steam run-off valves.

 (vi) Allow steam to issue from the steam cock for at least 5 minutes (or the time indicated on the instruction sheet).

 (vii) Observe the safety valve and adjust the power supply when steam begins to be blown. Watch the pressure gauge.

(viii) Allow the autoclave to operate for the required time and then turn off the heating.

 (ix) Before opening up the autoclave wait for the pressure gauge to read zero and then open up the steam cock.

  (x) Wait for 5 minutes or more and then open the lid.

It may be necessary to "dry off" the items within the autoclave before they are removed. This can be done by opening the steam run-off valve first before the steam cock is opened. This operation will take off water and steam which is at a high pressure within the autoclave. It is best run off by a tube leading into a sink or drain.

The above operating procedures may not be appropriate if the autoclave is an automatic model. In this case all phases of the sterilizing process are regulated automatically and commence at the push on a start button.

Testing the autoclave

It can happen that the autoclave is not working efficiently. The temperature and pressure gauges may not be giving a truthful reading. Some autoclaves may have pen recorder charts linked to temperature sensitive devices within the autoclave. These may be giving misleading records. Checks on the sterilization process may be carried out as follows:

Brown's tubes are small glass tubes containing an indicator liquid. They can be purchased for checking the efficiency of a hot-air oven, steamer, or autoclave. They need to be stored in cool conditions to prevent deterioration.

In use the liquid in the tubes will turn from red to green if the correct temperature/time ratio is employed.

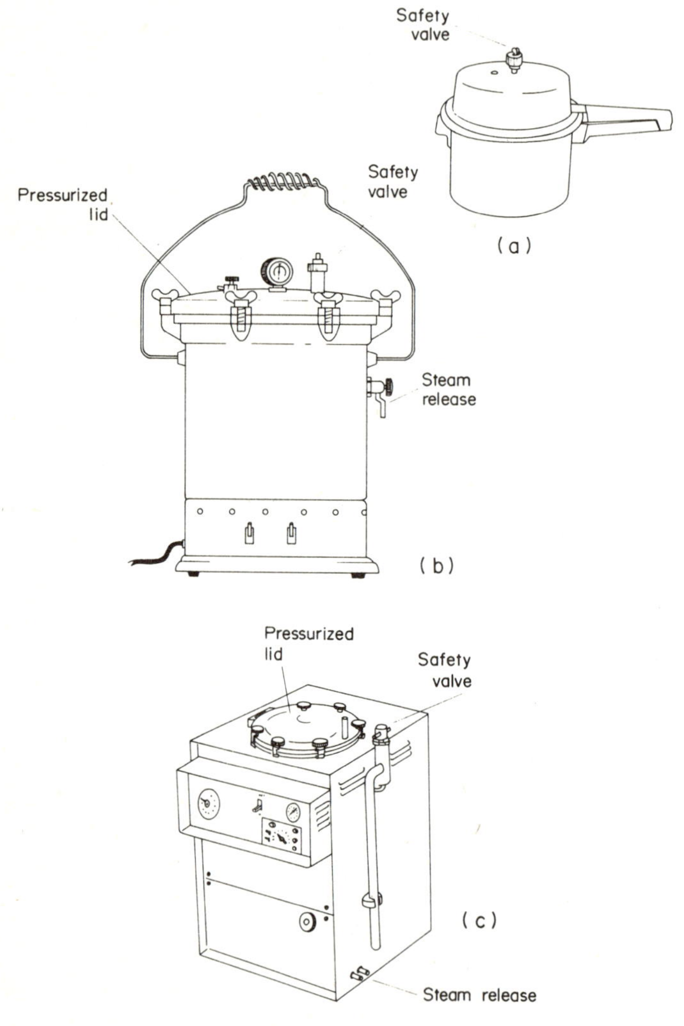

Fig. 35(*a–c*)

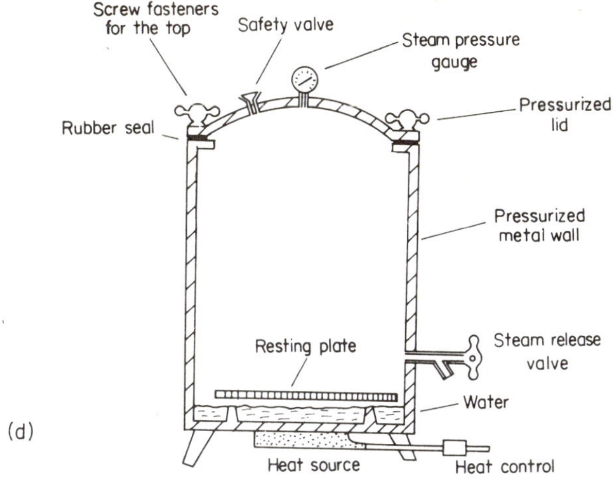

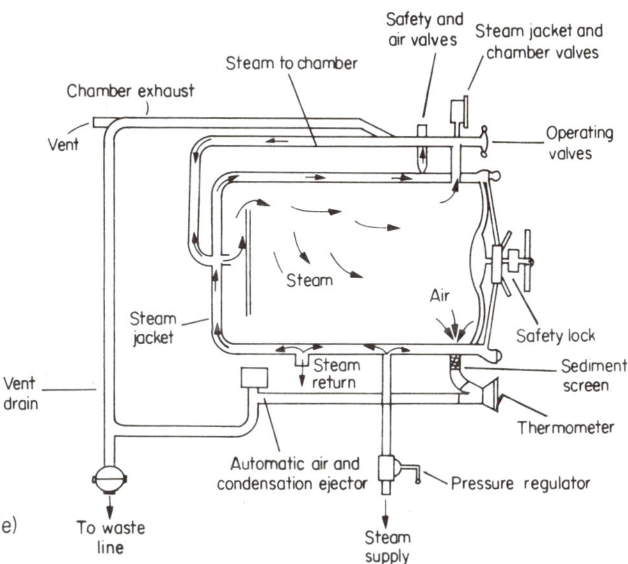

FIG. 35. Autoclaves (drawn by Ross MacKay, Oxford). (*a*) Pressure cooker. (*b*) Portable autoclave. (*c*) Floor-standing autoclave. (*d*) Section of the portable autoclave (adapted from *Practical Microbiology*, by Sirockin & Cullimore, 1969, McGraw-Hill)

Spore strips are small pieces of absorbent material soaked in a standardized suspension of *Bacillus stearothermophilus* spores. These spores are killed if subjected to 121°C (249.8°F) for about 12 minutes. The strips are put into the sterilization procedure, afterwards the strips are removed from their wrappers and put into a suitable culture medium and incubated for about a week to see if any growths occur. This technique is fair enough if time is available.

Autoclave tape can be purchased that is adhesive, and is attached to objects being autoclaved. A color change will indicate the progress of sterilization.

(*g*) *Filtration methods* of sterilization may be applied to liquids or gases. *Liquids* that contain heat sensitive substances such as vitamins or amino-acids may be treated by a filtration method that takes out any living matter. The filters used can be made of cellulose, asbestos, or porcelain. The size of the pores in these filters must be small enough to intercept all living particles larger than viruses.

Gases, such as the air in an aseptic work area, are usually filtered through wads of compressed glass wool, cotton wool, or special paper. All but the smallest particles will be intercepted.

(*h*) *Radiation techniques* of sterilization are commonly used for commercial animal diets. Radiations can be described by the following two terms: *ionizing radiations*—e.g. gamma radiation, beta radiation; *non-ionizing radiations*—e.g. ultraviolet radiation.

The ionizing radiations are highly lethal and their use requires very strict safety precautions. Gamma rays have a high energy penetration power and are obtainable from radioactive-isotopes such as cobalt 60. Ionizing radiations can be used to destroy microorganisms in animal foodstuffs but in doing so some chemical changes may occur in the food thereby changing its taste and odor.

Gamma rays are not only useful for the sterilization of animal diets but are also used to treat packaged petri-dishes, syringes, and catheters.

The non-ionizing radiations are lethal to microorganisms but have little effect on bacterial spores. *Ultraviolet radiations* are poor in their power of penetration. They are present in natural daylight, but concentrated rays may be obtained from special mercury vapor lamps.

Because of the poor penetrative powers of ultraviolet rays their use is generally restricted to treating the air, surface areas, and some liquids.

Ultraviolet rays can cause damage to the retina and so the eyes of humans and animals should be protected by shields.

Infra-red rays are lethal to microorganisms including bacterial spores because of the high temperatures produced. This sterilization procedure has limited use because of the high temperatures, and so generally metal objects only are exposed to these temperatures of 180°C (356°F) and more.

Pasteurization is *not* a sterilization technique but may for convenience be included here. This application of mild heat to substances that would spoil at higher temperatures reduces the microorganisms surviving within the pasteurized material. Many foods would spoil if sterilized at high temperatures or exposed to ionizing radiations (p. 137).

Pelleted diets can be exposed to temperatures of about 80°C (176°F) for a few minutes and the bacterial population will be considerably reduced. Bacterial spores are not destroyed, however.

3.4. Animal diseases

Laboratory animals are all prone to some disease or disorder no matter how efficient the hygiene measures in a conventional animal house.

This section presents an outline of some of the ailments found in a sample of the more common laboratory mammals and aquaria fish. The reason for confining our attention to these two groups is because their ailments are more easily recognized and treated. For each named mammal the information is presented in summary tabular form for ease of reference.

Some of the diseases are described as *acute*, others as *chronic*. These are technical words with the following definitions:

> *An acute disease* is one that has a rapid onset, a short course and presents definite symptoms.
> *A chronic disease* is one that persists over a longer period of time.

Students are recommended to back up this theoretical consideration of the signs of ill health with a practical examination of sick animals whenever the opportunity presents itself.

As a generalization, an experienced observer looking for indicators of ill health will note the following:

  (i) Change in feeding, drinking and sleeping routines
  (ii) Vomiting and diarrhoea
  (iii) Bleeding and discharges
  (iv) Breathing problems
  (v) Movement difficulties, unusual posture and behavior
  (vi) Irritation and scratching
  (vii) Abnormal temperature

A post-mortem will help to confirm any suspicions one has about cause of death due to disease.

3.4.1. AILMENTS OF MAMMALS

It is not really difficult to recognize that an animal is ill. It is a problem for the average person to say what is the disease or disorder that is causing the ill health. With the aid of the following charts the task should be a little easier, but there is no substitute for experience. By including this information it is not intended that the technician becomes a substitute for the veterinary surgeon, rather that an academic approach to disease is encouraged. The selected few diseases and disorders that are tabulated in the following pages can only be diagnosed accurately if backed up by laboratory studies of tissue or fluid samples. In this context the reader is referred to the valuable Medical Research Council publication noted in the Bibliography (*Microbiological Examination of Laboratory Animals for Purposes of Accreditation*).

The laboratory mouse

Salmonellosis

Cause of the ailment (bacterial infection)	*Salmonella enteritidis* *Salmonella typhimurium*
Acute disease symptoms	Rapid loss of condition Some diarrhoea and/or runny eyes
Chronic disease symptoms	Loss of condition over a period of time Some recovery in time but remaining infective
Source of infection	Possibly food and bedding that has been contaminated by animal carriers of *Salmonella*
Entry point of the infection	By way of the mouth or perhaps the eye membranes
Diagnostic aid (specimen—feces)	The *Salmonella* bacteria from the feces may be cultured and identified on media plates
Post-mortem indicators	*Acute disease* Generalized inflammation of the intestinal tract and some other viscera *Chronic disease* Inflammation of the intestines with fluid accumulations Liver enlarged with mottled patches of dead tissue Enlarged spleen
Control of the ailment	Monitor and quarantine incoming stock Destroy all infected animals and suspect carriers Sterilize all rooms and equipment where infective matter has been located

---

Mouse pox (Ectromelia)

Cause of the ailment (viral infection)	A pox virus
Acute disease symptoms	Few visible symptoms Death within hours
Chronic disease symptoms	Scaling of the tail and one or both hind limbs Poor coat condition Scab formation and some skin loss from toes Gangrene may affect the toes
Source of infection	By contact with infective animals or contaminated materials
Entry point of the infection	The virus enters through breaks in the skin of the tail or feet
Diagnostic aid (specimens—serum and live animals)	Injections of infective material from diseased animals into "disease-free" animals bring about noticeable changes in the liver and spleen, as seen in post-mortem Antibody and agglutinin studies may be carried out
Post-mortem indicators	*Acute disease* Some patches of dead tissue on the liver and spleen seen as white areas of fibres Some fluid accumulations around the intestines *Chronic disease* More extensive tissue death (necrosis) in the liver and spleen with the consequent patchy appearance
Control of the ailment	If an outbreak should occur the whole colony may have to be destroyed and all rooms and equipment sterilized; vaccination can be carried out

Tyzzers disease (liver degeneration)

Cause of the ailment (bacterial infection)	Probably associated with a bacterium *Bacillus piliformis*
Symptoms of the ailment	Some diarrhoea, loss of weight and condition Death follows the onset fairly quickly
Source of the infection	By contact with infected material, more pronounced in over-crowded conditions
Entry point of the infection	Probably by way of the mouth
Diagnostic aid (specimen—liver)	Liver sections show fibrous areas of dead tissue (lesions)
Post-mortem indicators	Intestinal inflammation Yellow and white patches on the enlarged liver
Control of the ailment	Destroy animals showing symptoms and all contacts Sterilize all contaminated materials especially any suspect bedding

Ringworm (skin and hair fungus)

Cause of the ailment (fungal infection)	*Trichophyton mentagrophytes* A ringworm fungus
Symptoms of the ailment	Hair loss and skin scaling
Source of the infection	By contact with infected materials contaminated with fungal spores
Entry point of the infection	The fungus establishes itself on the skin
Diagnostic aid (specimen—skin and hair)	Microscopic examination of the skin and hair taken from diseased areas Cultures of the fungus may be grown and identified
Control of the ailment	Infected stock may be destroyed as treatment can be lengthy and expensive Sterilize all areas of contact with the diseased animals
Hazards	This ringworm disease is infective to man

Pin-worms

Cause of the ailment (endoparasites)	*Aspicularis tetraptera* *Syphacia obvelata*
Symptoms of the ailment	No obvious symptoms and the animal shows little sign of ill health
Source of infestation	Fecal contaminated materials containing round-worm eggs
Entry point of infestation	By way of the mouth
Diagnostic aid (specimen—gut contents)	The microscopic examination of adult worms found in the large intestine
Control of the infestation	Avoid using bedding and food contaminated by feces of wild mice which may contain the worm eggs Difficult to eliminate the worms, possible by the long-term use of drugs added to the drinking water and food

The diseases just mentioned are not necessarily the most common. They are recorded to show the range of parasites that can affect the well-being of mice. The same applies to the examples that follow for rats and guinea pigs, etc. Other afflictions of mice that must not be overlooked because they do occur in some colonies are as follows: infantile diarrhoea; hepatitis; lymphocytic chorionmeningitis; and the sendai virus (pneumonitis).

The laboratory rat

	Salmonellosis
Cause of the ailment (bacterial infection)	*Salmonella enteritidis* *Salmonella typhimurium*
Acute disease symptoms	Rapid weight loss, poor ruffled coat and anemia Some diarrhoea Scabs around the nose
Chronic disease symptoms (more common—less severe)	No obvious signs of disease Animals may look "out of condition" Post-mortem indicators of disease
Source of infection	May be food or bedding contaminated by wild rat feces A carrier rat in the colony with infective feces
Entry point of infection	By way of the mouth or perhaps the eye membranes
Diagnostic aid (specimen—gut contents)	The Salmonella bacteria from the colon can be cultured on media plates and identified
Post-mortem indicators	*Acute disease* Liver pale and mottled with yellowish points of dead tissue Enlarged spleen with patches of dead tissue Intestinal swellings and ulcerations *Chronic disease* Enlarged and ulcerated caecum, some swellings in small intestine cysts on swollen glands in the mesenteries Spleen may be enlarged Liver appears normal
Control of the ailment	Monitor the colony for carriers Destroy all infected stock Sterilize all food, bedding, and equipment, avoid contaminations by carriers or wild rodents Up-grade all hygiene measures

Salmonellosis is described as being found less frequently in rat colonies in the UK. More often found among mice.

	Tapeworm cysts
Cause of the ailment (endoparasites)	Cysts of the cat tapeworm *Taenia crassicollis* (The rat is an intermediary host)
Symptoms of the ailment	Poor condition generally with some abdominal swelling Post-mortem indicators of the infestation
Source of infestation	Bedding materials contaminated by cat feces containing tapeworm cysts
Entry point of infestation	By way of the mouth
Diagnostic aid (specimen—gut contents)	Whitish cysts of the tapeworm seen in the liver
Control of the infestation	Avoid the use of possibly contaminated materials (i.e. sawdust) Sterilize all bedding

Scabies (Mange)

Cause of the ailment (ectoparasites)	Infestation by a species of mite, such as *Notoedres*
Symptoms of the ailment	Grey warty scabs around the root of the tail and near the ears and nose
Source of infestation	Other infested rats
Region of infestation	The mites infest the skin and hair
Diagnostic aid (specimen—whole animal)	Microscopic examination of hair and skin scrapings
Control of the ailment	Monitor the animals regularly for signs of scabies
	Apply chemical treatments such as gammaxane preparations to the skin (benzyl benzoate emulsions)
Hazards	The infestation can be transferred to man so care must be exercised when handling infested animals or cage contents

Middle-ear disease and labyrinthitis ("circling")

Cause of the ailment (bacterial infection)	Probably *Streptobacillus moniliformis* in association a "pleuropneumonia-like organism" (*Mycoplasma pulmonis*)
Symptoms of the ailment	Middle-ear and inner-ear disease symptoms associate to produce an animal with poor balance and tilted head
	Lack of coordination and circling movements when walking
Source of infection	Adults give the disease to young rats
Entry point of the infection	By way of the mouth or respiratory system
Diagnostic aid (specimen—nasopharyngeal swab inner ear culture)	The *Streptobacillus* bacteria from the middle ear pus can be cultured on media plates and identified
Post-mortem indicators	Inflammatory changes with pus present in the middle and inner ear
Control of the ailment	Monitor the colony for symptoms of the disease
	Destroy infected animals
	Maintain high standards of hygiene

Bronchopneumonia

Cause of the ailment	Probably a virus in association with a "pleuropneumonia-like organism"
	Often together with *Streptobacillus moniliformis* (perhaps a secondary infection)
Symptoms of the ailment	Poor coat condition, sneezing and coughing
	Chattering of the jaws
Source of infection	Diseased animals infect others by way of droplets from the respiratory system
Entry point of infection	By way of the respiratory system
Diagnostic aid (specimen—nasopharyngeal swab)	The examination of histological sections of the infected lung
	Inoculation tests on trial animals using infective material
Post-mortem indicators	Lung sections show microscopic grey translucent nodules
Control of the ailment	Difficult as all animals are likely to be carriers
	A "disease free" colony must be created under S.P.F. conditions

Rats are also susceptible to pin-worm infestations (as for mice), chronic respiratory disease ("chattering"), and leptospirosis.

The guinea-pig

Salmonellosis

Cause of the ailment (bacterial infection)	*Salmonella typhimurium* *Salmonella enteritidis*
Symptoms of the ailment	*Acute disease* No special symptoms other than maybe diarrhoea before one or two animals in a colony die *Chronic disease* Poor condition of the coat, wasting, coughing, and sneezing
Source of infection	Contaminated food and bedding
Entry point of the infection	By way of the mouth
Diagnostic aid (specimen—gut contents)	The bacteria may be isolated, diseased tissue and cultured for identification
Post-mortem indicators	*Acute cases* Slight enlargement of the spleen with some gut inflammation *Chronic cases* Enlarged spleen Areas of grey-white dead tissue on the spleen and liver
Control of the ailment	Destroy all diseased animals and their contacts Sterilize all rooms, cages, and equipment Improve all hygiene measures

---

Pseudo-tuberculosis

Cause of the ailment (bacterial infection)	*Pasteurella pseudo-tuberculosis*
Symptoms of the ailment	*Acute disease* Rapid death with no obvious symptoms *Chronic disease* Progressive wasting over a period of a month with death following although recovery is possible
Source of the infection	Food or bedding contaminated by feces containing the bacterium, birds have been suggested as carriers
Entry point of the infection	By way of the mouth
Diagnostic aid (specimen—nasopharyngeal swab)	The bacteria can be cultured and recognized
Post-mortem indicators	Intestinal glands (lymph tissue) show swellings of destroyed tissue which may contain pus The liver and spleen may be affected
Control of the ailment	Destroy diseased animals, sterilize all contacted materials and ensure no contaminated materials enter the animal area

---

Streptococcal pneumonia

Cause of the ailment (bacterial infection)	*Streptococcus pyogenes*
Symptoms of the ailment	Rapid loss of condition with coughing and sneezing
Source of the infection	Not clear, but the infection occurs most commonly in over-crowded, damp, and cold animal houses
Entry point of the infection	Most probably by way of the respiratory system
Diagnostic aid (specimen—nasopharyngeal swab)	The bacteria in the infected tissue can be cultured on a suitable media and identified
Post-mortem indicators	Inflammation of the gut, lung, and heart membranes Yellow pus-like accumulations in infected areas
Control of the ailment	Destroy the infected animals and sterilize all equipment Ensure that all bedding is sterilized

---

Some authors report *Bordetella* pneumonia as occurring amongst guinea-pigs kept in the same room as rabbits.

<table>
<tr><td colspan="2" align="center">Mucormycosis (mouldy-hay disease)</td></tr>
<tr><td>*Cause of the ailment*
    (fungus infection)</td><td>*Absidia ramosa*—a mucor fungus</td></tr>
<tr><td>*Symptoms of the ailment*</td><td>No obvious symptoms although some animals may die</td></tr>
<tr><td>*Source of the infection*</td><td>Mouldy hay</td></tr>
<tr><td>*Entry point of the infection*</td><td>By way of the mouth</td></tr>
<tr><td>*Diagnostic aid*
    (specimen—gut contents)</td><td>Microscopic examination of infected tissue will reveal the presence of recognizable fungal growths</td></tr>
<tr><td>*Post-mortem indicators*</td><td>Swollen glands within the gut membranes. (These swellings could be confused with those found in pseudo-tuberculosis)
Gut inflammation may be present in the cases of deaths due to diarrhoea complications</td></tr>
<tr><td>*Control of the ailment*</td><td>Avoid buying in hay that may be mouldy</td></tr>
<tr><td>*Hazards*</td><td>This ailment may be confused with more serious *Salmonellosis* or Pseudo-tuberculosis</td></tr>
</table>

<table>
<tr><td colspan="2" align="center">Scurvy</td></tr>
<tr><td>*Cause of the ailment*
    (dietary disorder)</td><td>Diet lacking fresh green vegetables or suitable source of vitamin C</td></tr>
<tr><td>*Symptoms of the ailment*</td><td>Loss of weight, unsteady walking, stiff joints, bleeding gums, and loose teeth</td></tr>
<tr><td>*Post-mortem indicators*</td><td>Emaciated carcass
Enlarged cartilage joints on the ribs</td></tr>
<tr><td>*Control of the ailment*</td><td>Feed the animals with green vegetables or suitable hay diet or vitamin C diet</td></tr>
</table>

Guinea-pigs kept in the same room as rats and mice may pick up a bacterial disorder called *cervical adenitis*. The swelling of neck glands in the animal with this disease do not necessarily impair the well-being of the animal.

The rabbit

<table>
<tr><td colspan="2" align="center">Intestinal coccidiosis</td></tr>
<tr><td>*Cause of the ailment*
    (parasitic infestation
       by protozoans)</td><td>*Eimeria perforans*
Eimeria magna</td></tr>
<tr><td>*Symptoms of the ailment*</td><td>Loss of condition
Diarrhoea frequently followed by death</td></tr>
<tr><td>*Source of the parasite*</td><td>Other rabbits or fecal contaminated material</td></tr>
<tr><td>*Entry point of the parasite*</td><td>By way of the mouth</td></tr>
<tr><td>*Diagnostic aid*
    (specimen—gut contents and feces)</td><td>Microscopic examination of fecal matter will reveal the egg cysts of this protozoan</td></tr>
<tr><td>*Post-mortem indicators*</td><td>Inflamed gut linings with localized areas of infection
White spots in the liver which examined microscopically reveal coccidia</td></tr>
<tr><td>*Control of the parasite*</td><td>Cautious use of hay or greens that may be contaminated by the droppings of wild rabbits
Regular removal of feces from the cage so as to prevent ingestion of infective matter
Regular cleansing operations
Correct weaning period and the separation of litters from the doe to reduce the chance of the young being infected by the mother
The use of drinking-water chemicals to kill the parasitic stages</td></tr>
</table>

Liver Coccidiosis

Cause of the ailment
  (parasitic infestation
    by protozoans)
	Eimeria steidoe

Symptoms of the ailment	Loss of condition with eventual death in severe cases
Source of the parasite	Other rabbits or fecal-contaminated material
Entry point of the parasite	By way of the mouth
Diagnostic aid (specimen—gut contents and feces)	Revealed in microscopic examinations of material taken from the gall bladder or liver The egg cysts are also found in the feces
Post-mortem indicators	In more advanced cases the white spots or lesions on the liver are seen to contain egg-cysts
Control of the parasite	As for the intestinal coccidiosis

Myxomatosis

Cause of the ailment (viral infection)	*Myxomatosis virus*
Symptoms of the ailment	Swollen eyelids which are often sealed together by a fluid exuding from them The nose, lips, anus, and genitals may be swollen Death usually follows the infection
Source of the infection	Infected wild rabbits with the flea or mosquito as carriers of the virus
Entry point of the virus	By way of the skin as the insect feeds on the blood of the rabbit
Diagnostic aid (specimen—serum)	The symptoms presented make diagnosis easier Laboratory tests can be made on the causative agent found in the fluids that exude from the eyes or skin tumors
Post-mortem indicators	Some swelling of the spleen and glands in the area of the genitalia Small tumors beneath the skin around the eyes and ears
Control of the infection	Barriers to insect vectors are essential High standards of hygiene amongst personnel Destroy and burn all infected animals, disinfect cages Isolate all contacts for at least one week A vaccine giving 6-month immunity is available

Snuffles—(nasal catarrh)

Cause of the ailment (bacterial infection)	*Pasteurella lepiseptica* *Brucella bronchiseptica*
Symptoms of the ailment	*Mild cases* merely exhibit repetitive sneezing but no loss of condition *The chronic cases* show nasal discharge and a more pronounced snuffling noise as the animal breaths Loss of condition Pneumonia may arise and the animal dies
Source of the infection	Other rabbits which are infected
Entry point of the infection	Most probably by way of the respiratory system
Diagnostic aid (specimen—nasopharyngeal swab)	The causal agent may be cultured on a suitable media and identified
Post-mortem indicators	Inflamed nasal passages Lung tissue damaged
Control of the infection	Isolation of infected rabbits Destroy chronic infected stock The use of chemical agents in the drinking water Injection methods of disease control do not always show success Stricter hygiene measures

Ear-canker (mange)

Cause of the ailment (ectoparasitic infestation)	Infestation by a species of mite *Chorioptes cuniculi, Psoroptes communis*
Symptoms of the ailment	Brown scabs within the ears causing itching The rabbit scratches and shakes the head
Source of the infestation	Other infested rabbits
Region of infestation	The skin of the ears
Diagnostic aid (specimen—whole animal)	The mites, the young stages and eggs can be examined under the microscope and identified They can be extracted from the scab areas
Control of the infestation	Regular examination of the ears Washing the affected ears with a suitable chemical agent such as the gammexane preparations Disinfecting all bedding and cage woodwork

Rabbit syphilis

Cause of the ailment (bacterial infection)	*Treponema cuniculi*—a Spirochaete resembling that causing syphilis in man. Harmless to man and other animals
Symptoms of the ailment	Ulceration of the skin in the region of the genitals Some ulceration also around the eyes, nose, lips, and anus
Source of the infection	Other infected rabbits
Entry point of the infection	The genital tracts as a result of sexual intercourse
Diagnostic aid (specimen—serum)	Bacteriological examinations of wet films show the presence of the causal agent
Control of the infection	Disinfect all caging by dry heat Examine all incoming animals Isolate and treat any infected animals with antibiotics

Rabbit pox

Cause of the ailment	*Pox virus*
Symptoms of the ailment	Inflammation and discharge from the eyes, nose, and mouth Rashes on the skin of the ears and genitalia High temperature and prostration
Source of the infection	Other infected rabbits
Diagnostic aid (specimen—serum)	Laboratory study of infective materials taken from the rabbit
Post-mortem indicators	Patchy grey liver which is swollen Damaged tissue in the testes, ovary, and uterus Glands enlarged in the neck
Control of the infection	Not so common in Europe, more so in the USA Some immunity given by vaccination

General enteric disorders that kill young rabbits in large numbers are described under such titles as: enteritis, mucoid enteritis; mucoid typhlitis; intestinal impaction; or simply diarrhoea.

The Cat

	Septicaemia
Cause of the ailment (bacterial infection)	A variety of bacteria such as *Streptococcus haemolyticus* *Haemophilus influenza*
Symptoms of the ailment	In young kittens no obvious symptoms before death in a few weeks Loss of appetite, raised temperature
Source of the infection	The mother may pass on the bacterium to the young Humans may pass on the bacterium
Entry point of the infection	The bacteria may enter through the ears as the kitten is being cleansed by the mother Entry through broken skin or by way of the nasal passage
Diagnostic aid (specimen—nasopharyngeal swab)	Few diagnostic features before death in acute cases Post-mortem examination aids in diagnosis
Post-mortem indicators	The liver, spleen, and intestinal membranes may show inflamation Microscopic examination of smears from infected tissue will reveal the identity of the causative agent
Control of the infection	Antibiotics can be administered to known infected lactating mothers

Cats are from time to time infested by fleas (*Ctenocephalides* species), lice, and ticks (*Ixodes* species). The fleas and lice are important as intermediate hosts for a tapeworm, *Dipylidium* species. Mites causing a variety of mange conditions are also found as ectoparasites on the cat; i.e. *Otodectes*—ear mite, *Cheyletiella*—fur mite. Cats are also subject to endoparasites such as the *Toxocara* species roundworm and the tapeworms *Dipylidium* and *Taenia*.

An enteric disorder caused by a virus (*feline infectious enteritis*) which can be typified by diarrhoea and wasting is not an uncommon cause of rapid death in cats.

Feline influenza, cat distemper, or simply cat respiratory virus disease are a group of names that cover a cat "sneezing disease" that can be controlled by precise attention to hygiene procedures.

The Dog

	Distemper and Hard Pad
Cause of the ailment (viral infection)	A virus usually affecting animals between 3 and 12 months of age
Symptoms of the ailment (*Distemper*)	Raised temperature with loss of appetite and an eye and nose discharge Ulceration of the tongue and mouth with breathing difficulties Diarrhoea and vomiting may be seen
(*Hard Pad*)	Thick, swollen foot pads, in addition to the above
Source of the infection	From infected or carrier animals by way of discharge from the eyes or nose
Entry point of the infection	By way of or mouth or respiratory system
Diagnostic aid (specimen serum)	The symptoms are clearly diagnostic of this ailment
Control of the infection	Vaccination doses given early, with boosters as advised

Dogs are also liable to infestation by fleas, lice, and ticks of the same species already mentioned for cats. The mange mites described for cats are from time to time found on dogs.

Endoparasitic "worms" such as *Toxocara* species are found in dogs. In excessive numbers they cause ill health in puppies. The larvae of these worms can invade the bodies of humans, particularly children, who may pick up the infestation from dog feces.

In some areas the dog heart-worm (*Dirofilaria immitis*) is a cause for concern as can be the hookworm (*Ancylostoma caninum*).

Illnesses affecting the liver in dogs can be Canine Virus Hepatitis or a bacterial disease caused by *Leptospira* species. Leptospirosis is infective to man and is known as Weil's disease. This disease is spread by way of rats' urine. Another type of leptospirosis in dogs is less serious to man but is a cause of kidney inflammation (nephritis) in the dog.

The Monkey

	B Virus disease
Cause of the ailment (viral infection)	Herpes virus (simian)
Symptoms of the ailment	Ulcerations of the lips and mouth cavity Some discharge from the nose and eyes
Source of the infection	Other infected monkeys
Entry point of the infection	Contaminated water or food, scratching or biting
Diagnostic aid	Typical ulceration features
Post-mortem indicators	Microscopic examination of liver, kidneys, and central nervous system shows damage to the tissue
Control of the infection	Monitor all imported monkeys for symptoms Reduce any handling of animals by special caging Staff should take appropriate action to avoid contact with possibly infected monkeys
Hazards	The ailment is communicable to man and is generally fatal

In addition to the above-mentioned serious disease there are the following:

Shigella—a bacterial dysentery.

Tuberculosis—a bacterial disease involving the lungs (*Mycobacterium hominis*).

Salmonellosis—a bacterial infection (*S. typhimurium*).

3.4.2. *AILMENTS OF FISH*

The diseases and disorders outlined here, it could be said, are rarely encountered if good aquarium management is in evidence. It is more than likely, however, that one or more of these ailments will be seen by any person keeping fish over a period of time.

When applying any of the treatments, handle the fish for short periods only and be sure to avoid drastic temperature changes.

	(i) White fungus—*Saprolegnia*
Cause of the ailment	A water-borne fungus that enters damaged skin
Symptoms of the ailment	Recognizable fluffy, white film of fungus growing on the fins, mouth, skin, and gills. The fins become frayed. Death may follow due to suffocation as the fungus attacks the gills
Control of the ailment	Avoid overcrowding and the fouling of water. Avoid rapid changes in the temperature of the water
Treatment	Trim away frayed areas of fins with a sharp scalpel pressing on to a hard surface. Do not use scissors because they can tear. Keep the gill region towel-damp. Clean the wound with a dilute antiseptic such as potassium permanganate To treat the body fungus immerse the fish in 3% sodium chloride solution for short periods (15 minutes) each day for 4 or 5 days Check that no distress is experienced by the animal. Reduce the salt strength if necessary. Disinfect all contact equipment, and the hands. Destroy severely diseased fish

---

	(ii) White Spot—*Ichthyophthirius*
Cause of the ailment	A ciliate protozoan parasite
Symptoms of the ailment	White spots over the skin, fins, and gills. Sluggish behavior and loss of appetite
Treatment	Immerse the fish in methylene blue solution made by adding four drops of $2\frac{1}{2}\%$ methylene blue to about 4 liters of water

---

TABLE (*contd*)

(iii) Swimbladder Disorders	
Cause of the ailment	It is unclear what causes this condition, but several suggestions have been put forward, including: digestive upsets due to excess dried foods and exposure to low temperatures.
Symptoms of the ailment	The fish swims to the water surface with difficulty and sinks downwards. It has an inability to swim upright. Floats on the top or bottom lying on its side.
Treatment	Isolate the fish in a shallow bath of salty water which is gently warmed and aerated. After 2 or 3 days of this treatment offer the fish live food. If there is no success the fish may as well be destroyed.

There are numerous other ailments that fish exhibit, such as tumors, which may or may not respond to surgical removal or chemical treatment. The above ailments are fairly common.

3.5. Signs of ill health

The selection of ailments previously outlined would hopefully be rare in animal houses. There are, however, more minor ailments that may well be found and need to be corrected. Ailments such as the sore hocks of rabbits or guinea-pigs living on grid floors, or the overgrown claws and teeth of rabbits and others. These are not conditions of ill health, but they could lead on to more serious disorders.

A technician would be well advised to make the following weekly routine observations looking for indicators of ill health.

(i) *Observe the animal in the cages* and answer the following questions.
   (*a*) What is the animal's posture? Is it alert, or is it curled up and inattentive?
   (*b*) How is the animal's breathing? Is it breathing with difficulty? Is it panting, or breathing slowly?
   (*c*) How is the animal's mobility? Does it move slowly and with difficulty? Does it sway as it walks? Does it have a stiff limb action, or drag any limb?
   (*d*) How is the animal behaving towards others? Is the animal aggressive or timid when approached by others?
   (*e*) What is the animal's reaction to food and drink? Does it eat, but with a labored chewing or swallowing action?

(ii) *Observe the animal when handled*

   (*a*) What is the reaction of the animal to being handled? Does the animal struggle and show fear? Do some parts of the body seem sensitive to touch?
   (*b*) What is the condition of the coat? Is the coat in good condition or dull and patchy? Is there any hair loss? Are there any damaged areas?
   (*c*) Is the body swollen or wasting? Are there any lumps? Are the legs swollen or sore? Are the claws normal?
   (*d*) Do the eyes seem bright and free from discharge?
   (*e*) Is the tail complete or broken and scab covered?

(*f*) Are the ears damaged? Does the animal shake its head? Do the ears seem to be blocked?

(*g*) Does the nose seem dry or wet? Is there any damage to the nose? Is there any nasal discharge?

(*h*) Are the lips swollen? Are the teeth normal?

(*i*) Is the urino-genital opening swollen and red? Is there any sign of bleeding?

(*j*) Is there any bleeding or discharge at the anus? Is the anus wet or dry? What is the nature of the feces?

An experienced observer will be able to tell the difference between temporary conditions and positive signs of ill health that may be more long lasting.

3.5.1. NOTIFIABLE DISEASES

There are several diseases affecting man that must be reported to the medical authorities in the UK. Such notifiable diseases include the following:

Smallpox	Tuberculosis	Pneumonia
Typhoid	Dysentery	Food poisoning
Typhus	Scarlet fever	Leprosy
Plague	Whooping cough	Encephalitis
Anthrax	Measles	Malaria

Diseases of animals that are notifiable include those listed below:

Sheep pox	Anthrax	Rabies
Sheep scab	Brucellosis	Fowl pest
Swine fever	Cattle plague	Glanders (in horses)
Tuberculosis (in cattle)	Foot and mouth disease	Pleuro-pneumonia (in cattle)

The legislation in this connection may be followed up by referring to J. L. Thomas in the Bibliography.

3.6. Zoonoses

A disease which can be transmitted from animal to man or from man to animal belongs to a group known as the *zoonoses*.

Zoonoses may be caused by viruses, bacteria, fungi, or protozoa. Non-microbial zoonoses are caused by helminths and arthropods.

Some selected zoonoses are tabulated below:

Disease	Agent	Animal reservoirs	Vectors, mode of spread
Bacteria			
Anthrax	*Bacillus anthracis*	Cattle, horses, sheep, goats, swine, wild animals, and birds	Contaminated wool, hair, hides, air, food, water
Brucellosis	*Brucella* species	Cattle, swine, goats, sheep, horses, mules, fowl, dogs, cats, deer, rabbits	Milk, meat, other contaminated food: contacts

Disease	Agent	Animal reservoirs	Vectors, mode of spread
Glanders	*Pseudomonas mallei*	Horses, mules, donkeys	Contact, inhalation
Leptospirosis	*Leptospira* species	Dogs, cattle, swine, rodents	Contact with infected animals, inhalation of contaminated matter
Plague	*Yersinia pestis*	Rodents	Flea bite
Salmonellosis	*Salmonella* species	Fowl, birds, cattle, swine, rodents, cats, dogs, turtles	Ingestion of contaminated matter
Q fever	*Coxiella burneti* (*Rickettsia*)	Rats, birds, domestic animals	Tick bite, inhalation
Rocky Mountain spotted fever	*Rickettsia rickettsii*	Wild rodents, dogs	Tick bites
Murine typhus fever	*R. typhi*	Rats	Flea bite
Psittacosis and Ornithosis	*Chlamydia* species	Birds of parrot type, fowl, pigeon, finches	Contact, inhalation
Viral			
Rabies	Rabies virus	Dogs and other canines, cats, bats, skunks	Animal bites
Yellow fever	Yellow fever virus	Monkeys, marmosets, lemurs	Mosquito bites
Herpes illness	Herpes virus	Monkeys	Monkey bites
Lymphocytic chorio-meningitis	LCM virus	Mice	Dust inhalation
Protozoan			
Trypanosomiasis	*Trypanosoma* species	Cats, dogs, rodents	Insect bites, tsetse fly
Malaria	*Plasmodium vivax*	Monkeys and humans	Mosquito bites
Fungal			
Ringworm	Several species of pathogenic skin fungi	Cats, dogs and other domestic animals	Contact

With the above diseases (zoonoses) in mind it becomes obvious that great care is necessary in the purchase, maintenance, and handling of laboratory animals.

3.7. Animal body defenses

The animal body is being constantly attacked by forces of the environment that will result in disease, disorder, or death if they are not checked. Animals have "built in" defense mechanisms against these invasions by viruses, bacteria, fungi, protozoans, and the chemical and physical agents of attack. The lines of defense are summarized below and later expanded a little. Many of our laboratory animals are used in research programs that are concerned with how the human body defends itself against disease.

First line of defense: The skin (epidermis); the gut-lining (mucosa).
Second line of defense: Blood clotting.
Third line of defense: Removing "foreign" chemicals, "neutralizing" viruses; reticulo-endothelial system; immunity processes.

The routes of infection (or equivalent) for any animal are across the outer covering of the body (the skin) or through the lining of the guts running through the body. Both these coverings offer a first line of defense.

If the attacking agent actually penetrates these covers of the body and gets into the blood there is a second line of defense—that is, the clotting mechanism.

In the event of the blood not being able to deal with the invaders, there is a series of final defenders of the body's vital organs. Foreign chemicals are converted and removed from the body. Viruses are prevented from reproducing inside the body, and antibodies are produced to inactivate invading foreign agents.

3.7.1. THE SKIN

This is the outer covering of the body and it is the first line of defense against biological, chemical, and physical attack.

The epidermis is the outermost layer of the skin and defends the body in the following manner. It has a horny, dead layer that is keratinized and offers some resistance to most chemicals and sharp objects. Bacteria may lodge themselves on the skin but are frequently shed with the discarded dead cells of the epidermis. Secretions produced by the skin (sebum and sweat) make it less than attractive for the growth of bacteria and fungi. Washing the skin too frequently can interfere with these important natural chemical defenders.

Finally, the skin acts as a barrier to the ultraviolet rays in sunlight by producing skin pigments (melanin).

3.7.2. THE MUCOSA (Internal surface cover)

This is the lining of the free surfaces of the body located internally. It has no hard, horny layer as does the skin. These surfaces or mucous membranes are covered in a protective, sticky mucus. Its function is to trap microorganisms and to lubricate the passageways preventing damage to tissue beneath.

In some cases this mucosa has a ciliated surface. These hair-like cell extensions beat back and forth to keep the mucus slime on the move. Such membranes are found in the windpipe (trachea), and so mucus trapped invaders are driven away from the lungs towards the back of the throat where they are swallowed from time to time.

3.7.3. BLOOD CLOTTING

If the outer coverings are damaged and invaders enter the tissues there are a series of steps taken by the body.

There is a reflex action that brings about a constriction of the skin capillaries (arterioles) so that blood flow to the injured area is reduced. Blood platelets tend to accumulate at the injury site and form a hemostatic plug to prevent further blood loss. A series of chemical actions now take place to produce a clot that plugs up the wound whilst tissue repair takes place.

3.7.4. REMOVING FOREIGN CHEMICALS (Xenometabolism)

When the body takes in foreign substances that are not used to produce energy nor are incorporated into the cell cytoplasm, they need to be disposed of in some manner.

This applies to materials like food additives, drugs, or air pollutants. Formaldehyde and carbolic acid, for instance, are taken across the skin and enter the blood. Once this happens, the foreign material may leave the body unchanged or, more commonly, it is metabolized in the liver. This xenometabolism converts the "foreigner" to a suitable product for excretion in the urine.

3.7.5. "NEUTRALIZING" VIRUSES (Interferon)

When cells are invaded by viruses (and other microorganisms) the challenged cells produce a protein called *interferon*. It does not destroy viruses, it is thought to interact with the nuclei of uninfected cells, so interfering with the ability of viruses to grow in those healthy cells. Normally viruses take over the cell nuclear material and multiply themselves. Interferon prevents this spread of viral infection.

3.7.6. RETICULO-ENDOTHELIAL SYSTEM

This system has two functions—it initiates the process of producing immunity and engages in *phagocytosis* without calling the immunological system into action.

The reticulo-endothelial system is a community of cells scattered around the body, present within or near all tissues. The cells are *phagocytes*, i.e. they can engulf and destroy foreign matter such as bacteria or worn out tissue due to be replaced, such as blood cells.

The phagocytic cells are of two types, fixed or mobile.

> *Fixed phagocytes*: within connective tissue (macrophages) abundant in thymus gland, spleen, lymph nodes, tonsils, Peyer's patches in the intestines, red and yellow bone marrow.
>
> *Mobile phagocytes*: in white blood cells in blood and lymph in connective tissue (mobile macrophages).

This community of cells is in constant daily action disposing of tissue debris and foreign invaders. If an infection is heavy, then the activities of the reticulo-endothelial system becomes noticeable as *inflammation*.

The inflammation response is characterized by the following:

(i) Redness, heat, and swelling of the tissues due to the increased blood supply to the area. These effects are produced by secretory materials such as histamine, serotonin, and heparin.

(ii) Pain is caused by pressure on nerve endings or changes in pH due to accumulation of acid materials.

(iii) Function may be reduced because of the local pain (as in joint inflammation—arthritis) or because of areas of tissue destruction (as in tonsillitis or hepatitis).

The inflammation process is characterized by a big increase in white blood-cell numbers and in some cases by the accumulation of dead tissue called *pus*.

3.7.7. IMMUNITY PROCESSES

The processes mentioned previously are defenses against foreign invasions, but they do not equip the animal with any future defense or immunity. When some foreign substances pass through the reticulo-endothelial system without being destroyed they initiate an immune response from the animal's immunological system. Substances that bring about this response are called *antigens*. Antigens provoke the synthesis of *antibodies* which subsequently destroy the antigen.

There are two kinds of immune systems:

(i) *Circulatory immune system* (humoral immunity) is immunity due to circulating antibodies in the blood plasma, in the gastrointestinal tract secretions, and in tears and nasal mucus. These antibodies are manufactured by cells in the reticulo-endothelial system, i.e. the B cells (bone marrow-derived lymphocytes). These react against some viruses and bacteria.

(ii) *Cell mediated system* (cellular immunity) is immunity due to antibody like molecules and proteins derived from T cells (thymus-derived lymphocytes). This form of immunity is dependent on the presence of the thymus in earlier life.

These cells react directly, as individual cells against certain viruses, bacteria, cancer cells, or foreign tissue transplants.

A young mammal has a short-term supply of antibodies at birth. These are derived from the mother's blood and are transmitted to the fetus before birth. In some mammals the colostrum (rich milk produced early in lactation) contains antibodies. This supply is of value for a limited period until the offspring begin to produce their own immunological requirements. In this context the reader is referred to a useful paper on the breeding and maintenance of *athymic nude* mice. These mice are devoid of the T cell lymphocytes because of the absence of a thymus (p. 301).

3.7.8. IMMUNITY

Immunity can be categorized as of the following types:

Natural Immunity is that type acquired from natural sources.

(i) It is described as *passive* if the animal does not itself take an active part in the production of the antibodies (e.g. temporary immunity given by the mother's antibodies before birth or provided in the colostrum after birth).

(ii) It is described as *active* immunity if the animal is itself actively producing antibodies as a result of an invasion of the tissue by an antigen. This produces a more long-lasting immunity.

Artificial Immunity is that type acquired as a result of the animal being given some assistance in their defense.

(i) It is described as *passive* if an injection of antibodies (immunization) is given. The antibodies are produced by another animal which has experienced the disease and manufactured its own serum antibodies.

This technique has limited value because the effects do not last, as the tissues of the animal being immunized have not "learnt" how to produce antibodies of their own. It is a short-term protection.

(ii) It is described as *active* if the animal itself produces antibodies as a result of an injection (vaccination) of relatively harmless modifications of the disease-causing microorganisms. There are various methods of rendering microbes innocuous whilst retaining their ability to stimulate antibody production.

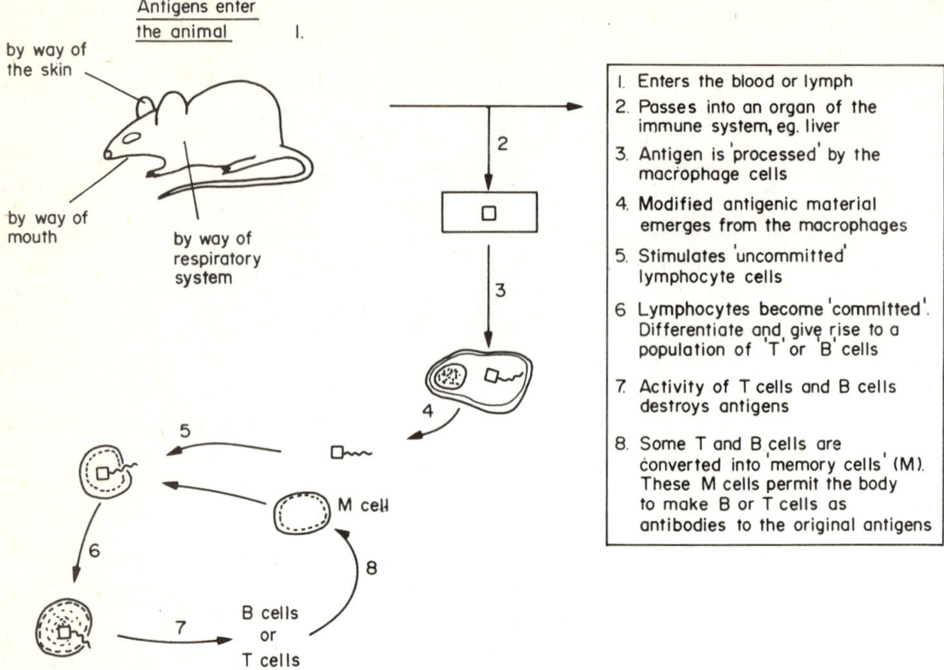

Fig. 36. The Immune Response

This technique is used to treat many of the major diseases of humans, including the following: tetanus, diphtheria, whooping cough, typhoid, paratyphoid, poliomyelitis, yellow fever, tuberculosis, and rabies.

In this area of study the name of Louis Pasteur (1822–95), the French chemist, is familiar. His name is incorporated into that technique of preventing disease transmission, *pasteurization*. This technique is used to kill microorganisms that cause tuberculosis, typhoid, etc., which may be present in untreated milk.

Milk pasteurization involves heating to 65°C (149°F) for 30 minutes and then immediately cooling it to not more than 10°C (50°F). Alternatively, it is heated to 72°C (162°F) for 15 seconds then rapidly cooled to below 10°C (50°F). This does not sterilize the milk, but it kills the disease-causing bacteria. For further information on antibodies and immunity the reader is referred to the Bibliography (p. 301).

3.8. Specific pathogen-free animals

From what has gone before it should be clear that obtaining a "sterile" animal, one free from microbiological parasites, is virtually impossible. It is possible, however, to make some attempt at producing a super-clean animal, but this will only be a reality if the "rules of hygiene" that have been discussed are observed in an almost fanatical manner. The super-clean animals which we use to start off a colony of S.P.F. animals cannot be born to mothers in the normal way otherwise she could pass on any microbiological organisms she possesses. Assuming we obtain a super-clean birth we must now maintain those animals in that exceptionally hygienic condition. We must set up a special animal unit, the "specific pathogen-free" unit.

This introduction to S.P.F. animals will be considered in two parts:

> (*a*) Obtaining an S.P.F. stock.
> (*b*) Maintaining an S.P.F. unit.

Whilst the student reads this portion of the study it is worth bearing in mind the rather special type of technician that needs to be employed in such work and the initial expensive outlay that is required. A breakdown in the whole system could result if the personnel are not of exceptional high quality and if the animal unit is poorly designed.

3.8.1. OBTAINING AN S.P.F. STOCK

An animal that is free of the majority of the commoner disease-causing organisms is in general classed as within the category S.P.F. This is no place to get involved in the debate about definitions, but some authorities prefer other titles for such animals, terms like: "pathogen-free", "disease-free", "clean-animals".

The question is, how do we obtain a "clean" animal from birth? The answer is, by caesarian birth (*hysterectomy*).

Hysterectomy techniques involve removing the uterus from the mother, when she is at "full term", containing the fully formed fetuses. This is carried out in the sterile conditions of an operating theatre where aseptic techniques are practised. This method involves the mother being humanely killed by a method that should not have any adverse effect on the unborn. This would exclude many chemical methods of killing.

Before summarizing this technique it might well be asked on what grounds one can assume the fetuses to be "disease-free" merely because they are extracted from the uterus surgically.

The developing fetuses obtain their nutrients and oxygen from the mother by way of the placenta. No blood flows across this placenta so the mother's microorganisms are isolated from her offspring. The young are therefore reared in a sterile condition within the uterus, because most known pathogens such as viruses, bacteria, worms, and so forth do not pass over the placenta. The structure of the placenta only permits nutrients and gases to diffuse across it. It should be mentioned that *some* disease-causing organisms can pass over the placenta but the majority do not.

Hysterectomy procedures

> (*a*) The animal is humanely killed by a method suitable to the animal, and in such a manner that the fetuses are not affected. Cervical dislocation can be applied to smaller animals.
>
> (*b*) The animal's abdomen is cleaned with a sterilizing fluid before the animal is laid out beneath surgical drapes in preparation for the operation.
>
> (*c*) The abdominal incision can be made by a scalpel, or by a hot needle (*diathermic needle*). The diathermic technique is useful because of the hygienic cut that it produces and the reduced blood loss.

(*d*) The pregnant (gravid) uterus is removed and placed into a vessel containing a sterilizing fluid at blood temperature.

(*e*) The container is passed into a sterile isolator or passed through a sterilizing barrier into an S.P.F. unit.

(*f*) Once in the S.P.F. unit the young animals are removed from the uterus and stimulated to breathing.

(*g*) The youngsters can be hand-reared, or more simply fostered to S.P.F. lactating mothers whose pregnancy has been timed to coincide with their reception of the foster group.

(For details of fostering and hand-rearing, see the Glossary p. 306).

The procedures just outlined are carried out by an experienced group of personnel whose attention to detail would hopefully eliminate any introduction of unwanted microorganisms to the initially sterile fetuses. These procedures provide the unit with the nucleus of the S.P.F. breeding stock.

3.8.2. MAINTAINING AN SPF UNIT

As a simplification, the techniques of maintaining an S.P.F. unit may be thought of as exaggerations of the routines used in a conventional animal house. The buildings will similarly be exaggerations of the conventional, with specialized additions.

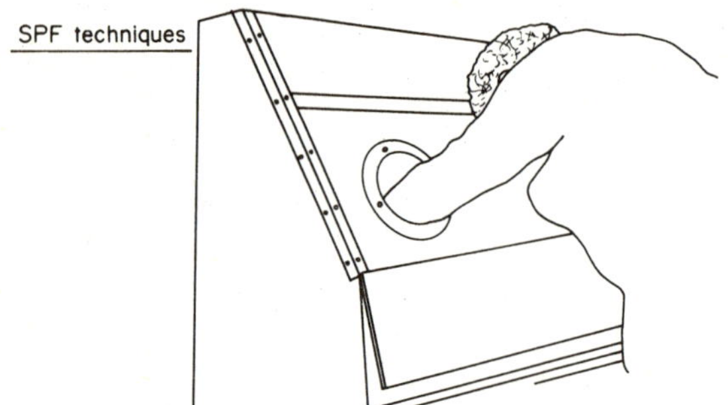

(a) Hysterectomy is carried out in a sterile glove box equipped with UV lamp.

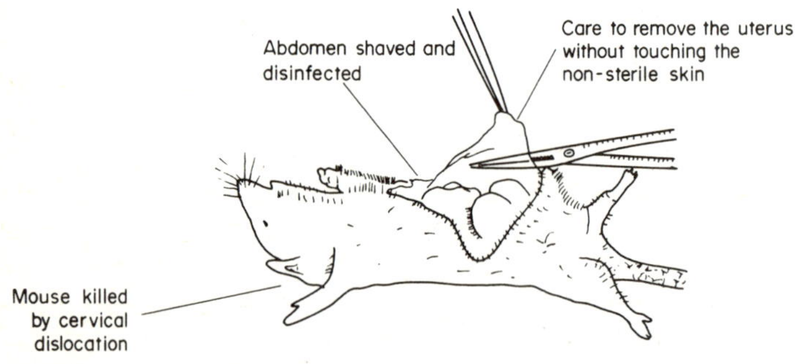

(b) The peritoneum is opened over about 3 cm (1 inch)

Fig. 37(*a–b*)

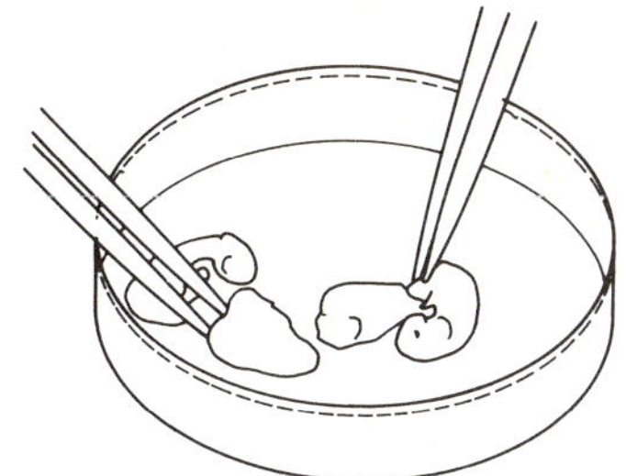

(c)   Remove the pups from the uterus rapidly in a
sterile petri-dish on a warming tray. Initiate
respiratory movements.

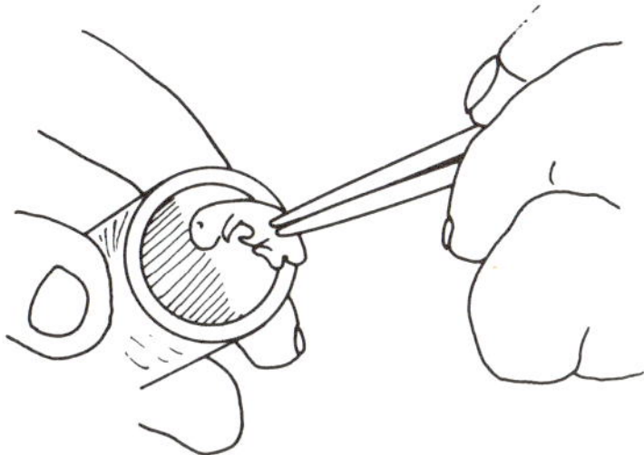

(d)   The young are removed from the sterile glove
box and passed through an entry lock (sterile)
to join the foster mother in an isolator.

FIG. 37. Hysterectomy procedures (drawn from photographs with permission from the editor
of the *Journal, Institute of Animal Technicians*, vol. 23, no. 4, December 1972)

The personnel working in an S.P.F. set-up are required to exercise special skills and
therefore must possess a more than average understanding of the importance of the
science of hygiene.

Entry to the unit by personnel will require that they first make a complete change of
clothing after showering. The building will be constructed in such a manner that
animal care persons must pass through a shower area in order to reach the animals.
Sterile gowns, masks, and other outer garments will be worn by personnel as they
enter the animal areas. These garments are passed into the S.P.F. unit directly from
sterilizing apparatus such as the autoclave.

Entry to the unit by personnel who are known to be carriers of microorganisms
must be prohibited. A technician with a cold or a septic throat would endanger the
S.P.F. conditions. Similarly, technicians who have had close contact with non-S.P.F.
animals will require to be excluded from the unit for a given period of time.

The buildings that go to make up the S.P.F. unit require to be specially planned and
constructed if microorganisms are to be excluded (reference, J. S. Patherson *IAT
Manual*).

Barriers to pests must be erected in the form of entry and exit locks through which
all items must pass in order to enter or leave the animal unit. There should be either
no windows or expertly sealed ones. There should be specially designed drains and
service facilities including air-circulatory devices with filters.

The barrier-type animal unit may be thought of in terms of the diagram in
Fig. 38.0b. A diagrammatic ground plan of a barrier-type animal house is shown in

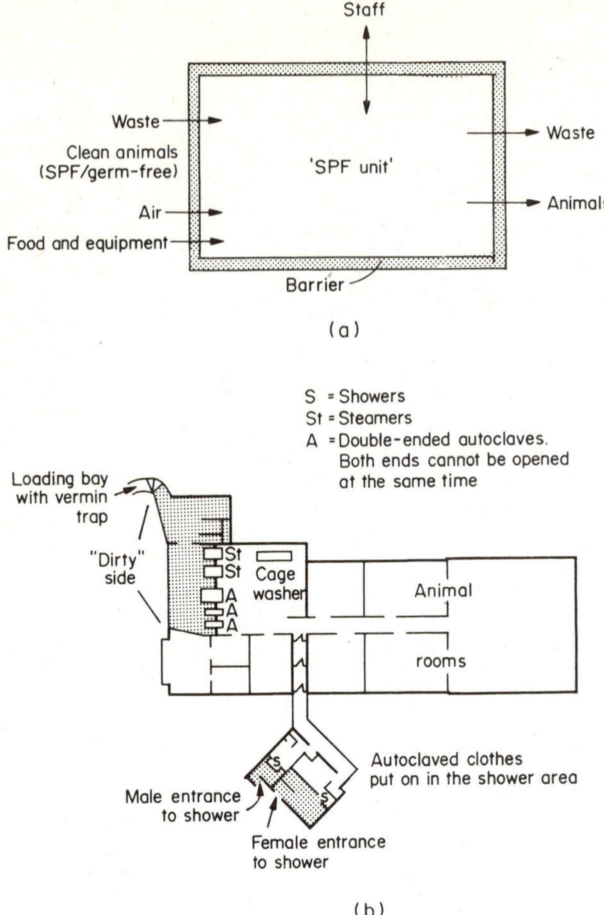

FIG. 38. Barrier type units. (*a*) The barrier type building used for disease-free animal breeding (adapted with permission from Bleby, *UFAW Handbook*, Churchill-Livingstone, 4th edn., 1972). (*b*) Plan of an SPF Breeding Unit (adapted from *LAC Collected Papers*, 1959)

Fig. 38.0a. It is not sufficient to merely have the facilities and the staff operating under a strict hygienic regime; the animals must also be subject to monitoring or quality control in order to ensure that they are as claimed, free of specific pathogens.

3.8.3. COLONY MAINTENANCE

A continuous process of preventing invasion across the barriers by microbial agents. For this reason work that is unconnected with the animals, such as repairing the services, like the ventilation or lighting, is best carried out on the outside of the barrier. This will necessitate a building design where the services are accessible on the outside, such as in corridors. In this connection it is worth mentioning that communications across the barrier should be made available because personnel confined behind the barrier for long periods of time may require conversation with the outside world. A telephone inside the barrier may be inadvertently overlooked at the design stage. A breakdown in the barrier can occur at the ports of entry outlined below. These portals of invasion have been stressed before in connection with general animal-house hygiene.

Air entering across the barrier for the purposes of ventilation needs to be filtered to remove air-borne particles. A filter removing all particles over 5 μm (micrometers) will substantially eradicate most infectious agents that are likely to be air-carried.

The number of air changes taking place per hour will depend upon the density of the animals in the room. The change will usually be 6–15 changes per hour. The air pressure in the animal rooms will be kept at a slightly higher pressure than the corridors so that any air movements through cracks in doors and so forth will be out of the room, not inwards. The air temperature and humidity should be controlled at a level suitable for the species being housed. This can be done by way of the ventilation system. For instance, laboratory rodents will require a temperature range of 19–21°C (67–70°F) and a relative humidity of 50%. These figures can be monitored by means of a max/min thermometer and a hygrometer.

Water supply from the mains is unlikely to be sterile but it should have a low bacterial count and therefore be of minimal concern to the maintenance of a disease-free colony. There are expensive methods of ensuring that the water is near sterile and this becomes necessary if the water is obtained from storage tanks, or is obtained from suspect sources. Some units add dilute hydrochloric acid or a proprietary additive to the drinking water. The water supply, as the air supply should be monitored regularly for its microbial content.

Drains and other water outlets should be sealed to prevent the entry of crawling pests.

Food in any form should be protected from contamination—in-transit, or in store from the invasion by rodents, microorganisms, or arthropods. Before entry to the S.P.F. unit all food should be sterilized in one of the following ways.

(*a*) High vacuum autoclaving.
(*b*) Ethylene-oxide gas chamber sterilization.
(*c*) Radiation by gamma-rays.

The last method has the advantage that diet is not affected adversely in terms of its nutrient value and it can be treated whilst packaged.

Bedding must be sterilized before entry to the S.P.F. unit. This can be done by high vacuum or by ethylene-oxide fumigation. This latter method can be a time-consuming operation.

Equipment brought into the S.P.F. unit must be sterilized by any method suitable to the materials concerned. The sterilization procedures available have been described in some detail earlier (p. 89). A feature of some animal houses is the *dunk-tank*, through which equipment is passed in order to be disinfected upon entry to the unit.

Exiting from the animal units needs to be regulated as well as the entering. It is a comparatively simple task to exit equipment or bedding. It merely has to be placed into a sterilizing autoclave or gas chamber. These have double openings, one inside the barrier, the other outside the barrier. The two doors are never opened at the same time and when the inside door is opened it will only follow upon a sterilization procedure to ensure that the air inside the apparatus is not contaminated.

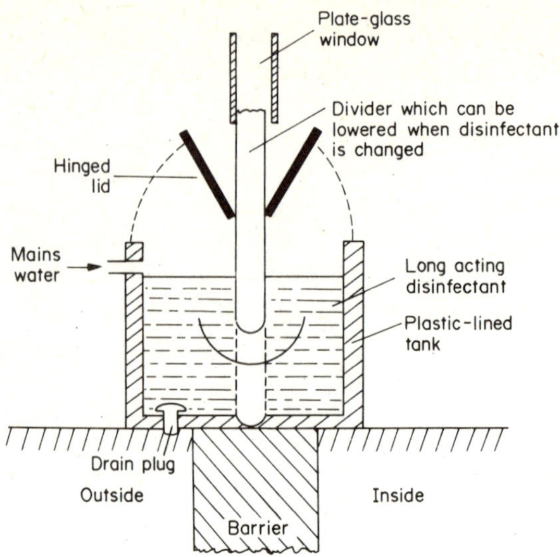

FIG. 39. Dunk Tank (adapted with permission from Bleby, *UFAW Handbook*, Churchill-Livingstone, 4th edn., 1972)

Exiting animals will be containerized in boxes with air filters on the side. The containers are put into ultraviolet locks communicating with the outside of the unit. Before the doors on the inside are opened again the air within the lock needs to be sterilized by the ultraviolet rays for at least 20 minutes.

Personnel will exit by way of the shower area where outside clothes will be exchanged for the working garments.

3.9. Gnotobiotic animals

Previously, animals free from disease-causing organisms have been described as S.P.F. Animals that are maintained in an isolated, sterile environment free from *any* microbial or other parasite, are referred to as *gnotobiotes*. If the environment is not completely free of living microorganisms, then in the case of gnotobiotes, the contaminating organisms are known and identified by the experimenter. Gnotobiotic animals are reared in *isolators* which are capable of holding one or more cages. Gnotobiotic animals are useful for a variety of reasons, not the least of which is that the experimenter has a complete inventory of the parasitic organisms likely to influence the outcome of his experiment.

The study of gnotobiotes will be approached in the following manner:

  (i) Obtaining a gnotobiotic mammal.
 (ii) Maintaining a gnotobiotic mammal.
(iii) Characteristics of a gnotobiotic mammal.

The information here is confined to a study of the mammal, which is not to suggest that other animal groups cannot be reared in the gnotobiotic condition. Birds can be reared similarly by treating the eggs to a sterilizing procedure prior to hatching in an isolator.

(i) *Obtaining a gnotobiotic mammal*

An animal free of contaminants can only be hoped for if that animal is born into sterile conditions, and it remains in a sterile environment. Mammals can be removed from the uterus of the mother by caesarian section in sterile conditions, as described for S.P.F. mammals. If the mother does not pass microbial organisms across her placenta to infect her young, then provided all aseptic techniques are followed we can use her offspring to produce a gnotobiotic stock. The caesarian delivered young are reared in an *isolator.* The cautionary point about some organisms passing across the mother's placenta and entering the fetus refers to some viruses like lymphocytic choriomeningitis (L.C.M.) in mice, or some larval stages of helminthes, like the roundworm Toxocara in dogs.

(ii) *Maintaining a gnotobiotic mammal*

The sterile young are housed in an isolator. This is a piece of apparatus that contains its own sterile micro-atmosphere. Isolators can range in size from housing for rats to larger mammals like pigs.

Isolators are generally transparent cocoons of firm or flexible plastic-like material enveloping cages, the accessories, and their living contents. The animals reared within this sterile cocoon are isolated from the outside world and so special structures have to be built on to the isolator in order that the contents may be serviced.

Air circulation and ventilation is achieved by having an opening through which air enters and exits. This opening is protected by a filter to remove any air suspended microorganisms.

The entry of equipment, animals, food, and so forth is achieved by having a sterile, two-ended lock. At no time will each end of the lock be open. This lock can be an autoclave, or a chemical treated cylinder containing the sterilizer 2% peracetic acid. Because of the acid nature of this chemical sterilizer the equipment has to be constructed from materials less likely to corrode, such as plastic, glass, or stainless steel. (Suggested hazard with peracetic acid—carcinogenic?)

The handling of animals and equipment within the isolator is carried out through disposable sleeve gloves. There can be two sets of such gloves to enable a pair of persons to manipulate an animal for injection or examination. If the animal bites the gloves, there is a danger of contaminants entering the isolator, so handling of animals must be done with care. In the case of mice they can be picked up by rubber-tipped forceps at the base of the tail.

The requirements of gnotobiotes are in the main the same as for conventional animals, with a few modifications here mentioned. One obvious difference between "germ-free" reared animals and conventional animals is in the region of diet. The gnotobiotic animal is fed a sterile diet. This diet can be high vacuum-autoclaved or radiation sterilized. The latter method appears to be advantageous because there are less nutrients lost and less heat and water affects on the diet. Even long-term storage of diet has been shown to result in some nutrient loss, and so nutrient or vitamin supplemented diets are recommended, as circumstances demand. Some workers have reported growth and reproductive difficulties with gnotobiotic animals fed on sterile diets in isolators.

The environmental requirements of temperature and humidity are the same for gnotobiotes as for the conventional equivalent, and need to be monitored by instruments located within the isolator. The frequency of air changes in the isolator will need some careful attention to prevent draughts or temperature and humidity changes.

(iii) *Characteristics of a gnotobiotic mammal*

An important feature of the gnotobiote is that it does not receive the usual compliment of microorganisms which in normal animals come to inhabit the intestinal tract. These microbial inhabitants of the gut aid in the digestive process and when themselves are digested contribute valuable additions to the diet. Another value to the animal is that the microbial presence in the body permits the tissues to manufacture some antibodies.

Gnotobiotic animals have been described as having distended large intestines. This can be serious enough to prevent reproduction in some animals.

3.10. Categories of laboratory reared mammals

An attempt to grade the mammals in terms of their level of hygiene is really a descriptive convenience. Attempts at accuracy must refer to specific pathogens and their presence or absence on the animal concerned.

The following is a classification used in the United Kingdom whereby an animal is given a star-grading, which equates it with a level of rearing practice. Breeders that are accredited by the Medical Research Council (Laboratory Animals Center) will provide a reliable source of graded laboratory animals.

Grade	Characteristics	Uses
xxxxx	Delivered by caesarian section and maintained in isolators. Free from all demonstrable organisms	Special research projects
xxxx	Delivered by caesarian section and maintained in barrier animal houses (S.P.F. animals). Free of all pathogenic protozoa and all helminths.	Breeding and research programs
xxx	Delivered by caesarian section and maintained under high standards of care. Free of all pathogenic helminths and all types of pneumococcus (guinea-pigs and rabbits). Free of *Streptobacillus moniliformis* (mice and rats) and *Corynebacterium murium* (mice).	Breeding and behavioral programs
xx	Reared under non-barrier, conventional conditions with a high standard of management. Free of all tapeworm stages and obligatory parasitic arthropods. Free of ectromelia (mice) and myxomytosis (rabbits).	Breeding and behavioral work
x	Reared under non-barrier conditions. Free from all infectious disease, particularly the zoonoses. Free from pathogenic skin fungi.	Dissection specimens

The higher grades of animal are not only expensive, but also unnecessarily "clean" for average laboratory purposes. The other grades of animal are described as being free from specific sorts of pathogenic organisms. The named pathogenic organisms are cataloged for each grade of animal. This can be checked out in the appropriate literature (M.R.C. Laboratory Animals Center, *Manual*, series 1, 1974).

3.11. Practical program

3.11.1. ELEMENTARY TECHNIQUES IN BACTERIOLOGY

The objectives of this laboratory-based program are as follows:

 (i) To instruct the students in aseptic techniques.
 (ii) To give the student an opportunity to apply the aseptic techniques in the preparation of a stained bacterial smear.
 (iii) To give the student further practice in microscope technique using the oil immersion lens.

3.11.2. GENERAL INFORMATION

For the purposes of the animal technology, bacteria are pests because they cause disease in both man and animals. This is only part of the truth because many bacteria are harmless, or are useful to man and animals.

The introductory techniques with bacteria presented here will employ harmless bacteria for the purposes of demonstration.

3.11.3. SPORE FORMATION

Under unfavorable conditions some bacteria can form spores. It is these spores that allow the bacteria to survive heat, drying, pH change and disinfectants. For this reason a sterilization process must take into account the possibility of spore survival.

The two most common spore-forming bacteria are bacilli—for example, *Clostridrium* (food poisoning) and *Bacillus anthracis* (anthrax).

As an indicator of the resistance of spores to heat it can be stated that *Bacillus stearothermophilus* requires a temperature of 121°C (249.8°F) for at least 5 minutes to kill them.

3.11.4. SAFETY IN THE BACTERIOLOGY LABORATORY

Before commencing any experimental work with bacteria, however elementary and harmless, it is necessary to adopt strict hygienic behavior.

Where bacteria are being handled it is important to have an efficient exhaust airflow and perhaps ultraviolet lamps in critical work areas. A supply of 5% phenol should be available to wipe down work surfaces and the outside of specimen containers. Suspect contaminated material should be immersed in this phenol solution before being autoclaved and disposed of.

All specimens sent for examination should be put into sterile, closed, and labelled containers. The origins, type of specimen, date of collection, and the name of the laboratory should at least be recorded on the container.

3.11.5. ASEPTIC TECHNIQUES (STERILE TECHNIQUES)

With careful attention to detail it is possible to prevent the atmosphere being contaminated by the bacteria we are handling (i.e. on the animal) and to prevent the environment contaminating the material we are handling.

It is unlikely that the majority of animals kept for laboratory purposes are to be found in aseptic conditions. Even so, it is the intention of the animal technician to approach asepsis—that is, to reduce the incidence of disease-causing agents in the animal area. Aseptic conditions can only really be achieved by following certain patterns of behavior. A sequence of aseptic techniques carried out in the confines of the laboratory are now described.

Aseptic procedures demand that all apparatus is sterilized because bacteria are to be found everywhere in the air, on the hands and clothes, in tap water, and on all surfaces.

Flaming is one method of sterilizing in order to destroy any bacteria that may be present on the wire loops that are used to transfer bacteria, or on the lips of tubes or bottles that are to be used.

3.11.6. OBSERVING STAINED BACTERIA

Bacteria must be smeared onto a sterile glass slide and then stained in order that they may be observed under the microscope.

Assuming that the technician has carried out the work in accordance with the rules of aseptic procedure, then the bacteria viewed with the microscope are the required ones, and not just bacteria found in the air or present on the metal transfer loop.

The following laboratory exercise will enable the student to prepare a stained bacterial smear. Saliva will be used as a source of the bacteria.

3.11.7. PREPARING THE BACTERIAL SMEAR

(i) Flame the loop and allow it to cool. Place the loop into a sample of saliva deposited in a watch-glass.

(ii) Spread out the saliva on a new sterile glass slide. Make sure the smear is of even thickness across the middle of the slide. Allow the smear to air dry for several minutes.

(iii) *Fix* the smear by passing the slide quickly three times through a flame. Test that the flaming is not excessive by touching the glass onto the back of the wrist.

   Fixing will kill the bacteria and help them stick to the slide so that they are not washed off in the staining procedure.

3.11.8. STAINING THE BACTERIAL SMEAR

(i) Put the prepared, fixed smear over a staining rack. Have the rack laid over a dish to collect excess stain. Flood the smear with *one* of the following stains: crystal violet, 10 seconds; carbol fuschin, 5 seconds; methylene blue, 30 seconds.

The equipment
in position before
and after use
Ensure the
loop is closed
as a circle
A A B B
(a)
Holding the
tubes in
one hand
Capalls
or other
closures
A
B
A-specimen
B-culture medium
(b)

Sterilizing the
inoculating loop
by red heat
(c)

A
B
(d)   Removing top from B
A
B
Removing top from A
Removing the enclosures with one hand

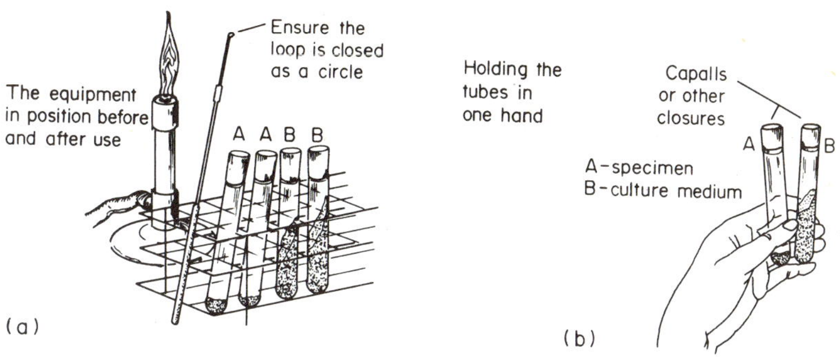

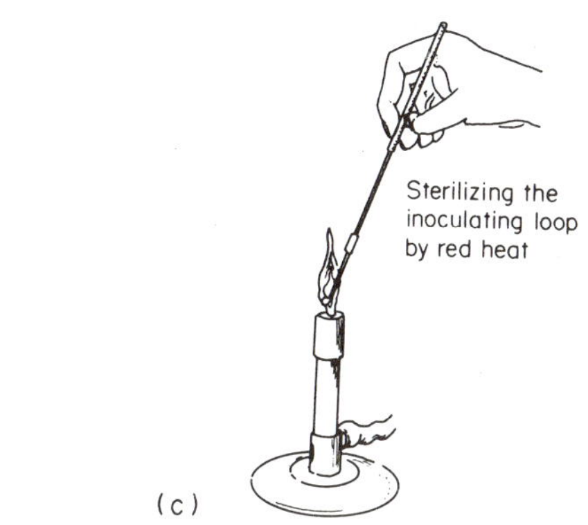

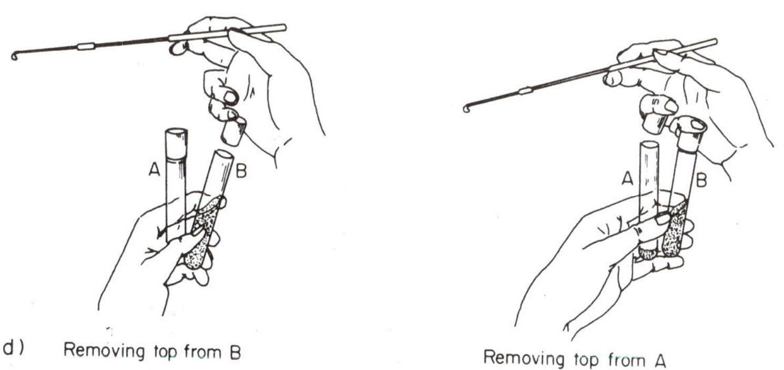

FIG. 40(*a–d*)

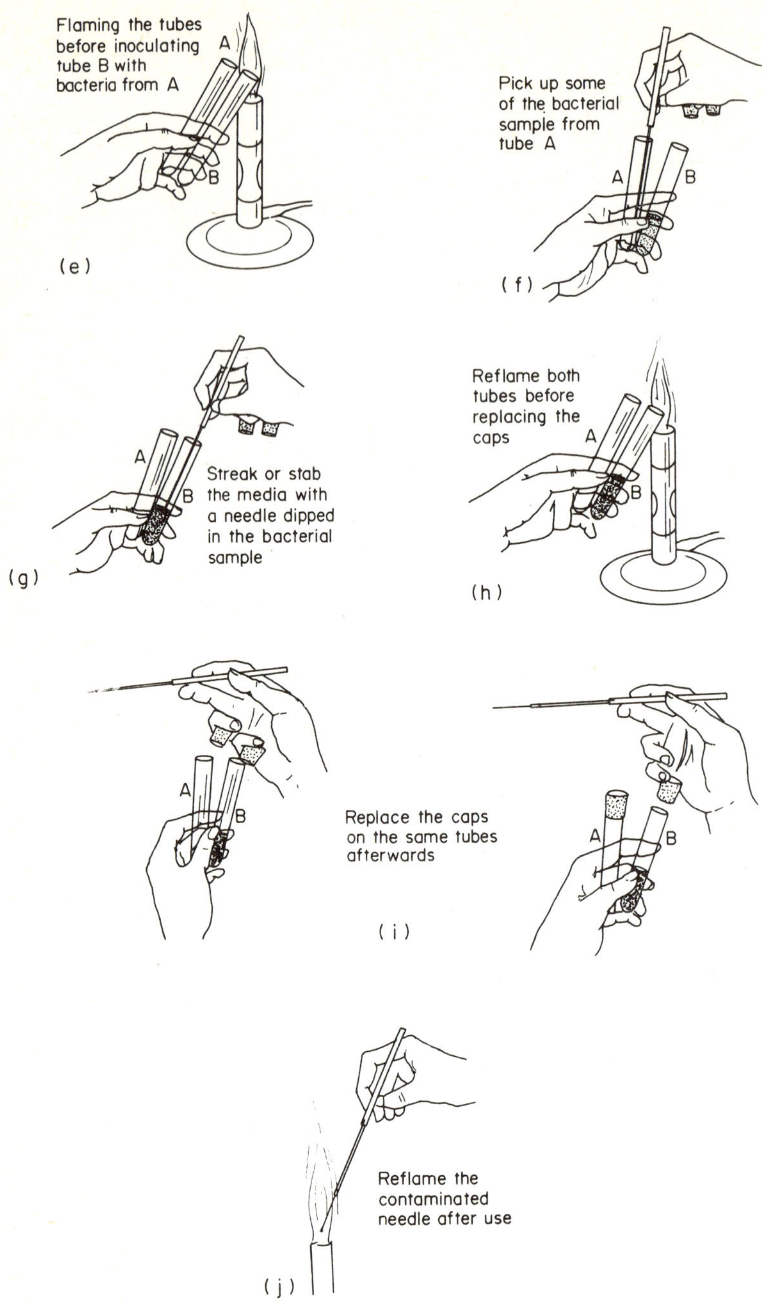

Fig. 40(*e–j*)

Fig. 40. Preparing a cultured specimen of a bacterial sample using aseptic techniques (drawn by J. Hinkley, College of Lake County, Illinois, USA)

(ii) Wash off the stain with a gentle stream of distilled water. Hold the slide firmly with a pair of forceps. Dry the slide carefully with clean tissue paper and allow it to air dry.

(iii) Now prepare the slide for microscopic examination by putting a drop of immersion oil directly on top of the stained smear. (The reason for doing this is explained in an earlier section describing the use of the microscope, p. 65).

(iv) Search the smear for epithelial cells (check lining cells) and note various types of bacteria which are darker staining smaller structures.

3.11.9. GRAM STAINING FOR BACTERIA IDENTIFICATION

A widely used staining technique which distinguishes between two types of bacteria, the Gram negative and the Gram positive.

The technique works like this. The colorless bacteria are all stained with crystal violet. An iodine solution is added to hold the violet in the cells (mordanting). The cells are then flooded with solvents to wash. Some of the bacteria lose the violet color, others remain violet. Another stain (safranin) is added and the colorless cells become red (Gram negative).

Method

(i) Prepare a bacterial smear and fix it as described previously.

(ii) Cover the smear with crystal violet stain (0.5% aqueous).

(iii) Flood with iodine (Lugol's) and leave (30 seconds). Wash away the iodine with distilled water.

(iv) Decolorize the smear with a solvent (50:50 acetone/absolute alcohol) for about 3 seconds until no more color comes away (not too much.)

(v) Cover the smear with safranin stain and leave (1 minute). Flood the slide with water to remove the stain.

(vi) Gently dry the slide with absorbent tissues.

Examine the smear for red cells (Gram negative) and violet cells (Gram positive).

Animal health and hygiene—summary

Parasites were described as ectoparasites and endoparasites.

Disease-causing (pathogenic) organisms outlined were viruses, bacteria, protozoa, fungi, platyhelminthes, nematoda, and arthropoda.

Methods of pest eradication and control included creating barriers to the pests and direct attacks upon the pests.

Animal-house personnel were described as potential hazards to laboratory animals.

Cleaning procedures were described for equipment and rooms.

Disinfectants were treated as having ideal characteristics.

Sterilization techniques included chemical and physical methods, such as treatment with phenolic compounds and vacuum steaming (autoclaving).

Diseases of some common laboratory animals were outlined with a description of symptoms of ill health.

Animal diseases communicable to man (zoonoses) were outlined.

The skin, the blood, and the immune response were described as three lines of defense for the animal body.

Animals maintained under barrier conditions were considered to include S.P.F. and gnotobiotic animals. Animals reared under barrier conditions are for convenience classified with "star categories".

Practical program of work on elementary bacteriological techniques.

STUDY OBJECTIVES

UNIT 4:

Diets, Feeding, and Drinking

(*a*) Describes the components of a balanced diet.
(*b*) Describes the nature of individual nutrients.
(*c*) Describes proprietary animal diets.
(*d*) Describes the composition of various diets.
(*e*) Discusses additional factors required in diets (i.e. vitamin C).
(*f*) Describes the requirements of efficient watering and feeding equipment.
(*g*) Describes food for fish.
(*h*) Practical program of animal diet analysis and an anatomical study of digestive and excretory systems. The use of metabolic cages outlined.

UNIT 4:

Diets, Feeding, and Drinking

ANIMALS confined to the laboratory are as a rule put into an abnormal feeding situation in as far as they are dependent upon their keepers for the selection and supply of their nutrients. In their natural states animals forage for food and presumably ingest the required balance of nutrients for healthy living.

Animal-care personnel, and researchers have made great strides into the creation of balanced convenience foods for animals. The ideal diet for a selected animal species does not exist and is unlikely to be produced. All that can be done here is to outline the characteristics of the ideal diet for the common laboratory animals and to say what use the animal makes of the nutrients. This study will deal with the following:

4.1. Nutritional requirements.
4.2. Diets.
4.3. The value of nutrients.
4.4. Water and drinking equipment.
4.5. Feeding equipment.
4.6. Fish foods and feeding.
4.7. Practical program (food testing, digestive and excretory systems, and the use of the metabolic cage).

Animals fall into feeding categories depending upon the nature of food that predominates in their diet.

Herbivores generally consume vegetation or plant materials.
Carnivores will usually eat animal flesh.
Omnivores will eat a mixture of plant materials and flesh.

The dietary preference of an animal species can generally be determined by examining the teeth. A carnivore has sharp flesh-tearing teeth as compared with the flat-grinding teeth of the herbivore. Whatever the animals' feeding category it will require a balanced diet made up from the nutrients to be considered here.

4.1. Nutritional requirements

All animals require an input from the environment so that they can make flesh and bone, and generate energy for movement and heat. This input we call food. Foods will provide the following seven essential *nutrients*.

Carbohydrates—in grains, cereals, fruits, root crops.
Proteins—in flesh, in some seeds.

Fats and oils—in flesh, in some seeds.
Mineral salts—in flesh, in some plant juices, in water.
Vitamins—in flesh, in plant juices.
Roughage—plant fibrous material.
Water—multiple sources.

These nutrients must be present in the diet in the correct balanced proportions otherwise the animal may be described as suffering from *malnutrition.*

The way in which the body makes use of these nutrients is the subject of the study of nutrition. The breakdown or digestion of the nutrients is considered elsewhere. Before digestion can be clearly understood it is necessary to know something about the various nutrients.

4.1.1. CARBOHYDRATES

These are compounds of carbon, hydrogen, and oxygen. The basic building unit of carbohydrates is the simple soluble sugar, glucose.

SOURCES   *sugars*—such as glucose, fructose, sucrose in plant juices.
starch—an energy store material found only in plants.
cellulose—a strengthening material found only in plants.

All these compounds yield energy when they are digested by enzymes in the alimentary canal. The final product of digestion (glucose) is absorbed into the blood-stream and taken to the tissues where energy is then released as a result of cell chemistry. Some of the excess glucose produced is converted to glycogen (animal starch) and stored in the liver or muscles. Other excesses of glucose are converted to fat. Consequently the obese animal which has been fed on excessive carbohydrate (starchy) foods.

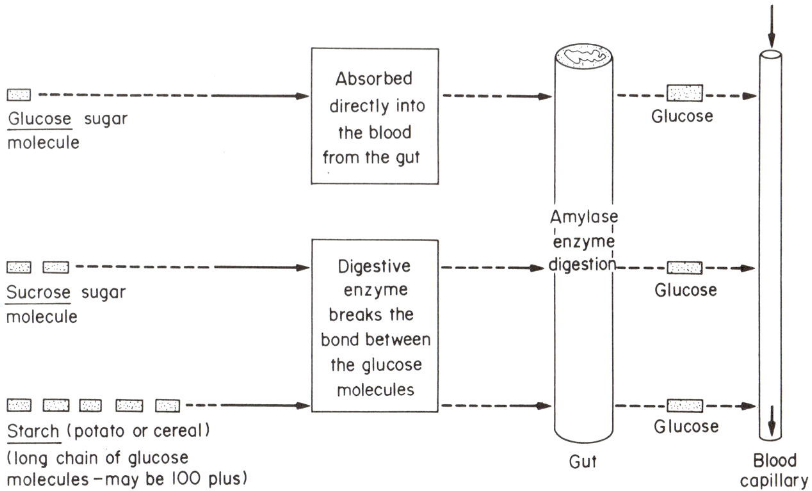

FIG. 41. Carbohydrates

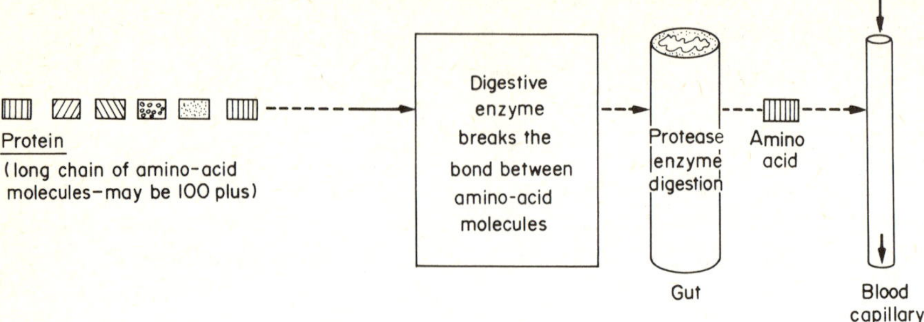

Fig. 42. Proteins

4.1.2. PROTEINS

These are nitrogen-containing compounds which are long chains of smaller molecules called *amino-acids*. It is these amino-acids that are needed by the animal because only they can be dissolved in blood and be carried to growth, or synthesis areas of the body. The amino-acids are used to build new proteins such as hormones, enzymes, or flesh.

Amino-acids are described as "essential" if they must be included in the diet, or "non-essential" if the tissues are able to synthesize them.

A diet must contain not only a quantity of protein, but more important a *quality* of protein that supplies the needed essential amino-acids.

SOURCES   *animal proteins*—meat, fish, milk, eggs.
               plant proteins—cereals, beans, nuts.

4.1.3. FATS

These are energy-rich compounds of carbon, hydrogen, and oxygen making up a fatty acid and glycerol in union.

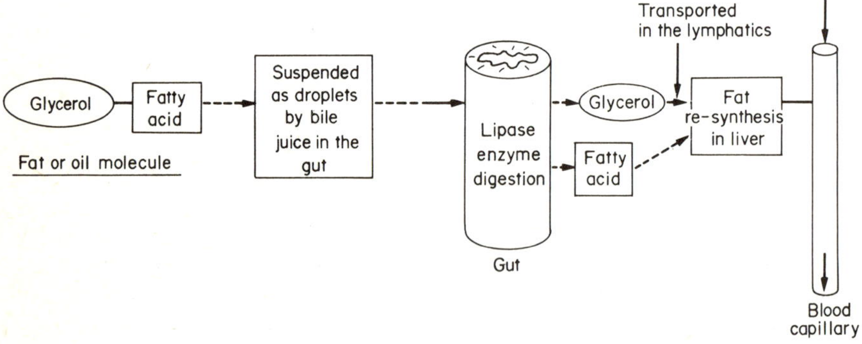

Fig. 43. Fats and oils

Oils are fats that are liquid at normal temperatures. Both supply energy when digested in the alimentary canal.

SOURCES: *animal fats and oils*—fish liver oils, milk.
            plant oils—soya bean, linseed, maize oil.

The vegetable oils are described as "unsaturated" as opposed to "saturated", which describes the hard animal fats.

4.1.4. MINERAL SALTS

Apart from the organic nutrients just mentioned, inorganic nutrients are also vital to the survival of the animal. They are generally required in small quantities. The list of required inorganic ions is considerable, some are required in such minute quantities that they are described as *trace elements*.

Calcium and phosphorus are required to aid in the building of bone and teeth. In order for calcium to be used correctly vitamin D also needs to be present in the diet. All these minerals are of much importance to the growing animal, the pregnant and lactating animal. Low levels of blood calcium interfere with the blood-clotting mechanism.

SOURCES: bone meal, dried milk, bran.

Sodium, potassium, and chlorine are required to make up the correct body-fluid compositions. Any abnormality in the salt balance will upset the water balance of the tissues because osmotic changes will take place (osmosis—see Glossary p. 305).

SOURCES: water will normally contain sodium chloride.
            Many foods contain sufficient potassium.

Iron is necessary for the construction of the red blood pigment, hemoglobin. A deficiency of this inorganic ion will result in an anaemic condition, because of the low production of respiratory pigment.

Iodine is necessary for the construction of the hormone thyroxine, which is secreted by the thyroid gland. A deficiency of iodine will result in an enlarged gland, or goitre, in adult humans. The infant animal will show cretinism or stunted growth.

SOURCES: drinking water or added iodide salts to food.

There are numerous other inorganic ions required in the diet including magnesium, manganese, copper, cobalt, zinc, and so forth.

4.1.5. VITAMINS

These organic molecules are essential in minute quantities in any diet if health is to be maintained. In many cases they act as *biological catalysts* in chemical sequences at cell level. The absence of a particular vitamin from the diet produces symptoms that result from the failure of normal cell chemical activity. A few vitamins will be outlined here using their chemical names as well as the alphabetical symbol. The deficiency resulting from the absence of that vitamin in the diet is indicated in order to show what functions the vitamins have. Vitamins can be grouped as follows:

Fat soluble—A D E K—fatty oily foods.
Water soluble—B C—fruit juices.

Vitamin	Alternative names	Deficiency effects	Food source
A	Retinol (carotene is a Vit. A precursor)	Retarded growth. Hardening of the corneal coat of the eye. Impaired reproductive ability. Visual problems in low illumination	Fish liver oils found in vegetables as carrots and spinach
B_1	Thiamine aneurin	Retarded growth. Reduced appetite. Polyneuritis causing odd movements in humans; Beri-beri.	Cereal husks, yeasts
B_2	Riboflavin	Retarded growth. Upset use of carbohydrates. Some hair loss.	Yeasts, milk.
B_5	Nicotinic acid Niacin Nicotinamide	Retarded growth. Skin disorders in humans; Pellagra	Yeasts, milk, liver
B_3	Pantothenic acid	Skin disorders and hair discoloration.	Animal tissues such as liver and heart.
B_6	Pyridoxine	Skin disorders. Reduced disease—defense.	Animal and plant tissues.
	Folic acid	Retarded growth. Anemia.	Yeasts, vegetables, liver, and kidney.
B_{12}	Cyanocobalamin	Retarded growth. Reproductive problems; pernicious anemia.	Liver
C	Ascorbic acid	Wounds fail to heal. Scurvy in guinea-pigs and primates. Loss of weight	Many animals can synthesize this. Essential, however, for guinea-pigs and primates. Fresh fruit and vegetables
D_2	Calciferol	Poor bones and teeth. Rickets. Defective use of calcium salts	Can be synthesized by the tissues exposed to ultra-violet rays (sunlight). Fish liver oils
E	Tocopherol	Sterility. Poor muscular action (muscular dystrophy)	Wheat germ, milk, egg yolk
K		Haemorrhages because of reduced blood-clotting ability	Can be synthesized in the guts of most species. Green vegetables

4.1.6. ROUGHAGE

That part of the diet which is undigestible, non-nutritive is usually plant-fibre, in the form of lignin and cellulose. The latter material cannot be digested by many animals because they have no cellulose enzyme. In those animals where cellulose can be used as a source of energy the enzyme is provided by some gut-dwelling microorganisms. It is worth mentioning at this point that excessive use of antibiotics can destroy gut bacteria and sometimes result in nutritional deficiencies. The roughage component of the diet acts as bulk to separate the nutritive components so that they are more effectively exposed to digestive enzymes. The bulk of undigested material acts as a stimulator of the bowel wall to empty (defecation).

4.1.7. WATER

A diet without water is useless. All the nutrients in a diet are reduced to small particles (molecules and ions) by digestive enzymes. These processes all take place in solution. Without water body chemistry will grind to a halt. The daily requirements will depend upon the animal species, its environment and activities.

4.2. Diets

It has been stressed earlier that it is not possible to point to a diet and call it, rat, mouse, guinea-pig diet, or whatever. There are different opinions about what constitutes the "correct" or "necessary" diet for a particular species. A diet will approach ideal if it has the following characteristics.

(*a*) Attractive so that the animal takes it readily.
(*b*) Balanced, with the correct proportion of nutrients for energy and tissue building.
(*c*) Quantity sufficient to supply the animal's needs.
(*d*) Contain additives that are especially required by the animal (e.g. ascorbic acid for guinea-pigs).

The provision of food for laboratory animals can be as a "home made" mash of the appropriate ingredients, or as a proprietary dry pellet form. The latter pellet or cube diet is for many reasons the most popular for the more common smaller laboratory mammals.

4.2.1. THE COMPOSITION OF DIETS

The simplest way in which to compose a diet must be to estimate the requirements of an animal related to its daily activities. This is easier said than done. How are the dietary requirements to be determined, by ad lib feeding observations, or studies of the animals in the wild? Assuming we can agree upon the ingredients that must be put together to make up the diet, can we guarantee that the components do not interact when they are mixed together to produce changes in their original condition? Will heating affect the quality of the dietary components as they are mass-produced into the commercial form?

If we decide that a commercial cube or pellet form is the most convenient for our purposes there are further problems. The diet must have a reasonable storage life without easily disintegrating or attracting mold growths.

As a starting-point in our discussion of the composition of diets, examine the tabulation below. The information presented is the result of *one* group of researchers on the quantities of daily food and water input (Table 2).

Another group of workers have constructed a more detailed breakdown of the average daily nutrient requirements as a percentage of the whole diet. An abbreviated form of this dietary study is summarized here (Table 3).

From this information it seems that the major component of an animal's diet, in many cases, is carbohydrate. This supplies most of the energy for the animal whereas the protein provides the building materials especially for the growing, pregnant, and lactating animal. It is the protein that is the most expensive item of the diet and thus

TABLE 2.

Average weight of commercially prepared balanced pelleted food consumed daily (in grams) and average daily water consumption (in cm³) for adult animals

Species	Grams of food	Cubic centimetres of water
Mouse	4–6	6
Rat	12–15	35
Guinea-pig	30	145
Rabbit	150	300
Hamster	10–15	15–20
Cat	100–200	320
Dog	300–500	350
Rhesus monkey	100–300	450
Ferrets	?	?

economy needs to be exercised here. Even so, the protein used must be a good-quality product providing the needed amino-acids.

At this point it will be useful to look at the make up of some commercially prepared diets to see in what form the various nutrients are supplied—cube or pellet.

In no way are we in a position to say which is the best diet for a particular species, this only comes with experience. From the diet charts, as shown, they do not tell us which food contains what food nutrient. In fact, some single foods contain many nutrients (Table 4).

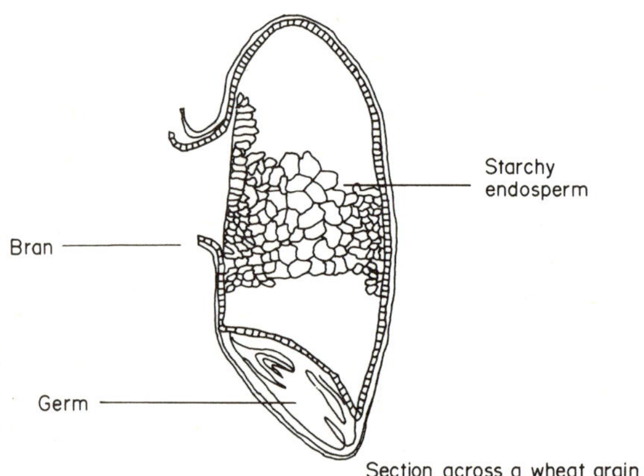

In milling the germ and bran are almost entirely removed leaving behind the less nutritious starchy endosperm (white flour). This wheat bran and germ, termed offal, is rich in protein and vitamins and makes good food for livestock.

Fig. 44. Wheat—a single food, many nutrients

TABLE 3.

Average daily nutrient requirement in percentage of whole diet or in grams (g), milligrams (mg), micrograms (mcg) per kilo of diet

Species	Carbohydrates (%)	Proteins (%)	Fats (%)	Calcium (%)	Phosphorous (%)	Sodium chloride (%)	Vitamins					Roughage (%)
							A	B_1	C	D	E	
Mouse	45–55	16–20	3–12	0.5	0.5	0.5–1.0	500 IU	3 mg	None	181 IU	40 IU	2.5–5.5
Rat	45–55	16–20	5.0	0.6	0.6	0.5	12 000 IU	4 mg	None	300 IU	20 mg	2.5–5.5
Guinea-pig	45–50	20	3–5	1.2	0.6	Needed	12 mg	16 mg	200 mg	None	60 mg	10
Rabbit	45–55	15	3–5	Needed	Needed	0.7	Needed		None needed		7 mg	15
Hamster	45–65	24	3.5	0.6	0.35	?	13 000 IU	6 mg	None	None	25 mg	2.5–5.5
Cat	25–35	30–40	20–30	Needed	Needed	Needed	25 000 IU	4 mg	None	1000 IU	136 IU	2.5–5.5
Dog	40–60	20–24	4–9	1.1	0.9	1.5	1.5 mg	7.3 mg	None	6.6 mcg	48 mg	2.5–5.5
Rhesus monkey	45–55	15–20	3–5	0.86	0.47	0.5	Needed	0.75 mg	25 mg	Needed	Needed	2.5–5.5

Vitamins and salts are summarized in terms of their requirements. This table is not intended to indicate total dietary needs. (IU = International Units). USA sources.)

TABLE 4.
Calculated compositions of laboratory animal diets

Diet	Digestible carbohydrates (%)	Digestible oils (%)	Digestible proteins (%)	Digestible fibre (%)	Gross energy	Calcium (%)	Phosphorous (%)	Sodium chloride (%)	Vitamin C
18 Guinea-pigs, rabbits	34.07	1.92	19.30	7.21	3730 Cals/kg	1.54	1.03	1.51	
41B Mice, rats, monkeys	49.67	2.42	12.73	1.41	3989 Cals/kg	0.61	0.64	1.20	
SG I Rabbits, guinea-pigs	43.81	2.38	14.48	4.50	3980 Cals/kg	0.97	0.98	0.37	
RGP Guinea-pigs, rabbits	42.69	2.58	13.67	4.44	3961 Cals/kg	1.04	0.85	0.73	1311 mg/kg
PRM Mice, rats	46.75	2.56	15.41	1.47	3740 Cals/kg	0.63	0.74	1.30	
86 Mice, rats	50.83	1.44	15.73	1.90	3942 Cals/kg	1.06	0.91	1.26	

(Adapted from commercial sources.) p. vi.

TABLE 5.
Foods and their nutrients

| Foods | Ingredients % average | | | Notes |
	Carbohydrates	Proteins	Fats	
Wheats (mixed)	62.0	11.5	2.0	Low fibre content. Nil vitamin C. Some vitamin B in whole grains
Oats/oatmeal	58.0	12.0	5.0–8.0	High fibre content in ground oats (11%). Nil vitamin C
Maize		10.0	4.0	Yellow maize contains some vitamin A (carotene) precursor. Low in two essential amino-acids
Barley		10.0	1.5	Fibre content about 5%
Rice		7.0		High carbohydrate content
Wheat-germ		30.0	9.0	Goes rancid on storage contains vitamin B and E
Wheat bran, offals				High fibre content (9–10%). The by-products of cereal milling
Grasses		22.0	18.0	Higher protein content in the budding Lucerne grasses. Protein value reduced as flowers develop. Good sources of carotene
Hays		7–15		Higher protein content before flowering and seeding. Late cut hays contain more fibre and less protein. Some vitamin D in the sun-exposed hays
Cabbage (vegetables)		0.5–2.5		Large quantities of water (90%). Good sources of vitamin C, raw, contains 20–100 mg/100 g
Fruits		1.0–2.0		Good vitamin C content, 25–60 mg/100 g
Potato		1.7		Low vitamin C content, 5–15 mg/100 g
Dried skim milk	50.0	35.0	1	Minerals. Water 5%
Whole milk powder	40.0	27.0	30.0	Minerals. Water 1% Some vitamin B, C, A, and D content
White fish meal	67.0	15.9	0.35	Minerals. Low oils. Important calcium content. Overheating may denature the proteins to reduce nutrient value
Whole milk (cow)	4.2	12.4	3.5	
Meat	—	53.0	28.2	
Eggs	1.0	38.8	11.6	
Soya-bean		35.0		
Oil-seed cake and meals				Made from linseed, ground-nut, cotton-seed, soya-bean, palm kernel. There can be problems with toxic products in the substance—these used to be removed.
i.e. ground-nut	7.7	28.2	49	
i.e. soya-flour	13.4	40.6	23.5	

(Adapted: *Manual of Nutrition*, H.M.S.O.)

4.3. The value of nutrients

A balanced diet can only be regarded as such if the components are digestible and yield to the body the required end-products.

Digestion changes insoluble, large molecules into soluble, smaller molecules. For instance, starch is a large insoluble molecule which when digested by an amylase enzyme changes into smaller, soluble molecules of glucose sugar. It has been mentioned earlier that these smaller digestive end-products are absorbed into the bloodstream and transported to various areas of the body for use.

The value of dietary protein is, for instance, judged by the amount and types of amino-acids yielded upon digestion. "Field trials" can be carried out to estimate weight gains of a growing animal being fed with known amounts of protein.

The value of carbohydrates and fats is judged by the amount of energy yielded upon digestion. This energy is generally measured as heat produced when a given quantity of the nutrient is used. The energy is measured as "Calorific value".

1 Calorie (kilocalorie) is the amount of heat required to raise the temperature of 1 liter of pure water 1° Celsius. (The calorie is no longer included in the S.I. units systems, it has been substituted by the joule. 1 calorie = 4185.5 joules, see Data Capsule p. 278.)

> 1 g of protein yields approximately—4 calories.
> 1 g of carbohydrate yields approximately—4 calories.
> 1 g of fat yields approximately—9 calories.

This *gross energy* yielded by the nutrients is measured in the laboratory, and does not necessarily represent the amount of energy available to the animal. The *metabolizable energy* is the *actual energy* yielded for use by the animal. In order to arrive at this more precise figure much laboratory experimentation must be carried out using known weights of food fed to a particular animal.

An in-depth study of the values of nutrients is outside the scope of this book. Those requiring further information are recommended to begin with the references found in the Bibliography (p. 299).

The value of fats as an energy source is closely associated with the amount of carbohydrate in the diet. The absorbed fat that is surplus to energy requirements is deposited in fat depots (adipose tissue) beneath the skin. When needed, fat is withdrawn from store and metabolized to produce energy.

Metabolism is a two-part process. *Catabolism* is the breakdown of a chemical material employing an enzyme action. *Anabolism* is the synthesis of chemical materials, also employing enzyme involvement. The reason why vitamins and mineral salts are so important in the diet is because many of them are needed in minute quantities in order to permit the continuing metabolic pathways.

4.3.1. ADDITIONAL FACTORS IN THE DIET

It should be made clear that commercially prepared diets are not available for use by all those that may be studying laboratory animal science. There are biomedical animal centres in less industrialized areas where feeding routines necessitate that the diet be made up fresh. This type of "mash" will be given to the animal in a variety of containers. The preparation of diets from the constituent ingredients is beyond the scope of this study. Reference should be made to the reading list. Experience has shown that diets, both self-prepared or commercial, very often require to be supplemented by additional nutrients. This applies particularly to the diets of some species, e.g. guinea-pig and animals that are hand-reared on liquid.

Vitamin C or ascorbic acid that occurs naturally in fresh vegetables and fruits is not essential as an ingredient in all animals diets because they are able to manufacture the vitamin in their own tissues. It may be an advisable component of the diet, but not essential. Guinea-pigs and primates require to have the vitamin added to their diet as an essential. This can be added to the drinking water or more advisably to the diet itself. Synthetic vitamin C added to drinking water can be an expensive item. Vitamin C is sensitive to heat and is liable to oxidize. Daily supplies of fresh green vegetables as a vitamin C source are an essential for guinea-pigs. It needs to be clean and provided in such a way that it cannot easily be fouled. Guinea-pig specialists have much to say on the types of vegetables offered to the stock because the animals are somewhat conservative, and dislike abrupt changes in their feeding routines. Some will eat roots more frequently than green leaves, or the reverse. They also appreciate hay, if only to keep them busy. Warnings are given about the use of grass cuttings because it can be fermenting and moldy, and therefore likely to create intestinal problems.

Primates, like the baboons and rhesus monkeys, also need to have their diets supplemented by vitamin C in a similar way to guinea-pigs. Unlike the guinea-pigs these primates are omniverous, as can be seen by examining their teeth. In the natural state these animals obtain their vitamins in a mixed plant; animal diet, as well as from sunlight that generates some skin vitamin D. In captivity the primates are fed commercial diets reinforced with stabilized vitamin C and daily fresh fruit.

In addition to the obvious need for supplementing vitamin sources in the diet there is the less obvious need to supply appropriate levels of mineral salts. Specialist studies in diet composition reveal that minerals are present and very necessary in all animal foods that have been analyzed. An obvious example is the need for mineral components in bird diets, that helps the female to manufacture egg shells.

4.4. Water and drinking equipment

Open bowls of water have limited use for most animals because of the obvious problems of spillage and the ease with which they may be fouled. Such open containers are used for animals that cannot obtain sufficient water from closed drinking bottles with tubes, such as dogs and cats. The provision of fresh water on demand presents problems. It can be provided in bottles or by an automatic watering device. The characteristics of a water provision system will include the features discussed below.

(*a*) The water must be easily available, fresh, and not open to contamination.

(*b*) The water containers, such as bottles, must be easy to remove and sterilize at regular intervals. They should be large enough singly or in multiples to provide sufficient water throughout periods when animals are unsupervised.

(*c*) Transparent glass bottles are liable to break if roughly handled, but are a practical choice in that it is possible to see the air bubbles rise as they function correctly. Any dirt or green plant growths (algae) that may be present in transparent glass bottles can be more easily located and removed.

(*d*) Bottles made of a translucent plastic are useful because they are less easily damaged unless subjected to extremely high temperatures during the sterilization process. The screw thread for the top may sometimes become damaged if treated roughly, and leaking will result.

(*e*) The bottle top and drinking tube are very important components of the watering system. They must deliver water and not permit leaking with the subsequent flooding of a solid bottom cage. There are all sorts of reasons for bottles leaking and most of them are associated with the type of bottle top and drinking tube being used.

(*f*) The bottle tops and tubes must be tough enough to resist gnawing animals. Stainless steel is a popular drinking tube. The diameter of the tube must be such that it permits water to flow only when the animal sucks or licks. Glass tubes are often used because they are cheap and deliver a good water flow. They are practical in the sense that any internal dirt can readily be seen. They can, however, easily get broken and the refitting of glass tubes to bottle tops is a hazardous occupation for those lacking caution.

(*g*) The positioning of the drinking tube has to be carefully thought out. It should not be too deep into the cage otherwise bedding may be heaped up to touch the tube end and cause it to drip. It must not be too high otherwise smaller animals cannot drink from it.

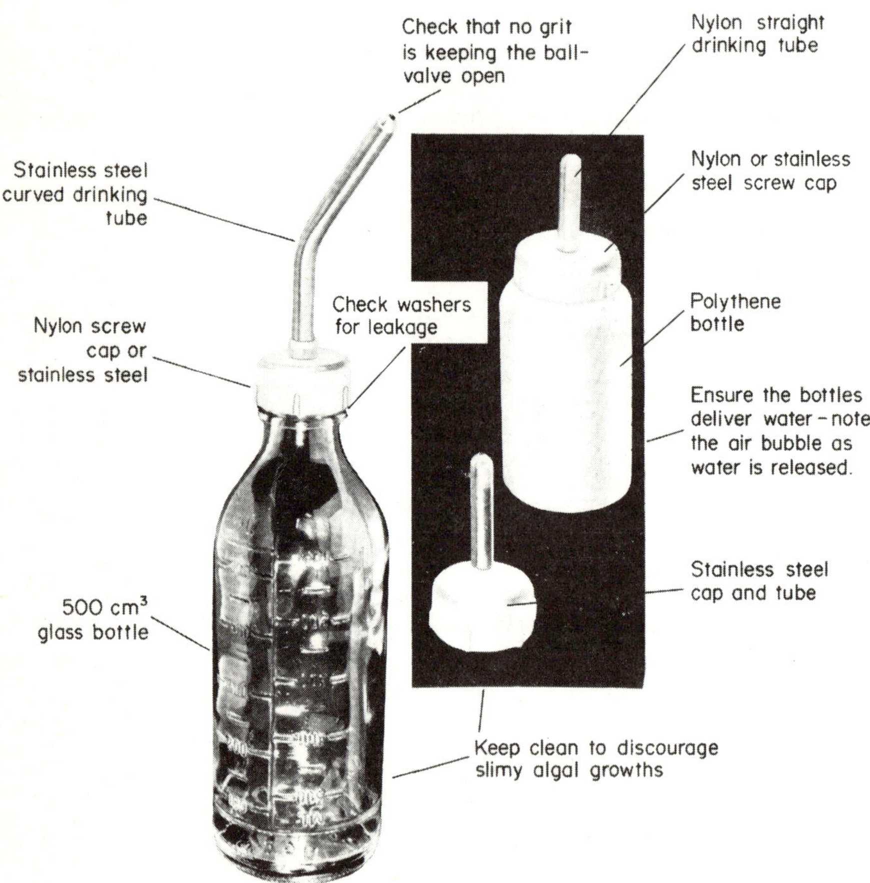

FIG. 45. Water-bottles and some drinking tubes

The routine labor of renewing water-bottles is very time consuming. If bottles are not to be sterilized when they are refilled from the tap (not from a static tank) then they should be replaced to the same cage from which it was taken. This will reduce the spread of possible disease.

The automatic watering devices are intended to reduce the routine labors and other problems connected with water-bottles. These systems are expensive to install and can only really be claimed to be of value if one is dealing with large numbers of animals. The conversion of an existing series of animal rooms to automatic water requires much planning and costing to make it a practical proposition. A poor automatic watering system can be more bother than water bottles.

Some of the practical problems associated with the automatic water system are as follows:

(*a*) The valve structure that delivers water to the small mammals licking the end of the tube has to be fairly sensitive to tongue pressures. Such sensitive valves can easily be stuck open if particles get trapped in them. Such dripping as results can be disastrous with solid-floor cages. The valve could also cut off water supplies.

(*b*) Hard water salts can sometimes encrust within the pipes and mechanisms causing difficulties. The water may need to be demineralized in order to get rid of these chalky salts.

(*c*) The maintenance of an automatic system can be more demanding than attending to bottles.

(*d*) The automatic water system does not give the animal care personnel a chance to check up on the water consumption of their stock. A half-full water bottle suggests either that it is leaking or that the animals are taking water. A full water bottle may suggest that the animals cannot get the water out or that they are not taking water. This kind of information is important. It is not easy to tell, at a glance, if the animals are taking or receiving water from an automatic set up.

4.5. Feeding equipment

Animals in captivity are supplied with their food, they generally do not have to search for it. This is a fairly abnormal circumstance and leaves the animals with virtually nothing to do! In the natural state animals spend most of their waking hours seeking for food. In the captive situation it is virtually impossible to imitate these circumstances.

Food can be presented to the animals in open, non-spilling basins that are attached to the side of the cage. In some cases these containers can be presented at floor level. This type of open-bowl feeding is only suitable if powdery foods or wet mashes are provided. The more common commercial diets such as the cubes or pellets are best presented in specially designed feeding hoppers. These food hoppers need to show some or all of the following characteristics.

(*a*) Food must be easily obtainable but not so easy as to enable the animal to overeat.

(b) Food must be provided so that some effort and gnawing is needed. Provide "near natural" conditions of feeding.

(c) There should be a minimum amount of waste. Pellets should not be easily extracted from the hopper to be dispersed around the cage by the animals.

(d) The hoppers should not be open so that animals can crawl inside and foul the food. Similarly they should be escape-proof for all sizes of animal using the cage, including the young.

(e) Where possible the food containers should be detachable for ease of cleaning and attending to the diet without disturbing the animals.

(f) The containers should be constructed of materials that resist corrosion and destruction by the animals. Galvanized iron is commonly used.

Feeding equipment used for non-mammals is described in Unit 5 and elsewhere.

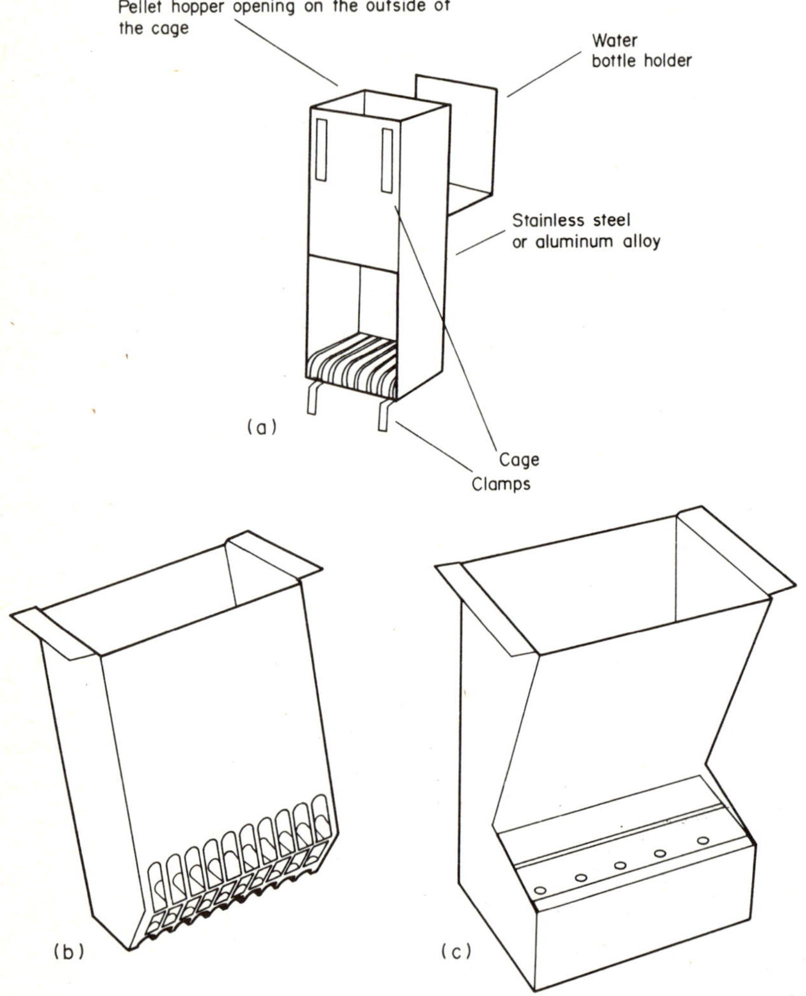

FIG. 46. Food hoppers (from Forth-Tech Services Ltd. catalog). (a) Hopper together with bottle-holder. (b) Stainless-steel rat hopper. (c) Stainless-steel guinea-pig hopper

4.6. Fish foods and feeding

Fish require a balanced diet as do all animals. They can be fed commercially prepared foods, bearing in mind that some fish are carnivorous, omnivorous, or herbivorous. If the dried foods are short on some ingredient, then this should be made up by supplementary feeding.

Fish can be bottom feeders, suspension feeders, or surface feeders, so the presentation of the food should be appropriate to the species.

Most of the fish kept in the laboratory are omnivorous, with the exception of trout and stickleback that are carnivorous. The inclusion of live food to supplement the dried diet is important for balance and for the "interest" of the fish. The large and small aquarist may find it economical to produce their own live food cultures. A selection of live foods and a brief outline of their culture methods are now described.

4.6.1. LIVE FISH FOODS

Infusoria cultures consist of myriad numbers of microscopic organisms, including paramecium and the distinctive "wheel animals", rotifers. They can be reared in the laboratory in many ways, the main point being to establish a nutritious stagnant water.

A beaker of pond water has added to it portions of pond debris and any vegetable matter like boiled hay, dried peas, chopped lettuce leaves, or pieces of soft potato. Leave the materials in the beaker to decompose in a warm place. As the growth of microorganisms increases, some can be sub-cultured to other beakers of pond water to generate a whole battery of culture food.

Before the beakers become too foul, samples of the infusorians can be net-strained and added to clean pond water of aquarium temperature. This live food can then be drip-fed to the young fish.

Water flea cultures are suitable for larger fry, and smaller adult fish. The term "water flea" covers the following small crustaceans: *Daphnia pulex*, *Cyclops*, *Cypris*, and *Simocephalus*. They can be cultured in the laboratory situation provided a sufficiently nutritious media is maintained at temperatures below 20°C (68°F). Details are to be found in the section on invertebrate animal cultures (p. 223).

Brine-shrimps (*Artemia salina*) are small salt-water crustacea that can be hatched out from commercially available dried eggs. They can be fed to baby fish 2 or 3 days after hatching in salt water at 20°C (68°F). Details are given in another section (p. 224).

Freshwater shrimps (*Gammarus pulex*) are larger scavenging crustaceans suitable as live food for bigger fish. They are found in shallow running water with a muddy, stony bottom. They can be removed with pond weed and kept in well-aerated pond water in the laboratory.

Freshwater louse (*Asellus aquaticus*) is a common pond and stream scavenger. It resembles the land-dwelling wood louse. It can be removed from water in handfuls of weed. This crustacean is a useful live food for larger fish.

Woodlouse (*Oniscus asellus*) can be used as a live food for larger fish. They can be reared in the laboratory in vessels with a high humidity at temperatures below 20°C (68°F). Details are available later (p. 223).

Whiteworms or Enchytraeid worms (*Enchytraeus albidus*) are found in damp humus or in rotting compost. They are useful live food for small fish. They can be fed in a pulped condition. Further details are recorded elsewhere (p. 226).

Earthworms (*Lumbricus species*) are a very good live food for larger fish. They can be cultured, as described in the next unit (p. 225), at temperatures below 10°C (50°F). can be found naturally on lawns after rain or beneath a piece of sacking left out on the grass.

Microworms (*Anguilluia silusia*) are minute nematoda that make good food for fish fry. They can be obtained commercially and cultured on cooked oatmeal at 21–26°C (69–78°F).

Mud worms (*Tubifex species*) are small red worms found in muddy freshwater. They can be obtained commercially and kept in muddy, well aerated water at temperatures below 20°C (68°F).

Insects, in the forms of larvae, pupae, and adults can be fed to fish. Included here are the life stages of the common gnat (*Culex pipiens*), mosquito larvae, blowfly (*Calliphora erythrocephala*) larvae and pupae. Mealworm beetle larvae are also useful fish food. Culturing of some of these insects is discussed in the next section on invertebrate animals.

4.6.2. DRIED FISH FOODS

The use of dried foods alone is not approved of by many aquarists. They suggest that it should always be supplemented by live foods each day. The dried flakes are best delivered through a floating feeding bowl obtainable commercially.

Proprietary dried foods are expensive and some aquarists prefer to make their own from powdered fish and shrimps mixed into a dough with whole-meal flour, egg, and water. To this mixture a little chalk (calcium) is added. For details of these home-made foods readers are recommended to check through the Bibliography.

As with all foods, the dried foods must be given in small enough quantities to be consumed by the fish, otherwise it will foul the water.

4.7. Practical program

ANIMAL DIET ANALYSIS—DIGESTION AND EXCRETION

Animals take in their food, digest it and some of the by-products are excreted. The practical work laid out here is concerned with an elementary study of diets, digestive, and excretory systems.

The objectives of this study are as follows:

(i) To enable the student to carry out simple biochemical tests on known and unknown foodstuffs as a technique in nutrient identification.

(ii) To examine the body structures associated with food usage in a variety of animals. To relate this knowledge to post-mortem examination of the digestive tract.

(iii) To examine the body structures associated with urinary excretion.

(iv) To demonstrate the use of a metabolic study cage.

(i) *Food testing*

This work will have to be carried out in a laboratory and the following items of equipment are required:

Requirements: Flame burner (e.g. Bunsen burner).

Rack of hard glass test tubes.

Spatula, "spoon-ended".

Beaker of water.

Watch glasses, two or three.

Test-tube holder.

Access to appropriate food testing chemicals (see below).

Access to sink with flowing water.

Protective glasses and coat.

The purpose of this practical work is to familiarize the student with the use of chemical and physical tests to identify unknown food nutrients. The best way to set out upon this task is to take *known* food nutrients and test them first to see how they react. Once we know what should happen then we can move on and test unknown food materials.

The tabulations below are tests and their results. Students are recommended to run through these tests first, before moving on to test the unknowns supplied by their instructor.

Nutrient	Test procedures	Reactions
Glucose sugar	*Ignition test:* Heat a small sample on a spatula until all is burnt off. Note any odors or carbon residues left over	Liquifies, bubbles, and burns. Large carbon residue left behind with "sweet" caromel odors
	Solubility test: Add a small sample to cold water in a test tube. Shake gently to dissolve. If unsuccessful, heat and shake	Glucose dissolves in cold water
	Benedict's test: Add a few drops of the reagent to a glucose solution and warm gently	A red through yellow coloration is produced
	Fehling's test: Add equal quantities of Fehling's reagents A and B to a glucose solution and warm gently	A red-brown coloration forms
Starch (potato or cereal)	*Ignition test:* Repeat as for glucose	Large carbon residue
	Solubility test: Shake in cold water and warm gently	Unsoluble in cold water. Becomes jelly-like upon heating
	Iodine test: Add two drops of iodine to the warmed starch "solution"	A blue-black coloration is produced

Nutrient	Test procedures	Reactions
Cellulose (woody fibers)	*Schultze's test:* Add a small quantity of the reagent to the cellulose in warm water	The cellulose "particles" show a blue coloration
Proteins	*Ignition test:* Repeat as for glucose	Some carbon residues and "burnt" flesh odors
	Solubility test: Shake in cold water and warm gently	Insoluble in cold water. Becomes jelly-like upon heating
	Million's test: Add a small quantity of the reagent (care—poisonous) to the protein in warm water. Heat to near boiling-point	A white precipitate turns red-brown upon heating
	Biuret test: Add a small quantity of the reagent to the protein in warm water	The protein particles turn a purple blue color
Fats and oils	*Ignition test:* Repeat as for glucose. Caution fats and oils are quite inflammable	Little carbon left behind. Smoky flame
	Solubility test: Shake in cold water and warm gently	Fats and oils do not dissolve in water. An oily layer floats on the water surface
	Grease-spot test: Shake the fat or oil in a small quantity of fat solvent such as ether or acetone (care—fire danger). Tip the dissolved fat onto a filter paper	A translucent spot is formed on a piece of filter paper
	Sudan III test: Shake up the fat or oil in warm water and add the reagent. Allow the mixture to stand for 10 minutes	A red-stained layer floating on water results

If any of the above chemicals are not available they may be purchased through any of the suppliers listed on p. 304.

These simple introductory tests may be used on "unknown" diet pellets in order to see if they show positive or negative reactions to these tests.

(ii) *Digestive systems*

The type of diet taken by an animal can be determined, very often, by looking at the following:

(*a*) Teeth and jaws.
(*b*) Alimentary canals.

Studies in biology stress the importance of "structures relating to function". This practical section will enable students to make comparative observations of some common laboratory animals in order to determine in what ways the animals are adapted to their feeding style.

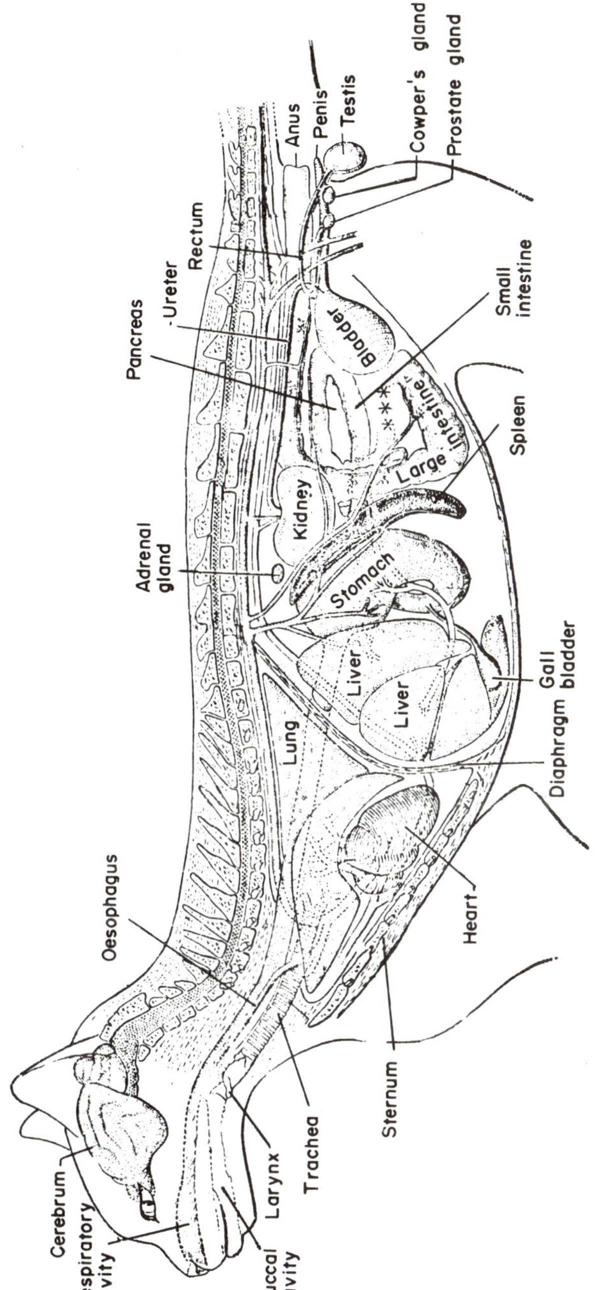

Fig. 47. Organs of the cat (adapted from *General Zoology*, Tracy Storer, McGraw-Hill, 1943)

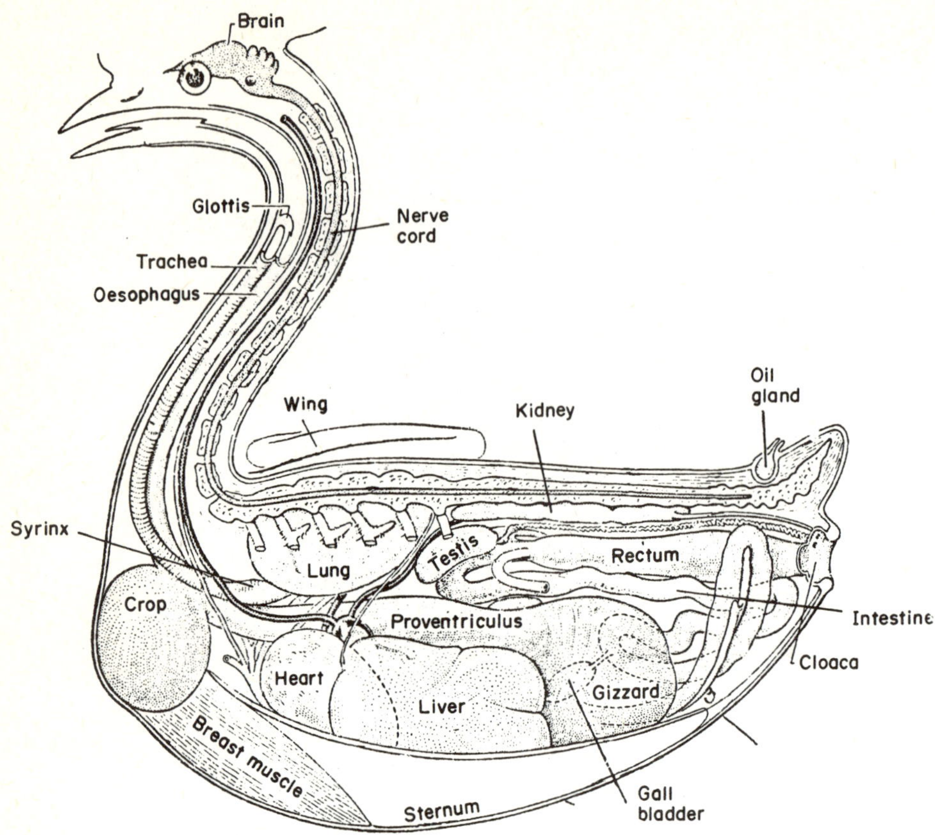

FIG. 48. Organs of the domestic fowl (adapted from *General Zoology*, Tracy Storer, McGraw-Hill, 1943)

(*a*) *Teeth and jaws.* Compare the teeth types and teeth distribution in the skulls of herbivores, carnivores, and omnivores. Notice the specialized cutting incisors of the gnawing animals such as rats and rabbits. Notice the diastema space into which the gnawed food is pushed by the tongue until it is subject to chewing by the pre-molars and molars. Notice the flesh-tearing teeth of the carnivores, such as cats and dogs. Notice the absence of upper and lower front teeth in the sheep. It uses its lips and tongue to pull up grass. Examine the skull and teeth differences in Fig. 49.

(*b*) *Alimentary canals.* The study of laboratory animals will include techniques in dissection in order to demonstrate the structure and locality of internal organs. An examination of the stomach, small intestine (ileum), caecum, colon, and rectum will reveal not only the state of health of the animal but also its feeding style. The "lower" animals have no roof to their mouths and thus cannot keep their food in the mouth without blocking off their respiratory organs. They do not chew their food, they swallow it whole. The long tubular stomach reflects this characteristic.

The herbivorous animals tend to have very much longer small intestines than the carnivorous types. They also have a much more developed caecum.

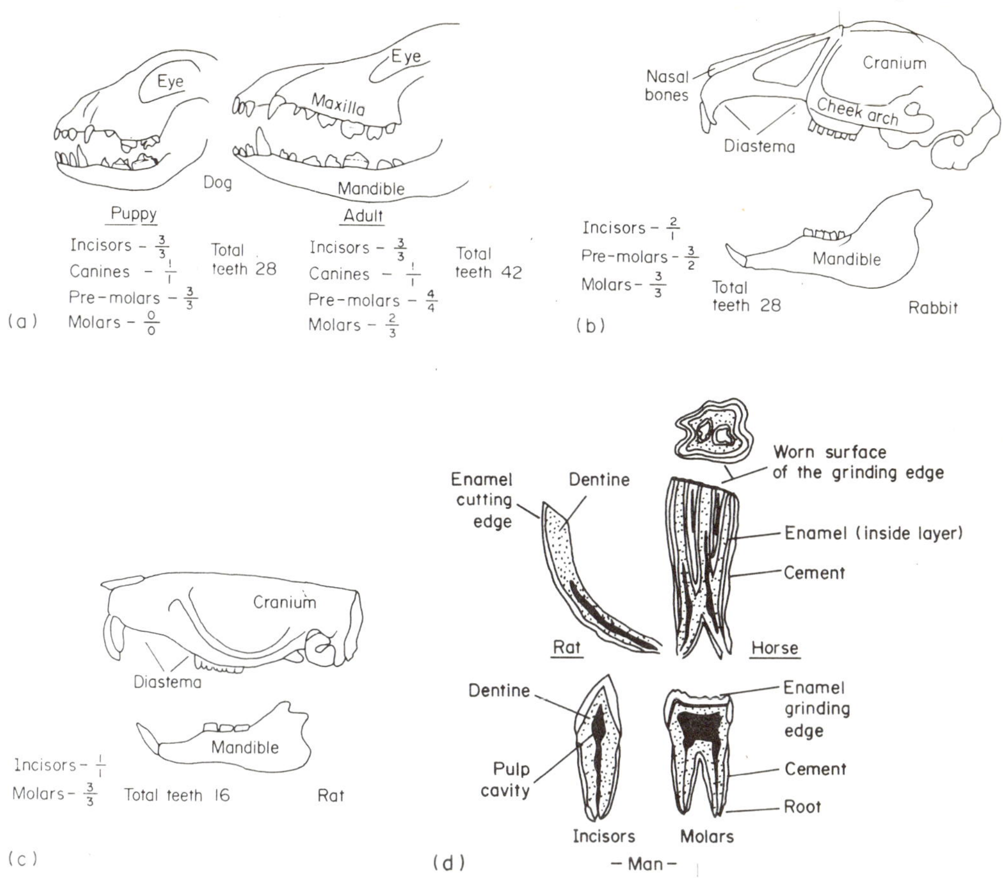

FIG. 49. Skulls and teeth. (*a*) Dog—puppy and adult. (*b*) Rabbit. (*c*) Rat. (*d*) Teeth of man, rat, and horse

A student is recommended to look for signs of inflammation along the length of the guts. The soft-formed feces or liquid feces in the rectum may represent some disordered condition.

(iii) *Urinary systems*

The by-products of the digestive process are either used, stored, or excreted. The potentially poisonous nitrogen compounds left over after the metabolism of proteins are converted to urea in the liver and passed out of the body in water, as urine. The urine is prepared by the kidneys.

Examine the urinary organs in the animals that have been opened up for digestive tract study. Cut a section across a kidney and examine it for the presence of chalky "stones" or any signs of swelling or inflammation.

Glandular areas
Nasal passage
Nostrils
Pancreas
Intestine
Stomach
Large intestine
Anus
Mouth
Buccal cavity
Salivary gland
Liver
Gall bladder
Caecum

(a)

Oesophagus
Duodenum
Stomach
Transverse colon
Caecum
Small intestine
Descending colon
Ascending colon
Guinea-pig
Gall bladder
Oesophagus
Liver
Appendix
Stomach
Pancreas
Duodenum
Ileum
Rectum
Colon
Caecum
Anus
Rabbit

Oesophagus
Stomach
Ileum
Duodenum
Colon
Rectum
Caecum
Rat
Anus
Stomach
Oesophagus
Duodenum
Ileum
Rectum
Caecum
Colon
Appendix
Mouse

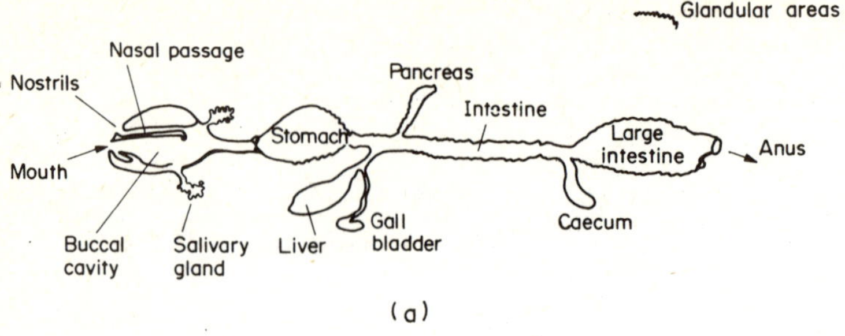

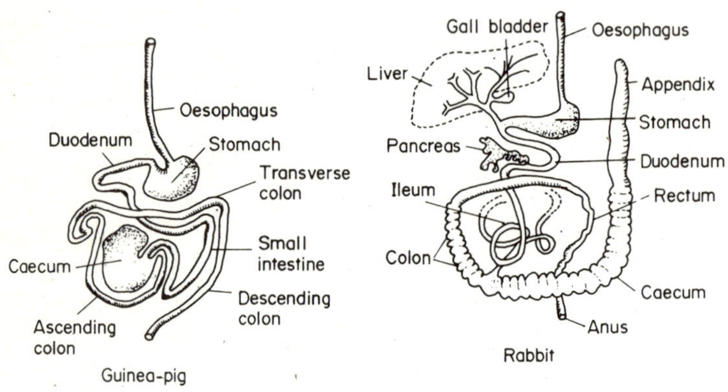

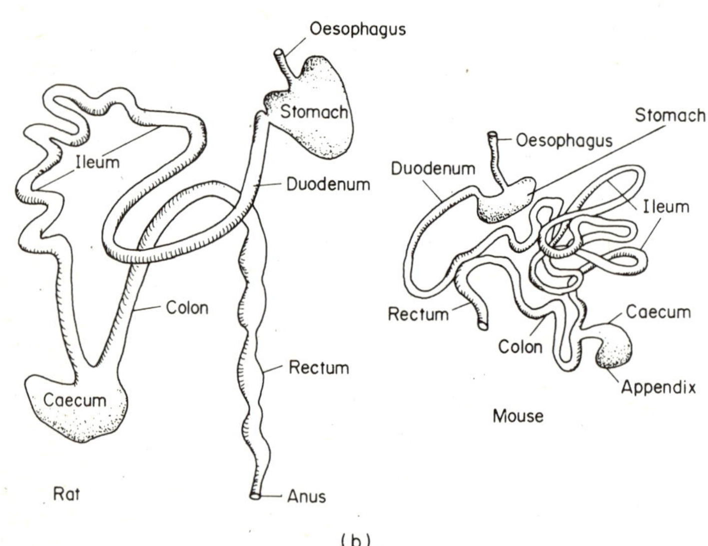

(b)

Fig. 50.

Animal type	Feeding type	Stomach structure
Fish – Shark	Mainly carnivorous Swallows food whole – no chewing	
Amphibia – Newt	Mainly carnivorous Swallows food whole – no chewing	
Reptile – Turtle	Mainly carnivorous Swallows food whole – no chewing	
Aves – Pigeon	Mainly seed feeder The bird has no teeth so the grinding takes place in the gizzard.	Muscular 'grinding' stomach-wall
Mammalia Rodent – Rabbit	Herbivore Food ground up by chewing action	
Ruminant – Cow	Herbivore Food regurgitated for chewing at later time	Four chamber stomach
Primate – Man	Omnivore Food chewed before swallow– ing	

(c)

FIG. 50. Mammal digestive tracts. (*a*) Alimentary canal (diagrammatic). (*b*) Comparative alimentary canals; guinea-pig, rabbit, rat, mouse. (*c*) Comparative stomach structures

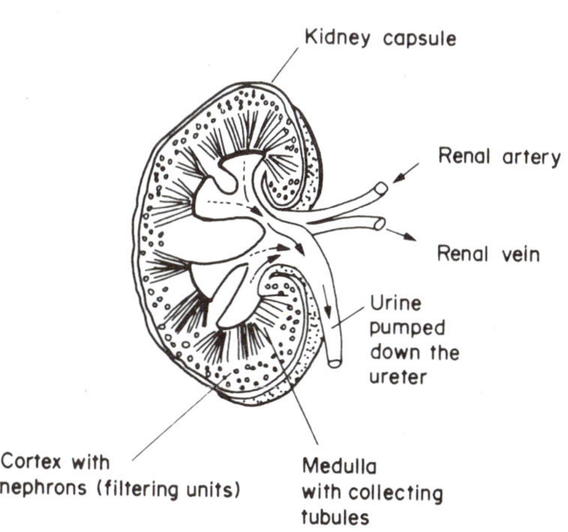

FIG. 51. Mammal kidney (diagrammatic section)

(iv) *Metabolic cages*

An interesting practical exercise connected with feeding and drinking is the use of the metabolic cage unit. The commercial set-up can be a fairly expensive item. An exercise in design and construction, to avoid this financial outlay, could be attempted by students.

In the illustration (Fig. 52) it can be seen that the weighed animal is kept in a cage resting on a collecting funnel that separates feces and urine. The procedure for use at this "beginners" level is as follows:

(*a*) Weigh the animal.
(*b*) Weigh the water bottle.
(*c*) Fill the water bottle and reweigh.
(*d*) Weigh food pellets in the food hopper.

Leave the animal in the cage for 24 hours. After this period:

(*e*) Reweigh the animal.
(*f*) Reweigh the water bottle.
(*g*) Reweigh the food hopper.
(*h*) Measure the volume of urine produced.
(*i*) Weigh the feces produced. (They may need separating from powdered food and animal fur.)

This type of exercise will give practice in accurate weighing and demonstrate the kind of technique used to determine daily food and water demands of laboratory animals.

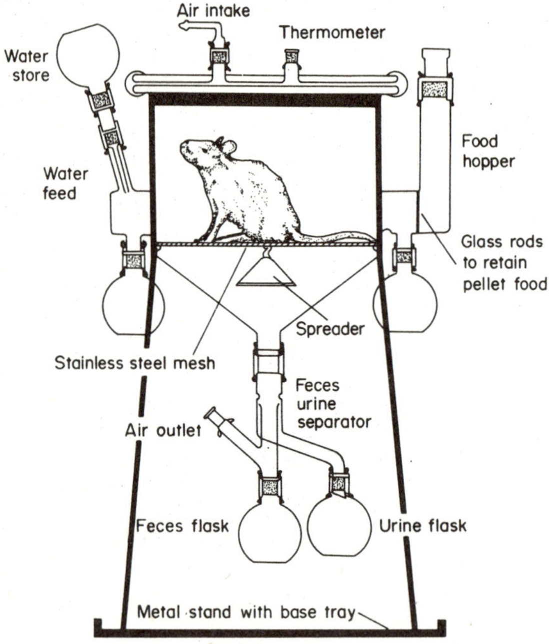

FIG. 52. Glass metabolism cage (drawn with permission from R. B. Radley & Co. Ltd.)

Diets, feeding, and drinking—Summary

The essential nutrients of a balanced diet were outlined as carbohydrates, proteins, fats, vitamins, salts, water, and roughage.

Diets for specific animals discussed with the suggestion that the ideal diet is difficult to achieve.

Commercially available diets discussed.

The function of dietary ingredients was discussed in terms of their energy contribution and the tissue-building as well as regulatory function.

Animal-house equipment for supplying food and water was described as requiring to satisfy certain characteristics, i.e. not easily fouled by the cage inhabitants.

Fresh, live, and dried foods for fish were described as needing to be fed, a little, often.

Practical work program was outlined to enable students to analyze foods.

A study of digestive systems and excretory systems was outlined to show the manner in which animals are adapted to their diets.

The use of the metabolic cage was described.

STUDY OBJECTIVES

UNIT 5:

Reproduction, Breeding, and Heredity

(*a*) Describes the structures of the male and female mammal reproductive organs.
(*b*) Describes the role of hormones in male and female mammal reproductive activity.
(*c*) Describes the events taking place during the oestrous cycle.
(*d*) Outlines the procedures recommended for successful animal mating.
(*e*) Discusses the influence of environmental factors upon biological rhythms.
(*f*) Describes the events taking place during gestation and pseudopregnancy.
(*g*) Describes the birth process, birth weights, and litter sizes.
(*h*) Describes lactation and the weaning process with relevant ages and weights.
(*i*) Reviews inbreeding and random breeding programs employing polygamous or monogamous methods.
(*j*) Describes chromosomes, genes, and cell division.
(*k*) Describes basic genetic principles of monohybrid and dihybrid inheritance.
(*l*) Describes rearing methods for a wide range of non-mammal laboratory animals.
(*m*) Suggests a practical work program.

UNIT 5:

Reproduction, Breeding, and Heredity

THIS unit is concerned with biological fact and methods of animal husbandry. There is little cause for discussion about the facts of reproduction and heredity, but considerable discussion arises from individual experiences of different methods of husbandry.

The newcomer to animal technology is recommended to use this material on breeding and husbandry as a starting-point for practical experience. The more experienced student will be aware how much of a problem it is to be precise about living organisms. A survey of the available literature will show how wide are the numerical descriptions of items such as optimum environmental conditions or the weights of animals at different times in their life. The newcomer may be impatient with this apparent lack of exactitude; the more experienced person will recognize that different authors have different sources for their information. The information being furnished from different circumstances of animal rearing.

The study is organized in the following manner:

5.1. Anatomy of the mammal reproductive organs.
5.2. Physiology of mammal reproduction.
5.3. Mating.
5.4. Breeding and the environment.
5.5. Pregnancy (gestation) and pseudopregnancy.
5.6. Parturition.
5.7. Lactation and weaning.
5.8. Breeding programs.
5.9. Basic animal genetics.
5.10. Rearing non-mammal laboratory animals.
5.11. Practical program (handling and sexing mammals).

The mammals referred to in this unit are as follows:

Mouse, Hamster, Guinea-pig, Rabbit, Cat, Monkey, Rat, Gerbil, Ferret, Dog.

The non-mammals included here are the following:

Vertebrates	*Invertebrates*
Zebra finch	Protozoans. Stick insects
Japanese quail	Hydra. Granary weevil
Chicken (eggs)	Planarians. Mediterranean Flour Moth
Goldfish	Worms. Mealworms
Guppies	Locusts. Flour beetles
Salamanders (axolotl)	Cockroaches. Water fleas

Newts	Drosophila. Woodlice
Frogs	Housefly
Toads (xenopus)	Brine shrimps

The reader is referred to the Bibliography for more detailed descriptions of rearing methods.

5.1. Anatomy of the mammal reproductive organs

The sex organs produce sex cells (gametes). The fusion of sex cells (fertilization) results in the developing foetus which is maintained by way of the female blood system through the *placenta* inside the uterus. The mammals to be considered here are described as *placental mammals*.

(i) *Male reproductive organs.* The paired *testes* produce *spermatozoa.* The production of sperm in most mammals takes place at temperatures that are lower than that of the abdominal cavity. For this reason most mammal testes descend to a position in the scrotal sac outside the body. The testes move into the scrotum seasonally in some animals (deer), in others they reside permanently within the abdominal cavity (elephant).

The testes also produce the male hormones (androgens) which influence the development of the secondary sexual characteristics (male behavior, male coat characteristics, and so forth).

The *male accessory sex organs* deliver the mature sperm to the female during copulation. These structures can be seen in the diagram and consist of the *penis* through which the urethra runs, the two sperm ducts and associated glands. The sperms travel down the urethra (which also conducts the urine) in a seminal fluid that is produced by the prostate and other glands.

(ii) *Female reproductive organs.* The paired *ovaries* are situated within the abdominal cavity and produce the *ova.* An ovum is released from the ovary (ovulation) at regular intervals, which is related to the periods of sexual activity or "heat". The ovaries also produce the female hormones (estrogens) which influence the development of the female secondary sexual characteristics (female behavior, female physical characteristics).

The *female accessory sex organs* consist of the *oviducts* (Fallopian tubes) which conduct the mature egg from the ovary to the *uterus* where the fertilized egg implants, and develops. The form that the uterus takes can differ from one group of animals to another. For instance, the duplex form is found among mice, rats, and rabbits, whereas the cow has the two arms of the uterus fused to form the bicornuate uterus. In humans the uterine "horns" are totally fused to produce the single uterus.

5.2. Physiology of mammal reproduction

This is not the place for a detailed study of reproductive physiology. It is sufficient to indicate the basics of sexuality as far as they influence the practical breeding of mammals.

(i) *Physiology of male reproduction.* Sexual activity in the male is related to the

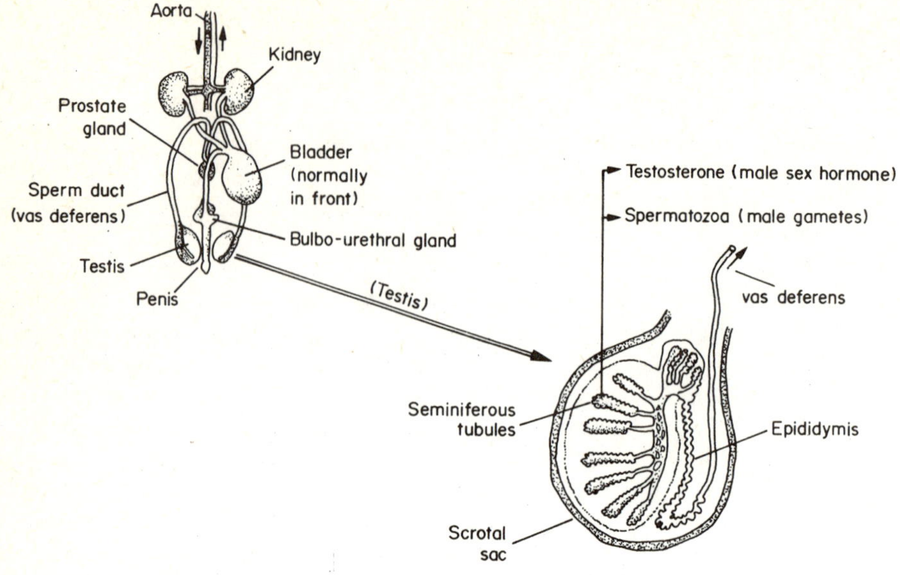

FIG. 53. Mammal reproductive system (male cat)

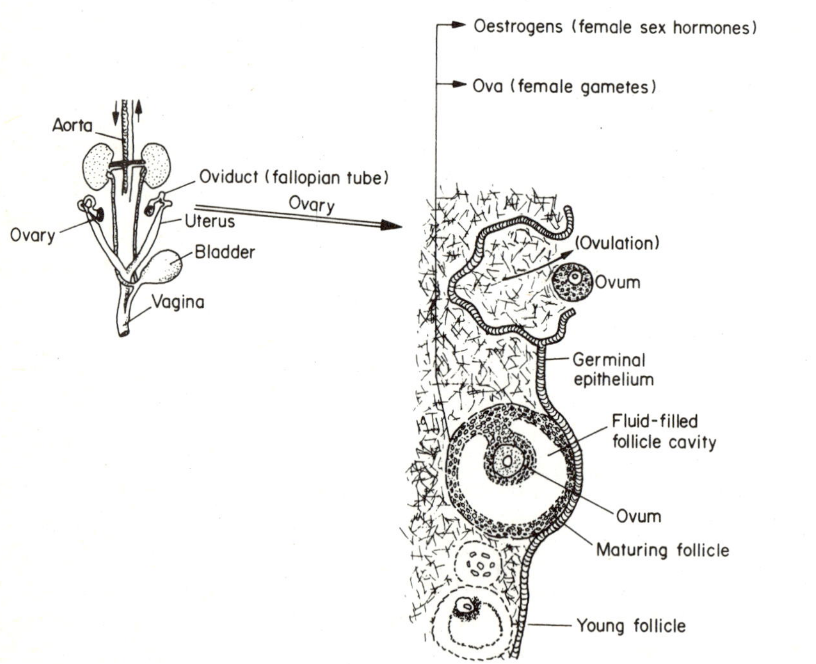

FIG. 54. Mammal reproductive system (female cat)

degree to which sex hormones, such as *testosterone*, are circulating in the bloodstream. Environmental factors such as temperature and daylight length also influence the onset of sexual activity and behavior. It can be said that environmental stimuli trigger off the release of sex hormones into the bloodstream in some animals.

The secondary sexual characteristics of the male, such as coat coloration and formation, body size, and perhaps aggression are all related to the physiological influences of the male sex hormones. The male has a relatively simple sexuality when compared with the female.

(ii) *Physiology of female reproduction*. Sexual activity in the female is also hormone related as well as environmentally influenced. The complexity of the female sexuality may be linked to the fact that she comes into sexual activity in a cyclical manner. The sex cycles of all female species are not the same. The main female sex hormone, *estradiol*, is responsible for the secondary sexual characteristics such as coat color, body size, development of mammary glands, mothering behavior and so forth. It is also responsible for the maintenance of the sexual cycles.

The estrus cycle is named after the word *estrus* ("heat"), when the female is sexually active and receives the male in copulation. This time of "heat" corresponds closely to the time, ovulation, when an egg(s) is released from the ovary into the Fallopian duct (oviduct). At this time the egg may be fertilized and pregnancy results. During the estrus cycle there are widespread changes in the physiology and behavior of the female. These changes result in the mature ovum being released into the oviduct and result in the preparation of the uterus to receive and nourish the fertilized egg. If pregnancy does not result then the cycle of physiological events is repeated. In the primates bleeding accompanies the breakdown of the uterine lining (menstruation) before the cycle recommences. If pregnancy does result, other physiological events occur in order to permit the foetus to develop until full term, and as a result the estrus cycle is suspended.

The frequency with which a species comes into "breeding condition" varies from one to another. The breeding season may consist of weeks or months and may occur many times or only once in a year. For instance, the silver fox has only one breeding season per year during which there is one estrus cycle (*monoestrus*). The dog has two such breeding seasons per year. The cat has two or three breeding seasons but within those seasons there are multiple estrus cycles (*polyestrus*). On the other hand, the mice, rats, and rabbits have no such breeding seasons and have repeated estrus cycles throughout the year.

The period of time covered by each estrus cycle is different for each different species of mammal. The small rodents have an estrus cycle which lasts only 4 to 6 days. The cat has an estrus cycle that lasts 2 weeks. Primate estrus cycles extend over 28 days.

There are occasions when the female is non-receptive to the male. This phase of sexual inactivity (*anoestrus*) is accompanied by lowered blood levels of female hormone (estradiol). There are methods of determining what phase of the estrus cycle an animal is experiencing, by a technique of examining stained vaginal smears. The technique will be outlined in the practical exercises at the end of this unit.

Ovulation is not the same for all mammal species. Some release the egg(s) *spontaneously* during the estrus cycle. This is true for the small rodents, guinea-pigs, dog, and primates, but ovulation needs to be *induced* by external stimuli for some other animals. The cat, rabbit, and ferret are induced to ovulate by mating.

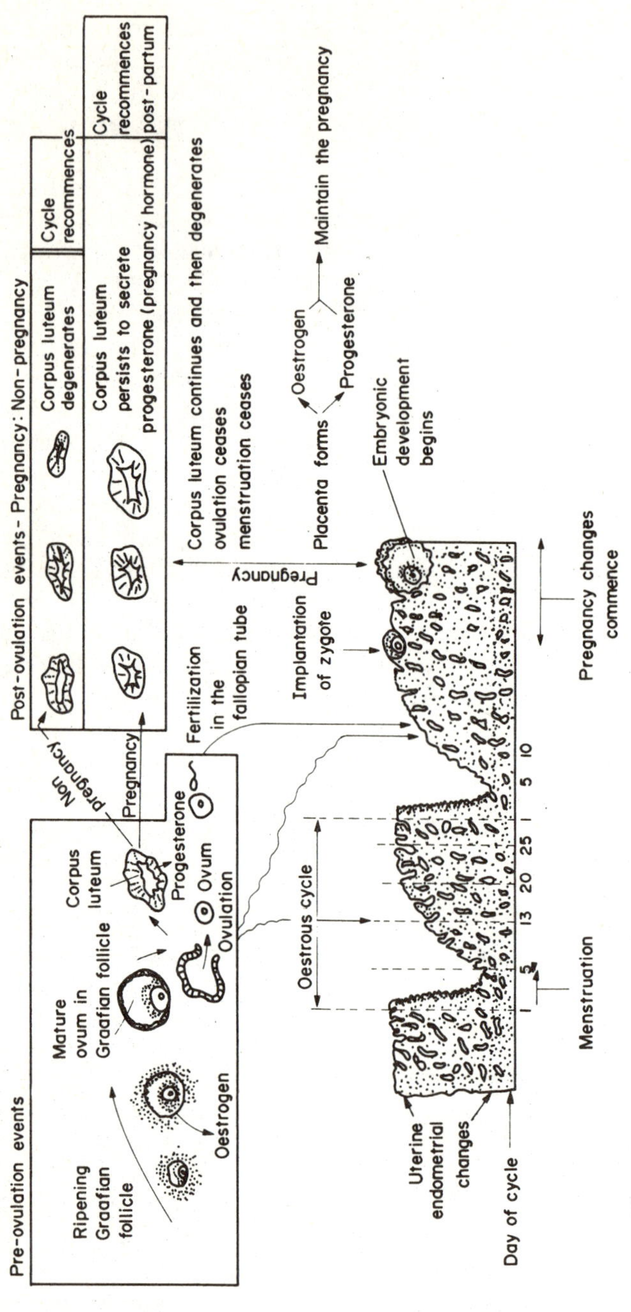

FIG. 55. Events during mammal estrus cycle

5.3. Mating

The physical act of mating (copulation or coitus) takes place only during the mating season for some species. As shown previously different species have different breeding seasons; some have no such season. These seasons are related to the environmental factors and to the female physiology, her estrus cycle. There are two mating seasons for the domestic dog in each year, but no such seasons for the small laboratory rodents, which mate at any time of the year. The male usually shows no breeding season and will mate given the correct circumstances. These circumstances are physical stimuli from the female such as odors, structure and color changes in the genital region, and behavioral postures or calls. It will be appreciated that copulation is not always guaranteed just because the female is on "heat". The female herself must select the male who presents himself for mating. As can be seen with hamsters, rabbits, and cats, for example, there are important preliminaries to the act of copulation which if absent may cause the male to be rejected.

For economic animal breeding it is necessary to select animals for mating that show good success in the mating and in the mothering situation, such that maximum litter sizes are weaned. In order that this becomes a reality, it is also necessary to keep precise records of the matings through to weaning so that the good breeders are usefully employed and the less successful excluded. This artificial selection on the part of the breeder will build up a production colony of animals.

It is not always an easy matter to judge whether an animal has mated successfully. It is not always practical to observe copulation. An act of copulation may not result in a pregnancy anyway. For the purposes of record keeping it is necessary to date and time the sex act in order to estimate birth expectations. In the case of rats some workers describe a method of observing the vaginal plug that is displaced from the female during copulation. This plug may for a variety of reasons not be found in the normal breeding cage so they set up a special observation cage which permits the plug to fall through the floor mesh on to paper beneath.

From what has just been said it becomes clear that successful animal breeding is not just putting male and female animals together. It is only by lengthy experience, accompanied by careful observation and record keeping, that one can make any meaningful utterances about a particular animal species. Then of course those utterances may only be true for the experience of the person making them. The environmental circumstances with regard to disturbance, work routines, health, and so forth are specific to a particular animal house. So how does one start to learn how to breed animals? Learn from the experience of others is the obvious answer. Try out schemes recommended by experienced people and determine if the recommendations work for your circumstances. If not, modify, or try another scheme.

At this point we will outline the recommendations that have been made by experienced animal technicians concerning successful mating of a variety of species. One thing that becomes obvious on reading the literature on animal breeding is the wide variety of timings given by different authors. One has to tolerate this irritation of apparent indecision and use the figures as approximations.

(i) *Mating mice.* Observers tend to agree that female mice kept *singly* come into "heat" irregularly, and that their estrus cycle is lengthened to 5 or 6 days. *Groups* of females kept without male contact tend to reduce their estrus cycles, and may even

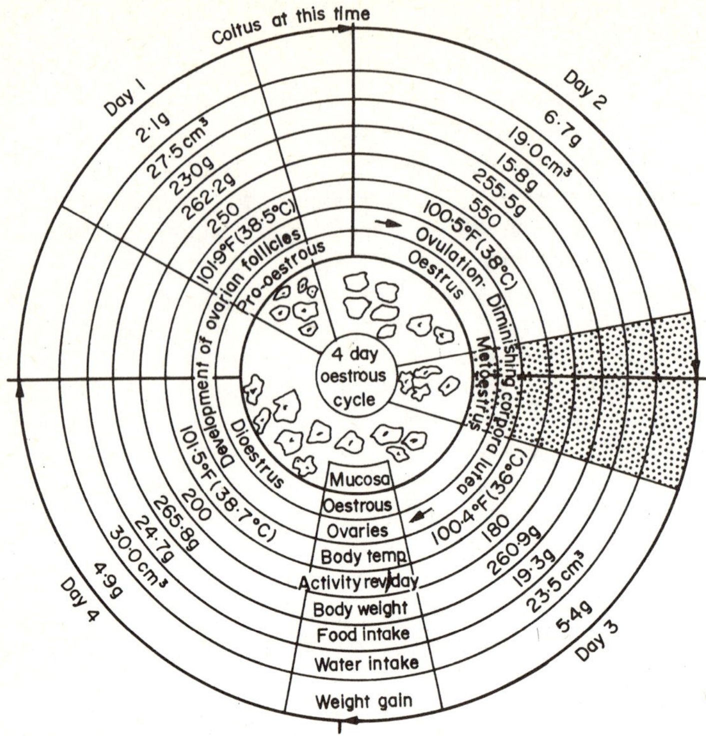

FIG. 56. The estrus cycle and related effects—rat (K. B. Simpson and D. May, *IAT Journal*, vol. 24, no. 1, March 1973)

exhibit "mock pregnancies". It seems that when a male is introduced into the group then the females come into estrus, more or less at the same time. In order to be sure that mating takes place at a given time when mice are paired, then the following procedures have been suggested.

Previous to putting the male in with the grouped females for mating, they should be brought into estrus by 2 days or so exposure to a male who is cage-confined within their midst. After this preliminary, the male can be run with the grouped females and a confident estimate that most or all will be mated by about the second or third night.

Other specialists will report that good results, in terms of litter numbers, are achieved by running a male with grouped females all the time. She is removed to a litter box with other pregnant females. This of course does not permit post-partum mating which will be achieved by permitting the female to litter in the box with the male in a monogamous or polygamous situation.

(ii) *Mating rats.* Keeping monogamous pairs of rats is not very economical because equal numbers of males land females need to be caged and fed. It has the advantage, as mentioned for mice, that post-partum mating can take place.

Rats kept in polygamous groups of one male to five or more females will breed well. The suckling mother rat seems to be less tolerant of other suckling mothers unless there is much nesting space, and so she may have to be removed to a litter box until the young are weaned. Her place in the harem can be taken by another young female.

Where it is necessary to have exact records of the stage of estrus for a rat the vaginal smear technique may be employed. The timed matings can be assessed by observations of the vaginal plug, as described earlier. Changes in body weight are also recorded at the time of estrus, as shown in Fig. 57.

(iii) *Mating guinea-pigs.* Monogamous breeding of these animals is expensive in space and labor, but it is a good method for post-partum mating.

Polygamous breeding of guinea-pigs is most commonly employed with one boar servicing a harem of five to twenty or so sows. The pregnant sows are removed to a communal litter pen. Alternatively, the pregnant sows may be littered separately if accurate breeding records are needed.

These mating methods have produced ten to twelve young annually for each sow.

(iv) *Mating rabbits.* Rabbits are comparatively easy to breed both naturally and by artificial insemination. Accurate records are possible as the copulation will generally be observed. The does appear to remain in estrus for long periods of time in the absence of the male and are therefore prepared to mate at most times. The stage of estrus may be determined by an examination of the genitalia. The swollen vulva region is generally highly colored when the doe is in "heat".

The recommended procedure for mating the buck to the doe is to assess, as far as possible, from physical signs whether she is in estrus. Transfer the doe to the buck's cage and note the behavior of the doe. If she is in "heat" she generally hollows her back and the buck mounts. If problems in mating are obvious for one reason or another, it is wise to remove the buck and use another. She may mate with the alternative buck.

The doe is kept separately to litter. In common with mouse and rat she can be mated post-partum and a pregnancy results whilst she is suckling her litter.

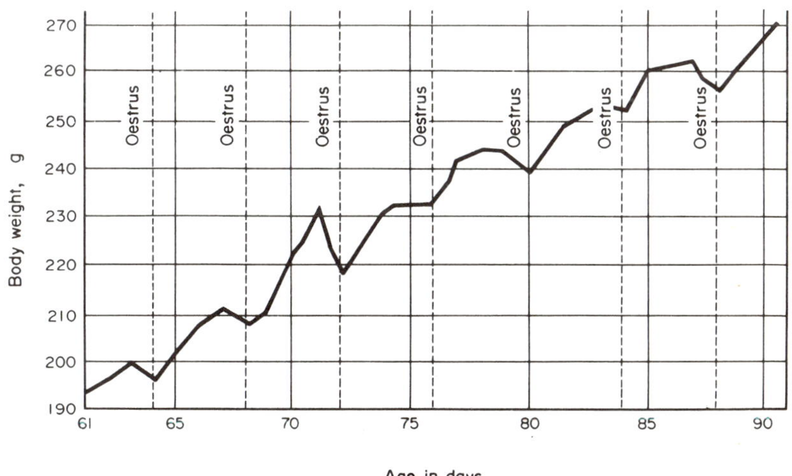

Fig. 57. Daily body-weight fluctuation in relation to estrus cycle—rat (drawn with permission from *IAT Journal*, vol. 24, no. 1, March 1973)

The pregnant doe will permit mating, but she produces no eggs at this time due to the action of the hormone of pregnancy (progesterone).

(v) *Mating hamsters*. Syrian hamsters may show a breeding season that extends from spring to autumn. They can be bred as monogamous pairs, provided they are paired up just after their weaning. This early pairing is necessary because hamsters tend to be aggressive little creatures. The mating of these animals should be observed if accurate records are to be kept. They tend to mate at dusk when they revive their activity after a day sleeping. They are noctural animals. The readiness of a female hamster to mate will be indicated by her behavior towards the male. She will fight off any male attentions to the extent of serious damage in some cases. This fighting behavior is more common in pairs where they are strangers and the female is on new ground having been introduced to the male for mating.

The female hamster ready for mating can be recognized by the presence of a waxy vaginal plug, or the absence of such, with little or no vaginal discharge. After the period of "heat", a vaginal discharge becomes more obvious and this develops into a more waxy plug over 3 or 4 days.

The harem-breeding method is more common for hamsters where a ratio of one male to five females can be adopted. The pregnant female is removed to litter and a young female replaced in the harem. The introduction of new members or even previous weaned-off mothers to the harem may cause much fighting amongst the members. This has to be monitored and action taken if needed, as the males can be seriously damaged around the genitalia.

(vi) *Mating ferrets*. These animals have a breeding season which extends from spring to autumn. The females tend to come into season in March/April, a little later than the males. The readiness to mate is accompanied, in the hob (male), by the testes descending into the scrotum and enlarging. The gill (female) shows an enlargement of the vulva region. About 3–4 weeks after this swelling appears, the females come into estrus and can be mated. If the swollen area has not reduced after 1 week of mating another attempt should be made. This area shrinks after ovulation, induced by the mating behavior. This is no guarantee that the mating was successful.

If ferrets are required to be mated at periods throughout the year, then they need to be reared under special conditions. Ferrets reared with short daylight lengths in the animal house have been described as being in breeding condition throughout the year. For the practical details relating to this work, refer to the Bibliography.

Ferrets are usually mated in a monogamous manner. The female is placed into the male quarters and left there for about 24 hours. The mated gills are then removed and caged with other mated females. Experience suggests that it is better not to put the mated females straight back into a cage with non-mated estrus females because they may be induced to a pseudopregnancy. The mated gills can be returned to the colony of estrus females about 6–7 days after mating. This kind of procedure can be adjusted in the light of practical experience and requirements.

The pregnant gills are separated a day or so before giving birth (as are any females exhibiting pseudopregnancy) because their maternal behavior can be disturbing to other cage inhabitants.

(vii) *Mating cats*. Cats have multiple estrus cycles (polyestrus) during their breeding seasons which are two or three in number between January and October. The queen (female) comes into heat every 14 days for about 3–6 days. During this period she will exhibit special forms of behavior rather than obvious physical signs. She tends to be "touch hungry" and shows pleasure in physical contact. She will tend to be more alert and more noisy. The typical arched back and raised tail can suggest the queen is in estrus. The vaginal smear technique can be used to confirm the stages of the estrus cycle that may be unclear from behavioral evidence.

Cats for breeding are generally kept in polygamous groups with one or more males present. Alternatively, the tom may be kept near to, but separated from, the queens. The queens are then taken individually to the tom's territory for mating. It can happen that these "visits" do not result in pregnancy. The reasons for these unsuccessful matings could be related to the fact that ovulation in cats is not spontaneous but related to not only the sex act but also the premating encounters such as the displays of various males. It must not be forgotten that many mammals are territorial creatures, claiming various dimensions of the environment as their own. They tend to mark out this area by means of urine odors and so forth. They are protective of their territory. The captive animal may have much of this "natural aspect" of their life omitted and their fertility may be affected. Breeders stress the importance of the attendants who care for the cat colony and their part in the success of the breeding program. It seems that cats are particularly sensitive to their environment and to those that attend them. They must also have adequate vitamin A and mineral supplements in their diet for successful breeding. Queens can be made pregnant by techniques of artificial insemination at any time of the year.

(viii) *Mating dogs.* Bitches generally have two seasons in a year at 6-month intervals. During each season there is one estrus cycle (monoestrus). The duration of estrus is about 21 days. The bitch accepts the dog for mating at about the 10th day after the beginning of the estrus. This period of "heat" lasts for a week or so and successful matings are reported at this time.

The bitch shows swelling in the area of the vulva as she moves into the early stages of the estrus cycle. Dogs begin to take notice, but mating is not permitted until a later stage when some signs of discharge are present at the vulval region. The stages of the estrus cycle can be determined by the vaginal smear technique as for rats and cats, if doubts exist.

The dog and bitch can be introduced to one another whilst both are restrained, in case aggression results. If the bitch shows receptive behavior, the two are left together in sufficient space for them to go through their courtship activities. After the mating both dog and bitch remain in a back to back "tie" because the male organ cannot be quickly removed from the female. When the mating phase has passed, the vulva swelling reduces and there is no further vaginal discharge. Ovulation is spontaneous and occurs during the early stages of estrus.

If the mating does not result in fertilization, the bitch may show a "phantom pregnancy" (pseudopregnancy) with the accompanying maternal behavior. This can even be exhibited by bitches that have not been mated.

(ix) *Mating primates.* Monkeys may be kept in polygamous groups if space permits. Reports suggest that fertility rates are higher with harem groups in open-air colonies as compared to the indoor monogamous pairs. There are distinct disadvantages with the larger colonies, however, in terms of record keeping for breeding purposes and in terms of health care. The mating of monogamous pairs requires that the technician document precise details of the female's menstrual cycle.

Successful mating is reported to be most likely on day 11 or 12 of the menstrual cycle. Day 1 is defined as the first day of bleeding. The female is left with the male for a period of 2 or 3 days commencing on day 11 of the menstrual cycle. The timings quoted here are to be regarded as generalizations because the menstrual-cycle lengths vary from female to female and so individual records are an indicator of when the female should be taken to the male. Some workers pair the animals for a longer period of time either side of day 11 in order to ensure fertilization. The monkey does not usually show a breeding season in captivity and has repeated estrus (menstrual) cycles throughout the year. The average duration of the menstrual cycle is 28 days. Ovulation occurs between days 8 and 21 of the cycle. The most common recorded times being on day 12 and 13. This is a spontaneous ovulation.

In most of the species considered so far, there has been some period of the sexual cycle of the female known as estrus (heat). This does not appear to be the case with primates as they mate at most periods of the menstrual cycle. This being the case, it is difficult to estimate when ovulation is taking place and thus difficult to be precise about conception times. It is common to assume that conception takes place when ovulation occurs if the female has been mated because sperms will be present in the oviducts. (Note: not all species of monkeys have the same type of sexual cycle.)

A combination of techniques are employed to estimate the time of ovulation in the monkey. The vaginal smear test and a study of genital mucus for the increased presence of chloride ions previous to ovulation are two useful aids. Another method is to palpate the genital tract by way of the rectum when noticeable changes in the uterus and ovaries are obvious at ovulation time.

5.4. Breeding and the environment

It has been stressed previously that some animals are ready to breed at any time of the year, others have definite breeding seasons. It is reasonable to ask why this is so.

We are all familiar with the birds coming into reproductive activity during the spring. This is at least true for many birds, and many mammals, in the cooler Northern Hemisphere. In the warmer, southern areas, there may be no equivalent of spring when new plant growths become available as a food supply. For this reason the breeding behavior is not linked to that particular time of the year.

Wherever the animals live, it is clear that to bring up young in unfavorable circumstances with low temperatures and poor food supplies is going to be avoided.

Researchers have shown that many animals have evolved a type of "internal biological clock" that is sensitive to environmental factors, such as daylight length and temperature. Linked with increasing daylight and temperatures is the growth of new vegetation, and the insect life that feeds upon it. So springtime in temperate zones seems to be the best time to bring up the young because of the readily available source of food.

The biological rhythms of animals of the Northern Hemisphere will be altered if they are transported to the Southern Hemisphere. In this way they adapt to the availability of food. On the other hand, it has been shown that Southern Hemisphere animals transferred to the temperate Northern Hemisphere do not change their activities, as they appear to be unresponsive to the changes in daylight length. This can create problems as the animals are likely to breed during unfavorable environmental conditions. This "unresponsiveness" has been reported for the Java deer imported to northern lands.

5.5. Pregnancy (gestation) and pseudopregnancy

The estrus cycle is suspended when a fertilized egg is implanted into the uterine lining, and develops through embryo and fetus to become a young animal. The events leading up to and during pregnancy may be summarized as follows.

Firstly, the egg membrane is penetrated by a sperm, up in the Fallopian tube. Spermatozoa are introduced into the female in their millions during *coitus*, or as a result of artificial insemination techniques. It is not possible to cross unrelated species because the egg will not be penetrated by "foreign" sperms. The fertilized egg, or *zygote*, begins to develop up in the Fallopian tube, and by the time it implants into the prepared uterine lining, it is already a ball of cells.

This ball of cells receives its nutrients for development by way of the increasing blood supply at the point of implantation in the uterine lining. This specialized area of the uterus builds up to become the placenta (after birth). It might be pointed out at this stage that the bloodstreams of the mother and young are not joined across the placenta. The placenta is a selective diffusion barrier.

Whilst these events of early pregnancy are in motion, there are changes taking place in the female body. In the ovary the remaining structure of the Graafian follicle, from which the egg was expelled, becomes glandular in nature. It secretes the hormone *progesterone*. This glandular structure is called the *corpus luteum*.

As gestation advances and the embryo develops to become a recognizable foetal animal, so the corpus luteum degenerates and the production of the progesterone secretion is taken over by the placental tissue. Estrogens are also produced by the placenta which inhibit continued follicle growth and ovulation. These quantities of hormones produce the changes associated with pregnancy, and the later birth (parturition). The changes of pregnancy occur in the vagina, the uterus, and the mammary glands. The non-pregnant animal does not show these changes. The corpus luteum degenerates and the estrus cycle usually recommences after a certain time. However, there are occasions when the non-pregnant animal will show physical and behavioral changes that suggest that the animal is pregnant. This *pseudopregnancy* is noticeable in some, not all mammals. Some brief descriptions of pseudopregnancy and its effects are outlined below where the gestations of individual species are discussed.

Mouse

The female laboratory mouse has a useful reproductive life of about 9 months. The young female is sexually mature between 4 and 6 weeks, but workers suggest the optimum age for mating is between 6 and 8 weeks.

The female will come into "heat" or estrus every 4–5 days in the presence of the male. She ovulates spontaneously. She is receptive in estrus for a period of 9–20 hours. After a fertile mating the female will go into a gestation period of about 19–20 days. After a sterile mating, the female may go into a period of pseudopregnancy which lasts 12 days, or about the same length of time as a normal pregnancy, 19–20 days.

A longer gestation period will result if the female is mated post-partum—that is, in the short period of estrus 20–24 hours after giving birth. The pregnancy may be extended by a week or more in this case. The pregnancy duration depends upon the strain of mouse and the demands of her suckling young.

Rat

The female laboratory rat also has a useful reproductive life of about 9 months to 1 year. The young females are sexually mature between 6 and 8 weeks, but are generally mated at between 12 and 14 weeks.

The female comes into estrus every 4–5 days. She is receptive to the male for about 12 hours, and ovulates spontaneously. Reports have been made of males force-mating a female, which is not in estrus, thereby inducing her to ovulate and to become pregnant.

After a fertile mating, the female is pregnant for a period of between 21 and 22 days. If the mating is post-partum then the gestation period is lengthened, by a week or more. This post-partum estrus is 20–24 hours after parturition.

Pseudopregnancy is reported to be less common in rats than in mice after mating with a sterile male. The male may be sterile for dietary reasons or because of high cage temperatures. Some workers suggest an animal that has been used for multiple matings over a short period of time may be less fertile, if not given sufficient rest between matings.

Guinea-pig

The female guinea-pig has a useful reproductive life span which varies between 2 and 3 years, depending upon the method of breeding employed. The young female is sexually mature between 4 and 5 weeks of age when she weighs about 200 g, but she is generally mated at about 12 weeks of age when she weighs 450 g. The males are slower to mature than the females, becoming fertile between 8 and 10 weeks of age. The males are generally presented for mating at about 12 weeks when they weigh 500 g. The figures quoted will vary for different breeding establishments. The female comes into

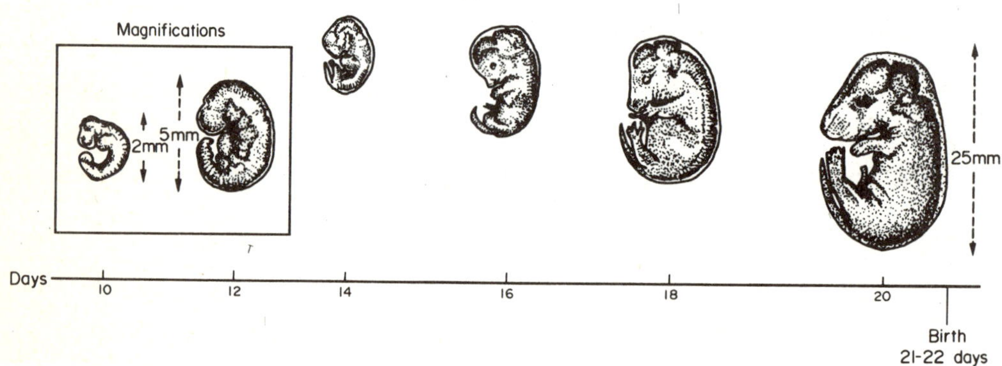

FIG. 58. Rat gestation

estrus every 13–20 days. The period of estrus lasts for 9–20 hours and ovulation is spontaneous.

After a fertile mating, the female is pregnant for a period of between 59 and 72 days. If the mating is post-partum then the gestation period is lengthened from an average of 63 days to 68 days. The post-partum oestrus is 20–24 hours after parturition. There are breeding advantages in terms of litters produced over a period of time in all cases of post-partum mating. This is particularly true for the guinea-pig when the interlude between succeeding litters is reduced from 15 weeks to about 10 weeks.

Pseudopregnancy as seen in mouse and rat is not reported to occur in guinea-pigs.

Rabbit

The useful reproductive life of the rabbit depends upon the type of animal and the way in which it is employed for breeding. Bucks have been described as breeding for 6–7 years and does for 5 years.

The young does reach sexual maturity between 6 and 9 months. The smaller the breeds, the earlier they mature. The does are generally mated when they reach sexual maturity rather than later. Their weights can vary between 2.5 and 3.0 kg, when mated.

The does have no estrus cycle in the same way as the female mice and rats. There is some suggestion that they show changes in sexual response which appear to be related to environmental factors, like food availability or daylight length. The female is generally on "heat" for prolonged periods in the absence of the buck. She is induced to ovulate by the males mating behavior or even by being mounted by a female.

After a fertile mating, the female is pregnant for a period of between 28 and 32 days. She can be palpated as early as 9–12 days to feel for the developing young. She will generally accept the male in mating whilst pregnant, after a period of time has elapsed (33–39 hours).

The doe shows a post-partum estrus when she readily accepts the male, and may become pregnant. This post-partum mating is most effective for breeding purposes, 21 days after the birth of a litter. The litter will thus have been sucked for 3 weeks before the mother is remated.

A sterile mating will generally result in a pseudopregnancy. This can happen even when a doe is mounted by another female. The duration of pseudopregnancy can range from 14 to 16 days during which the doe shows nest-building behavior. She can be remated after a period of time has elapsed (33–39 hours) if the previous mating is thought to have been a sterile one.

Hamster (*Syrian*)

The useful reproductive life of the Syrian hamster is about 9 months to a year. The young reach sexual maturity at about 45–60 days and are mated between 6 and 8 weeks of age.

The females come into estrus every 4 days. The estrus cycles are repeated throughout the year (polyestrus), but there are reasons to suggest that a breeding season exists between February and October. The period of estrus lasts between 4 and 24 hours and ovulation is spontaneous.

After a fertile mating, the female is pregnant for a period of between 15 and 18 days. The suckling mother usually does not permit post-partum mating. She permits mating at the end of her lactation.

After a sterile mating hamsters show a period of pseudopregnancy with a duration of 8 to 10 days.

Ferret

The useful reproductive life of the female (gill) ferret is about 3 years (3 seasons). The females become sexually mature earlier if reared under conditions of short daylight length. The short light periods can be 2, 6, and 9 hours, causing both male and female to remain in breeding condition. The female first comes into estrus at about 7–8 months under short daylight conditions. If she is kept under daylight conditions entirely, it takes about a year before she reaches maturity.

The hob and gill are first mated between 9 and 12 months when they weigh about 700 to 800 g.

The females are in estrus for prolonged periods when no male is present. She is generally in season from March to August. The ovulation is induced by mating. After a fertile mating, the female is pregnant for a period of between 41 and 43 days. The pregnancy can be checked out by palpating at about 21 days. The suckling mother may come into estrus early in lactation when she may be mated. If this is not achieved, mating can be made 16 days after she weans off her litter.

After a sterile mating, the female ferret may enter a period of pseudopregnancy lasting 42 days. This condition may also result, if unmated estrus gills are put with those that have been mated in the last 4 or 5 days.

Cat

The useful reproductive life of the queen is about 8 years and the tom about 6 years. The females reach sexual maturity at between 5 and 8 months of age. The males take longer to reach puberty, about 1 year. These times are not precise because animals born in the spring may not become sexually active until the year following. The queens are 2.5 kg or more when mated at between 8 months and 8 years of age. The toms weigh about 3.5 kg or more when mated at between 1 and 6 years of age.

The queen has recurrent periods of "heat" every 14 days. The duration of this estrus is 3–6 days, or even longer if no mating has occurred. The queens are polyestrus throughout the year showing no definite breeding season, despite the fact that breeding is likely to be confined to the first 10 months of the year unless daylight lengths are increased artificially.

After a fertile mating, the queen is pregnant for a period of 64–65 days. Ovulation is generally induced by the mating. The ovulation occurs some 25–27 hours after the mating. If the pregnancy is successful, the abdomen can be palpated 21–35 days later feeling for the presence of developing fetuses, no larger than the size of a garden pea.

A suckling mother may show a period of estrus during later weeks of lactation (4th or 5th). She may be mated at this time.

After a sterile mating, a pseudopregnancy has been described as lasting for about 23 days.

Dog

The bitch may attain sexual maturity when she is between 8 and 14 months of age. The weights of the mature animal will depend upon the breed. They are generally mated in their second season.

Mating is permitted during estrus which occurs within the two seasons annually. The bitch is monoestrus with an estrus period lasting 2 or 3 weeks in each season. She will attract the dog during this period, but will usually only permit mating at a particular time within her period of "true estrus" that lasts 4–10 days. This happens about 11–17 days after she commences the full estrus period. Ovulation in the bitch is spontaneous and commences during the early part of estrus.

After a fertile mating, the bitch is pregnant for a period of about 60 to 66 days. The pregnancy can be checked out by abdominal palpating around 3 weeks to 30 days. This technique of pregnancy diagnosis becomes more difficult as the bitch goes further into gestation.

After a sterile mating or in the absence of a mating, the bitch may display a pseudopregnancy when she behaves as if pregnant. She may even produce milk. This can last for a week or two, or longer, up to full gestation time of about 63 days.

The nursing mother or pseudopregnant animal is next mated in the following season which may be spring or autumn.

Primate (macaque monkey)

The rhesus female becomes sexually mature earlier than the male. The female is generally showing menstrual (estrus) activity at about $4\frac{1}{2}$ years of age and can be successfully mated later that year. The older the male ($5\frac{1}{2}$–15 years) used for the mating, the more success is reported. The weights of the paired animals are recorded as being between 4.5 and 9 kg.

The female imported from the Southern Hemisphere may show signs of a breeding season related to her natural environment, but this disappears after a period of residence in the north. This applies particularly to cage confined animals. These matters are the subject of discussion amongst the experts and generalizations are unwise.

The sex cycle of the female is described as the menstrual cycle and extends over 28 days, although this timing is subject to variation. Ovulation is spontaneous and occurs most frequently between days 12 and 13 of the cycle. It is at this time that mating is most likely to result in pregnancy. This period is not generally accompanied by an estrus or "heat" in the Macaque. They can be sexually receptive throughout the menstrual cycle.

After a fertile mating, the female enters a period of pregnancy lasting between 150 and 174 days. The average times are recorded as 164–168 days. The pregnancy can be diagnosed by a session of bleeding that occurs on days 15 to 21 after mating and lasts for longer than the usual menstrual bleeding. This implantation bleeding has been described as lasting 10–30 days. As a confirmation that pregnancy is underway a palpation of the gravid uterus may be carried out 3–4 weeks after the conception time. The palpation technique is carried out by way of the rectum.

The female returns to normal estrus activity 2–3 months after giving birth. Pseudopregnancies are not recorded for the Rhesus monkeys.

5.6. Parturition

Pregnancy comes to "full term" with the birth or parturition of the developed fetus. The degree to which the new born are independent varies with species. Some, like rats and mice, are born naked and helpless and require extensive mothering. Others, like guinea-pigs are born with fur, their eyes open and are able to move around.

At the end of gestation, the fetus generally assumes a position with the head pointing towards the birth canal (vagina). The foetus is contained within the fluid filled amniotic sac. This sac is pushed towards the vagina by contractions of the uterine muscles. These "birth contractions" are influenced by the pituitary hormone oxytocin. More powerful uterine contractions can be induced by an injection of oxytocin.

At birth the link between the mother and the young is severed. The placenta and umbilical cord are expelled: the mother generally eats this "after-birth". From time to time the mother may eat part of her litter if there is some confusion in recognition of the placenta and offspring. Sometimes a mother may eat her young if she is disturbed during parturition.

The mother may need to be separated from other animals at parturition in order to ensure good litter sizes, this can be true for hamsters and guinea-pigs under some breeding programs.

If the offspring is presenting itself in a breach position (posterior first) it may be necessary to perform a caesarian delivery. This can be true for monkeys, but not for dogs which do sometimes deliver tail first.

Species	Average gestation (days)	Average litter sizes	Average birth weight (g)
Mouse	19–21	4–12	1.0–1.5
Rat	21–22	6–12	4.0–7.0
Hamster	15–18	5–7	1.5–2.5
Guinea-pig	59–72	3–5	85 (female) 90 (male)
Rabbit	31–32	5–10	30–70
Ferret	41–43	3–15	6–12
Cat	64–65	3–6	90–140
Dog	60–66	4–8	350–50
Monkey (rhesus)	150–174	1	465 (female) 490 (male)

In the interests of economics, those mothers that cannabalize their young more than twice or have small litters, for whatever reason, must be excluded from future breeding programs.

5.7. Lactation and weaning

Lactation is the production of milk by mammary glands. The glands are brought into functional activity by the sex hormones oestrogen and progesterone secreted by the placenta during gestation. Towards the end of pregnancy, the oestrogen secretion by the placenta decreases and the "milk producing hormone" (lactogenic hormone) from the pituitary gland initiates lactation after parturition. The first milk to be secreted is rich in proteins and in some species it contains antibodies. This first secretion is called the *colostrum*.

The glandular cells of the mammary gland secrete milk continuously. The milk only flows along the ducts if there is suckling. The period of lactation is different for different species. This is related to the state of development of the offspring at birth. The rat requires a relatively longer period on milk than does the guinea-pig.

Weaning is the stage of development when an animal changes from a dependence on the mother as a source of a balanced diet, by way of milk, and takes to using solid foods and water.

Species	Average weaning age	Average weaning weight (g)
Mouse	18–21 days	10–12
Rat	20–22 days	35–50
Hamster	20–21 days	25 (female)
		30 (male)
Guinea-pig	14 days	180–200
Rabbit	6–8 weeks	800–1500
		(variable)
Ferret	7–8 weeks	300–400
Cat	4–6 weeks	500–1000
		(variable)
Dog	7–8 weeks	1000–1500
		(variable)
Monkey (rhesus)	3–6 months	700–1000

If attempts are made to wean animals too early, their growth may be retarded. Similar effects may be seen if animals are brought up in crowded conditions.

Mice weaned as early as 14 to 16 days may not grow as well as those that stay with the mother until 20–21 days. If young mice are kept at a density of thirty plus in a cage there can be a tendency for them to show less weight gain. Similar effects are also noticed if the density falls below a certain optimum group size. Experience only can lead one to determine the best conditions for rearing the "optimum" mouse in terms of weight gain. Mice at this age are very active and require skilled handling.

Rats weaned too early (17–19 days) also show reduced weight gain. Experience shows that the weaned young thrive better if kept in groups up to thirty. They tend to heap, as do mice, and thereby conserve body heat.

As the animals get older and bigger, the density of the grouping is reduced until they show weight gains consistent with the requirements of the breeder. This density figure will be determined by experience, under the prevailing conditions.

Hamsters are generally weaned by about 20–21 days and the sexes separated early in order to prevent any unwanted matings. If mating is intended at a later date, it is an advantage to have the sexes paired up or harem grouped just after weaning. As has been mentioned elsewhere, hamsters will fight vigorously when strangers are placed together. Incapacitating attacks on the male sex organs are a common feature of the hamster in conflict.

Guinea-pigs are weaned between 21 and 28 days when they weigh about 180–200 g. The sexes are separated and reared to the required weight for experiment or mating.

TABLE 6.
Mouse: Development Profile

Time (days)	Development features
0–1	Eyes closed (color discernible through the eye lids)
	Ears closed
	Short whiskers
	Sexing possible—greater genito-anal distance in the male
	Bright pink and naked
2–3	Ears open
	Longer whiskers becoming visible
5–16	Skin becomes increasingly pigmented
8–10	Hair color becoming visible in half-grown hair
	Mammary teats becoming visible
11–14	Eyes open
	Teeth begin to erupt
	Weaning begins
	First moult begins
	Elongation of the face area and reduced fatty tissue
	Greater activity
18	Outer ear pinna increases in size
21–23	Adult appearance achieved
	Weaning complete
	Animals very active
	Sexing more difficult because of hair covering
	(Remove those wanted for breeding to separate cages)
35–46	Sexuality clearly observed
	Vagina open
	Testes descended and visible

The body weight of a guinea-pig can vary with the conditions under which it is reared. A change of environment, for instance, can cause the animal to lose weight temporarily. The amount of free run available to the animal also has effects upon its weight gain.

Rabbits weaned early (4th week) show retarded growth and benefit by longer periods with the doe, up to 6 and 8 weeks. Once the young are weaned, they should be separated into groups by sex. The males become particularly aggressive to one another by 3 months. Their attacks on the genitalia resemble the behavior of hamsters.

Those animals to be mated can be grouped, if space permits, into colonies of 15 to 20. The alternative is to cage pair the bucks and does.

Ferrets can be started on wet food pellets mashed in warm water as early as 3 weeks. When the young weigh around 300 to 450 g at 7–8 weeks, they tend to take on dry food pellets.

Cats commence weaning at around 4 weeks when they may take cow's milk in addition to mother's milk. As time passes their body weight increases as much as 80–100 g per week. The weaning period is between 4 and 5 weeks.

Dogs may lick up fluid from a dish at around 3 weeks of age. If soft solids are added to a meat broth, this can also be taken during the 4th week. As time passes the pup begins to take more solid offerings, such as minced meat in week 5. At 6 weeks meat, milk, and fresh water is taken. Weaning takes place between 7 and 8 weeks.

Primates such as the rhesus monkey may make the transition from milk to solid foods, like a mash of fruits or egg and cereal products, at 2 to 3 months. The weaning can be at the age of 3–6 months.

5.8. Breeding programs

Breeding animals in a controlled situation can follow two possible pathways. The offspring can be selected for some desirable characteristics, and then mated with close relatives in order to perpetuate the wanted characteristics. Alternatively, the offspring can be mated with remote relatives, or newly introduced animals, so that "new blood" is put into the stock.

The study of breeding programs is to be considered under the following titles:

5.8.1. *Inbreeding programs*

5.8.2. *Random breeding programs.*

The choice of breeding program will depend upon a variety of factors, including those below.

(a) The species of animal, and its mating characteristics (i.e. frequent or infrequent estrus, passive or aggressive intersex behavior).

(b) The genetic characteristics required (i.e. "pure lines", with genetic similarities, "mixed-lines" with genetic variety).

(c) The age of the animals required (i.e. new born, weanlings, adults of given weight).

(d) The numbers of animals required over a given period of time.

The mating method chosen to produce inbred or random-bred animals will depend upon some of the factors mentioned above.

The mating methods may be described as:

(i) Polygamous (harem)

(ii) Monogamous.

The advantages and disadvantages of these methods will be described later after a closer look at the genetic implications of inbreeding and random-breeding methods.

5.8.1. INBREEDING PROGRAMS

If animals are produced by a program of inbreeding then they will become more and more "pure bred", or homozygous for particular gene-pairs (p. 181). If brother–sister matings (sib matings) are continued for twenty or more generations, then the degree to which the animals are inbred increases to a figure of about 98.6% (*coefficient of inbreeding*). If matings are between more distant relatives, such as cousins, then a high degree of inbreeding is less rapidly achieved.

This type of breeding program requires accurate records of animal identity, physical and behavioral characteristics as well as mating records.

To return to the beginning again. If we intend to set up an inbreeding program we must commence with a healthy accredited nucleus of animals. In this respect we begin

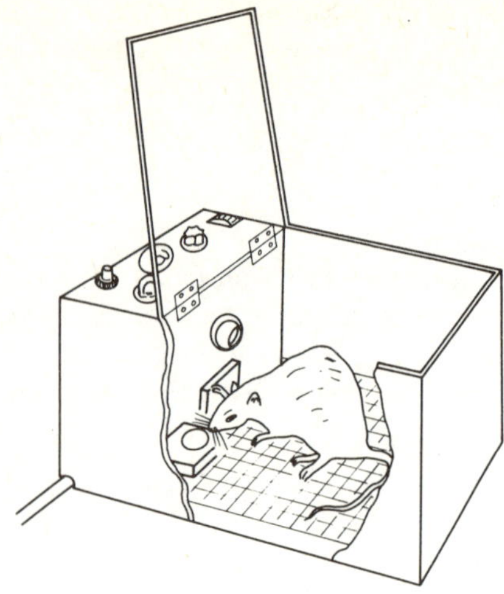

FIG. 59. Skinner-box experiment (adapted from S. A. Barnett, *A Study in Behaviour*, Methuen)

by selecting our breeding stock for certain desirable characteristics such as productivity, tameness, physical condition, and so forth. They will perhaps be "brought in" from specialist breeders. So the program begins with a selection process on the part of the breeder and it continues throughout the breeding program. It could be said that it is the selection process that reflects the skill and experience of the personnel. If these qualities are absent in the personnel then the breeding program may be less successful. It needs to be said at the outset that inbred animals are required by researchers because they are genetically similar and make it easier in the experimental situation to ignore, or give low priority to heredity variation as a relevant factor in the researcher's results. If he uses genetically similar animals throughout his experiments, then he can more easily ignore the hereditary differences.

Inbreeding for genetic uniformity

There are different methods of producing inbred strains of animals. Only one of the two described here produce the required genetically uniform population for use in experiments.

"Parallel line" programs of inbreeding involve mating the selected parents, followed by a brother–sister mating of the offspring. For convenience, let us say, six sib-matings take place. From each of the families of these sib-matings *one* brother–sister mating is carried out. If this continues, then six parallel lines of pairings are produced as shown in Fig. 60.

The result of this parallel line breeding program are six genetically different groupings, even though the matings are brother sister.

Single line programs of inbreeding involve mating the selected parents, as previously, but in this case only *one* pair of the offspring is selected for sib-mating. From their

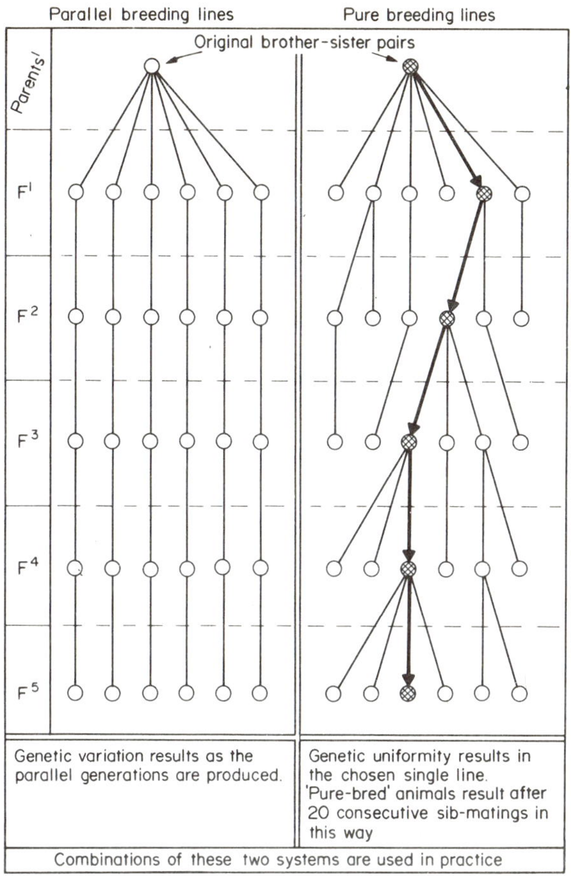

FIG. 60. Breeding programs

litter only one pair is again selected for brother–sister mating. This procedure continuing for many generations produces "pure bred", genetically uniform animals. However, if carried out exactly as described, there might be a danger of the line dying out if breeding did not progress as hoped. For this reason a compromise is made and breeding programs may include elements of both the "parallel" and "single" line breeding programs.

The effects of inbreeding are to increase the animal's "purity" (homozygosity) for certain gene-pairs. Amongst these gene pairs could be "disease tendency" or "low productivity" and in the pure-bred animal there is a greater showing of any "pure" hereditary characteristic. Inbred animals exhibit *inbreeding depression*. They lack "vigor", which means they show reduced fertility and reduced disease resistance. These unwanted characteristics of the inbred animal can be avoided by carefully controlling the size of the breeding population in a *closed colony* so that it does not become too small. The introduction of an "outsider" to the inbreeding closed colony will also in time revive the vigor of the stock (hybrid vigor). This "outcrossing" with new stock will break the closed colony, but it is necessary if the original closed colony shows inbreeding depression as a result of a single line breeding program.

5.8.2. RANDOM BREEDING PROGRAMS

If animals are produced by a program of random breeding they will not always be mated with close relatives. They will be mated with animals in the closed colony that are also more distant relatives. In this manner genetic differences of the individuals may be perpetuated and inbreeding depression will be avoided. This type of program does not exclude the possibility of inbreeding, what it does is to avoid established lines of ancestry such as results from inbreeding programs.

If one is interested in reducing the degree of inbreeding, that even becomes obvious within a random breeding program, then planning of the matings is necessary. A system of planned mating to produce offspring with *minimal inbreeding* requires careful consideration at both theoretical and practical levels. The reader is referred to the Bibliography for further reading in this area.

5.8.3. POLYGAMOUS MATING (HAREMS)

This method of breeding animals is where more than one male runs with more than one female. The numbers (ratios) of females to one male will depend upon various factors, such as the species of animal, and the available space. There are many species (hamsters) where this method is difficult because of the aggressive behavior of the partners or other harem members. Problems can arise with this method because litters may be disturbed by other members of the harem or they may be suffocated by overlying. It becomes problematic to keep accurate records of the matings and litters by this method.

An alternative to the above is to remove the pregnant females a few days before she is due to give birth. She is then transferred to a litter-box where the young will be kept until weaning. This permits the young to be monitored during their growth. After weaning the mother is returned to the breeding colony (harem). If the pregnant females are littered as groups, rather than singly, it then becomes more difficult to keep individual records as the young may become mixed up.

A disadvantage of the separation of pregnant females from the harem is that they cannot be mated in the short period of "heat" after birth (post-partum estrus). In the interests of productivity post-partum-mating (where it occurs in a species) is important because the suckling mother is carrying developing fetuses and there is no time lost waiting for her to be mated after she has weaned her litter.

5.8.4. TRIOS MATING

This method has one male to two females. There are larger litter numbers occupying one cage, which is an advantage. It can be more difficult to keep exact records of the offspring, especially if the litters are produced by both females at the same time.

5.8.5. MONOGAMOUS MATING

This method allows one male and female to remain together throughout their breeding life. The female is in some cases removed to a separate litter-box, because the presence of the male is disturbing to the mothering process. In other cases, the male stays with the female throughout, and can therefore fertilize her during the brief period of post partum estrus. This method is easier for record keeping.

The monogamous method is one way of producing inbred animals, but it presents problems in terms of space and extra labor. Keeping one male to one female clearly means more cages to be maintained, and more mouths to feed.

The most suitable method of mating for a particular species of animal may be determined by looking through the mammal data capsules (p. 255).

5.9. Basic animal genetics

Genetics, or the science of heredity, is concerned with the fact that parents and offspring resemble one another in some respects, and differ in others.

An individual animal originates from the fusion of a sperm with an egg at coitus. This fusion of a male gamete (spermatozoan) with a female gamete (ovum) is termed *fertilization*, and results in a structure called a *zygote*. An *embryo* is formed from the zygote as cell multiplication (growth mitosis) continues. Eventually the embryo takes on the characteristics of the adult and is described as the *fetus*.

These statements apply mainly to mammals. From this brief outline of reproductive biology it can be seen that the major link between parent and offspring is the sex cells or gametes. The heredity information is transmitted by way of these cells. The study of genetics examines the way in which this information is transmitted and the effects that it produces in the offspring.

The adult animal characteristics are not a sum total of the inherited information because the environment also plays a part. For instance, the quantity and quality of food available is important to the animal's adult body weight and size.

This study of heredity will overlap to some extent with material generally found in a biology course book. The extent to which this overlap occurs will be limited to a sufficient background to enable the reader to understand basic genetics.

The study is organized as follows:

5.9.1. Chromosomes and genes.
5.9.2. Cells and cell division (mitosis and meiosis).
5.9.3. Monohybrid inheritance.
5.9.4. Dihybrid inheritance.
5.9.5. Sex determination and sex linkage.
5.9.6. Lethal genes and gene interaction.
5.9.7. Mutants and inbreeding.

For a more detailed study of the genetics of animal breeding, the references quoted in the Bibliography (p. 299) will serve as an introduction.

5.9.1. CHROMOSOMES AND GENES

The nuclei of animal body cells contain thread-like structures that become more obvious at cell division, called *chromosomes*. These chromosomes are paired (diploid) reflecting the fact that all body cells originate from the zygote, a fusion of two cells, the sperm and egg. The chromosomes can be described as "paternal" or "maternal" suggesting their origins from the sperm or egg. This duplication is represented symbolically as 2n.

Different species of animals have different diploid chromosome numbers. The chromosome numbers for the laboratory species are tabulated in the data capsule (p. 255). Some examples of the different chromosome numbers are as follows:

Species	Diploid chromosome number (2n)	Haploid chromosome number (n)
Mouse	40	20 (pairs)
Rat	42	21
Guinea-pig	64	32
Gerbil (Mongolian)	44	22
Hamster (Syrian)	44	22
Rabbit	44	22
Cat	38	19
Dog	78	39

Chromosomes have located in them structures that determine our inherited characteristics. These structures we call *genes*. If a chromosome is confined to a particular gender then it is called a *sex chromosome*. If genes are located on such a chromosome then the characteristics are referred to as sex linked. For instance, color blindness is a sex-linked characteristic linked to maleness in humans.

At fertilization, the genes contributed by male and female are united in the zygote and the offspring's characteristics are an expression of this gene mixture.

Genes are represented by alphabetical symbols for convenience of discussion. Some genes encountered in laboratory animals are listed below:

Dominant gene	Phenotype	Recessive genes	Phenotype
A	Agouti pattern of hair	as	Non agouti pattern of hair
B	Black hair pigment	bb	Brown hair
C	Colored hair of any sort	cc	Albino or colorless hair and eyes
D	Non-dilute color of hair	dd	Dilute color of hair

The wild type, agouti mouse, has black hairs that have a yellow sub-terminal band. Genes ABP and C need to be present. The convention is that of capital letters for *dominant* characteristics and small letters for those that are *recessive*. If the characteristic is found in the wild-type condition this is indicated by the symbol +. The wild characteristics are adaptations to the natural environment and any mutation of normality can be a disadvantage to survival. Some interesting color mutants in mice have been produced with the following names.

Phenotype	Genotype
Chocolate	bbaa
Cinnamon	bbAA
Lilac	ppaa
Champagne	bbppaa

A gene occupies a particular position (*locus*) on a chromosome. When chromosomes pair up at fertilization, the genes at each of the loci or positions along their lengths come into opposition (see Fig. 61). These paired positions of genes, in their different versions, are known as *alleles*. Sometimes one of the genes in the pair dominates and masks the effects of the other. Sometimes the two genes interact and produce an intermediary characteristic or a new characteristic, as can be seen with comb characteristics of the domestic fowl (Fig. 69).

If the paired chromosomes carry different genes for a particular characteristic (i.e. coat color) then the condition is described as *heterozygous*. If the paired genes are the same with respect to coat color then the condition is described as *homozygous*.

5.9.2. CELLS AND CELL DIVISION (MITOSIS AND MEIOSIS)

The chromosomes, with their genes, are distributed throughout the body in all cells. It is the genes that "inform" the body chemistry in which way to proceed; to produce skin cells or sex cells and so forth. The distribution of chromosomes into daughter (offspring) cells takes place at cell division. There are two types of cell division, mitosis and meiosis.

Mitosis is a process of cell division that takes place in growth areas of the body producing new cells or to replace dying cells. The daughter cells are identical to the parent cells. The process is a continuous one, lasting about an hour, and for convenience is described as having four stages (prophase, metaphase, anaphase, and telophase). The mechanics of how the chromosomes are distributed during the division stages is not relevant here, but can be checked out in any introductory biology text.

Meiosis is a process of cell division that is confined to sex organs (testis, ovary). Sex cells (gametes) are produced as a result of meiosis. These cells unlike those produced

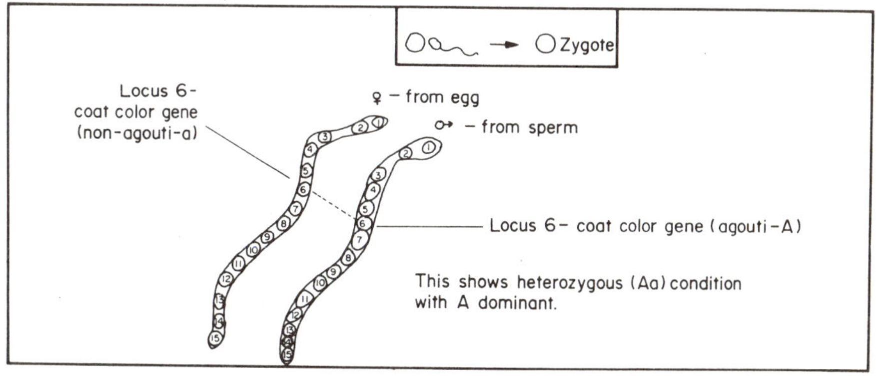

Fig. 61. A hypothetical pair of chromosomes

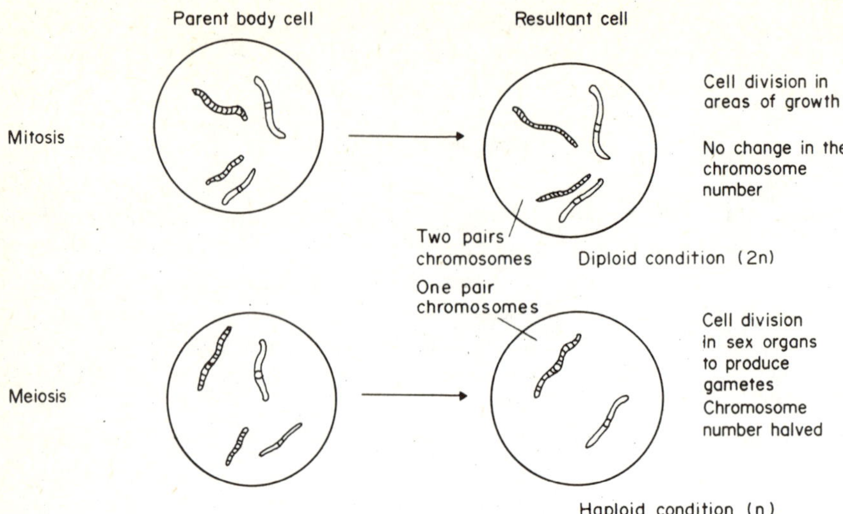

Fig. 62. Mitosis and meiosis

by the growth divisions, are not identical to the parent cells. The chromosome number is halved (haploid). This for good reason, because sex cells combine at fertilization to return to the original parent duplicate (diploid) number.

5.9.3. MONOHYBRID INHERITANCE

The earliest studies of heredity were carried out by Mendel using garden peas. He demonstrated that when two plants (which differ only with respect to one characteristic—such as size) are crossed, then one characteristic appears to be dominant over another. The pea plants he used were pure breeding for the selected characteristic. That is they were homozygous. This feature of dominance, and its variations, can be demonstrated with animals provided the animals are known to be "pure" for the characteristic (such as coat color) chosen for this demonstration.

A demonstration of dominance: recessiveness in mouse crosses

If a pure breeding wild type (agouti-AA) mouse is crossed with a pure breeding black (aa) mouse the result will be all agouti mice (Aa). These are known as *hybrids.* They are heterozygous, having a "mixed" genotype.

If these hybrids from the F1 (first generation offspring) are "selfed" (brother–sister crossed) then the F2 (second generation) will statistically result in three agouti mice to one black mouse. This will not necessarily be seen in each cross, but it is the pattern after a series of such crosses. The agouti mice are all physically similar (same phenotype) but they do differ in their genotype. Two of the agoutis are hybrids, one of them is pure breeding. The only way to determine which is which is to carry out a "test cross" or "back cross".

The "test cross" involves mating the pure recessive (in this case, the black mouse) with the animals (the agoutis) under question. If the agouti animal has a pure, homo-

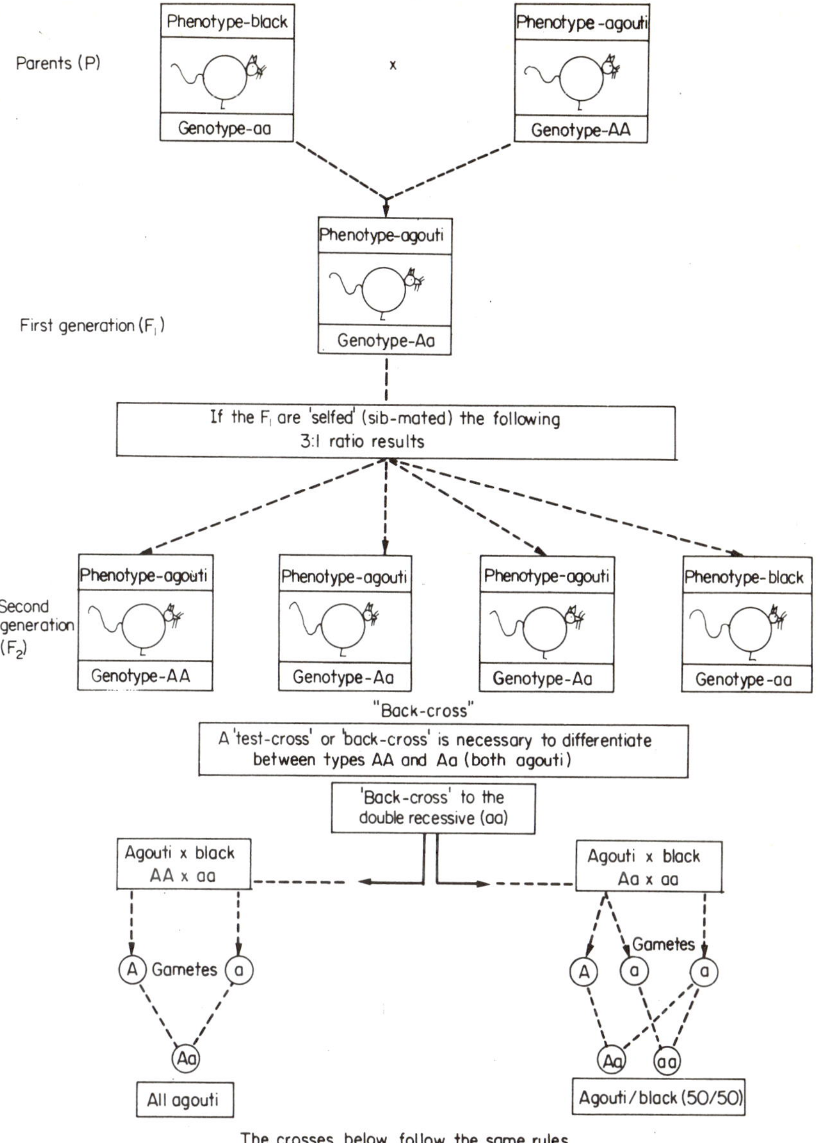

FIG. 63. Monohybrid crossing of a mouse

zygous genotype then all the offspring from the test cross will show the dominant agouti. If the agouti animal is not pure, but heterozygous for the coat color, then the offspring will show 50:50 agouti and black, as shown in Fig. 63.

When mice are "sib mated" (brother–sister mated) for at least twenty generations they are described as being inbred and homozygous for a particular gene (i.e. coat color). There are in excess of eighty inbred strains of mice available for research purposes. Some of the strains are bred specifically because they show a high incidence of different types of cancer. There are some inbred strains tabulated below:

Strain	Phenotype	Genotype	Cancer type	Others
BALB/c	Albino	AA bb cc DD	Plasmacytomas (induced)	Hypertension Heart defects
A	Albino	aa bb cc DD	Lung; mammary in female	Cleft palate
$C_{57}BL$	Black	aa BB CC DD	Low incidence of lung and mammary tumors	Alcohol (10%) preference
$C_{57}BR$	Brown	aa bb CC DD		
DBA	Dilute brown	aa bb cc dd	Mammary	Hypertension Heart defects
C_3H	Agouti	AA BB CC DD	Mammary: some liver tumors	

Some inbred strains of mice. Coat colour phenotypes and genotypes. (Source—see Bibliography.)

Inbred strains of mice obey the "law of dominance" in the manner indicated earlier. The genetic results of some crosses are shown here.

	Offspring	
Strains mated	Phenotype	Genotype
$C_3H \times C_{57}BL$	All agouti	Aa BB CC DD
$C_3H \times DBA$	All agouti	Aa Bb CC Dd
$C_3H \times BALB/c$	All agouti	AA Bb Cc DD
$C_3H \times A$	All agouti	Aa Bb Cc DD
$C_{57}BL \times DBA$	All black	aa Bb CC Dd
$C_{57}BR \times A$	All chocolate brown	aa bb Cc DD

Any albino mated with any colored mouse will produce only colored offspring. However, the colors will be different depending on the genotype of the albino parent.

The agouti pattern is dominant over any other color pattern. Black is dominant over brown. Inbred strains are increasingly being used in biomedical research because they represent a "genetic standard animal" that is then subject to a variety of experimental procedures. The reader is referred to the writings of Michael Festing (see Bibliography, p. 300).

A demonstration of dominance/recessiveness in guinea-pig crosses

The principles that governed monohybrid crosses of mice may also be demonstrated using guinea-pigs. This is seen in Fig. 64.

5.9.4. DIHYBRID INHERITANCE

The previous section has been concerned with crossing animals that differ by only one characteristic. In reality animals differ in a whole host of characteristics, but for

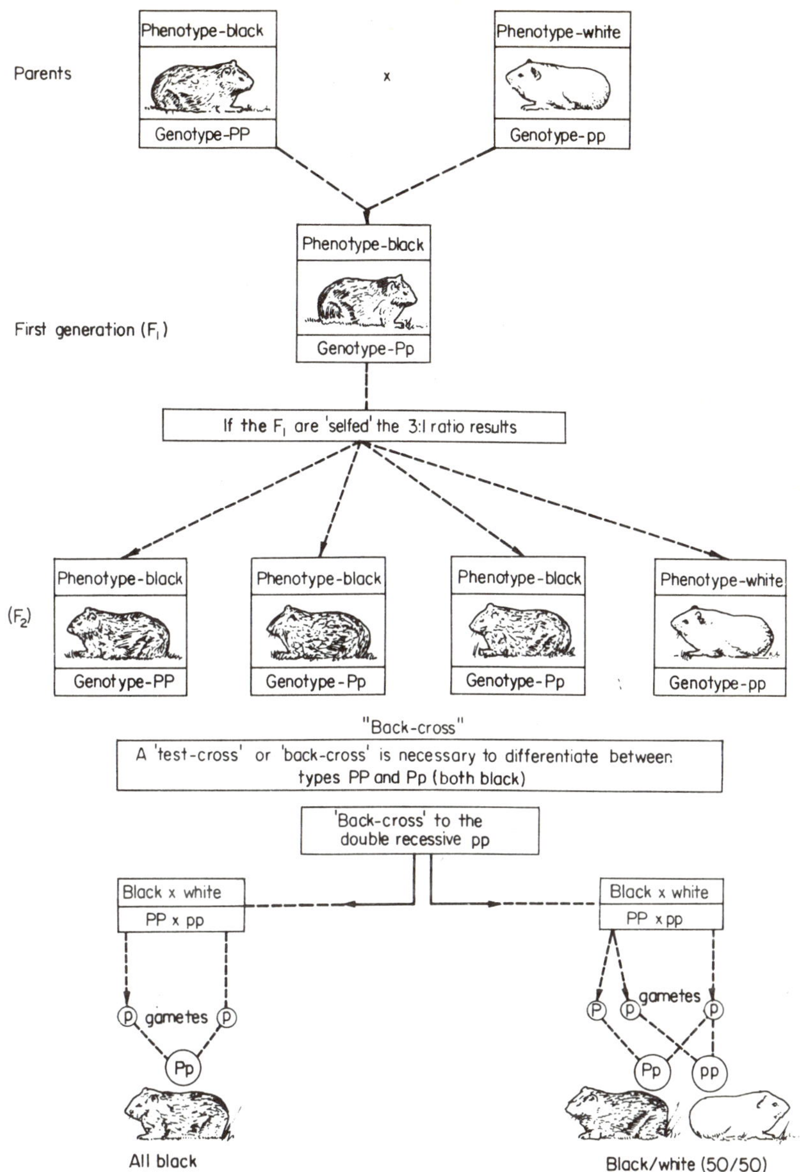

FIG. 64. Monohybrid crossing of a guinea-pig

simplicity in explanation we limit the differences being observed to small numbers. Dihybrid inheritance is concerned with the study of two character differences when pure-bred animals are crossed. The two characters may be hair color and hair length.

The way in which genetic characters segregate themselves out and are present in the offspring can be seen in Fig. 65. The dihybrid situation results in an F_2 with the described 9:3:3:1 ratio.

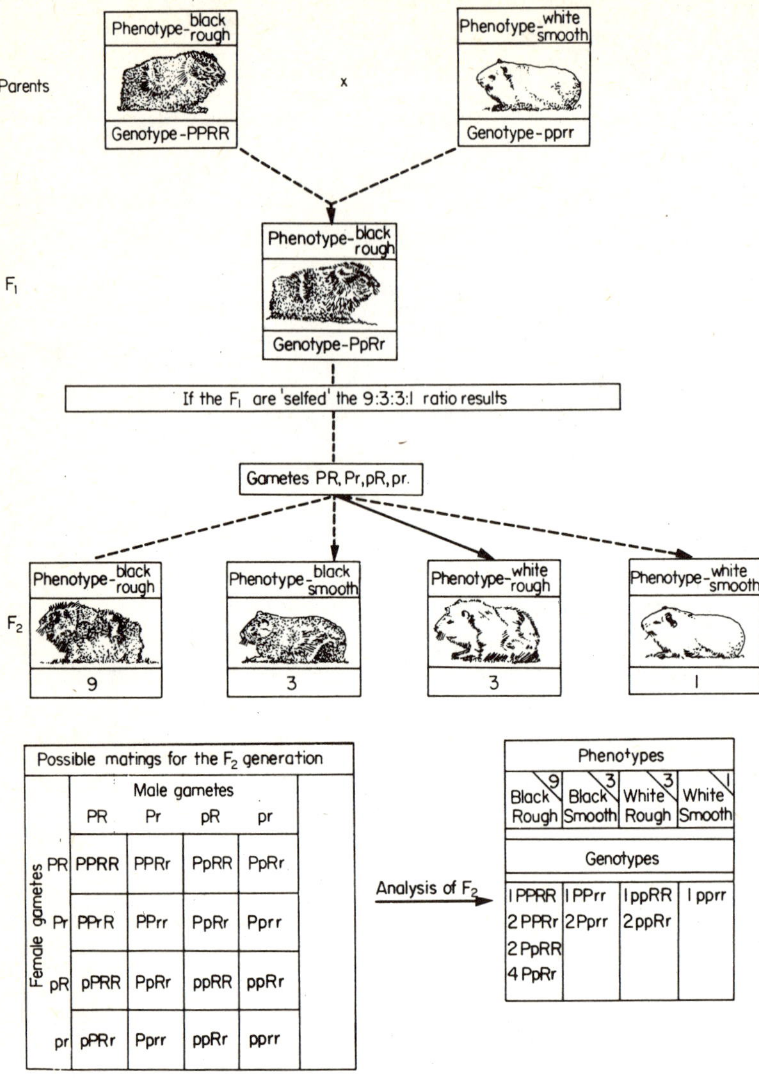

Fig. 65. Dihybrid crossing of a guinea-pig

Both monohybrid and dihybrid inheritance experiments may be set up for educational purposes using Drosophila fruit-flies. Provided the routines described on p. 214 are carried out efficiently then these animals are more suitable for short-term demonstrations. Suitable crosses are as follows:

Monohybrid crosses using Drosophila

    Yellow body (yy) × wild type (+ +)
    Ebony body (ee) × wild type (+ +)
    Dumpy wings (dp dp) × wild type (+ +)
    Vestigial wings (vg vg) × wild type (+ +).

Dihybrid crosses using Drosophila

Dumpy wing (dp dp, ee) × wild type (+ +, + +)
Ebony body
Vestigial wing (vg vg, ee) × wild type (+ +, + +)
Ebony body.

Genetics experiments with these flies is "fiddling", but they are considerably cheaper in terms of cash and time than the use of fish, birds, or mammals.

5.9.5. SEX DETERMINATION AND SEX LINKAGE

The gender of an animal is determined by the presence in the cell nuclei of sex chromosomes. Each cell has paired sex chromosomes, XX or XY. In most mammals the pair XX produces the female, and XY produces the male gender.

Males and females are approximately equal in an animal population for the reasons shown in Fig. 66.

Sex-linked characteristics are those related to the sex chromosomes. For instance, eye color in the Drosophila fruit-fly is determined by a gene located on the X chromosome. Color-blindness and hemophilia are examples of sex-linked disorders in man. They are caused by a recessive gene carried on the X chromosome. The male cannot receive any of the sex-linked characteristics from the father (because he contributes only the Y chromosome) but he does receive the X-carried characters contributed by the mother. The females tend to be carriers of sex-linked traits rather than express them, because her other X chromosome "covers up" any recessive genes.

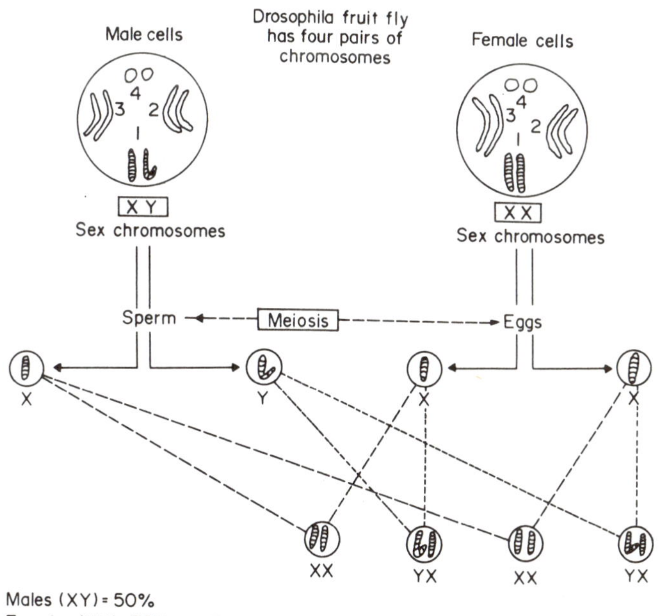

FIG. 66. Sex determination as demonstrated by fruit-flies

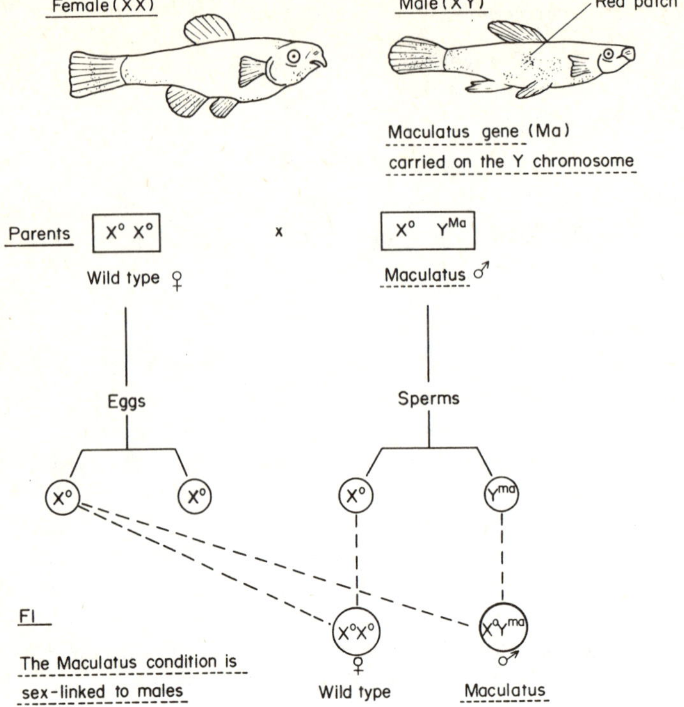

Fig. 67. Sex linkage demonstration using guppies

An explanation of sex-linkage is best achieved by running through an example. For laboratory demonstrations fruit-fly or guppy breeding experiments can be set up. Refer to Fig. 67 for a demonstration of sex-linkage in guppies.

5.9.6. LETHAL GENES AND GENE INTERACTION

There are occasions when a monohybrid cross will produce a homozygous recessive genotype which does not survive. Pure yellow ($A^y A^y$) mice die before birth. The presence of this lethal condition reduces the 3:1 ratio of F_2 to 2:1. Refer to Fig. 68 for a demonstration of this condition.

Genes do not always segregate out as if separate entities having no effects on other genes. Genes do interact in most complex ways. A good demonstration of this situation can be seen in the example shown in Fig. 69.

5.9.7. MUTANTS AND INBREEDING

Many of the rodents used in biomedical research are homozygous recessives which survive and breed well. The mutant albino type rodent is an example (cc). This stock is continued by mating always with other albinos.

There are occasions when the stock may need to be given more "vigor" by out-crossing with a non-albino to produce a non-albino hybrid.

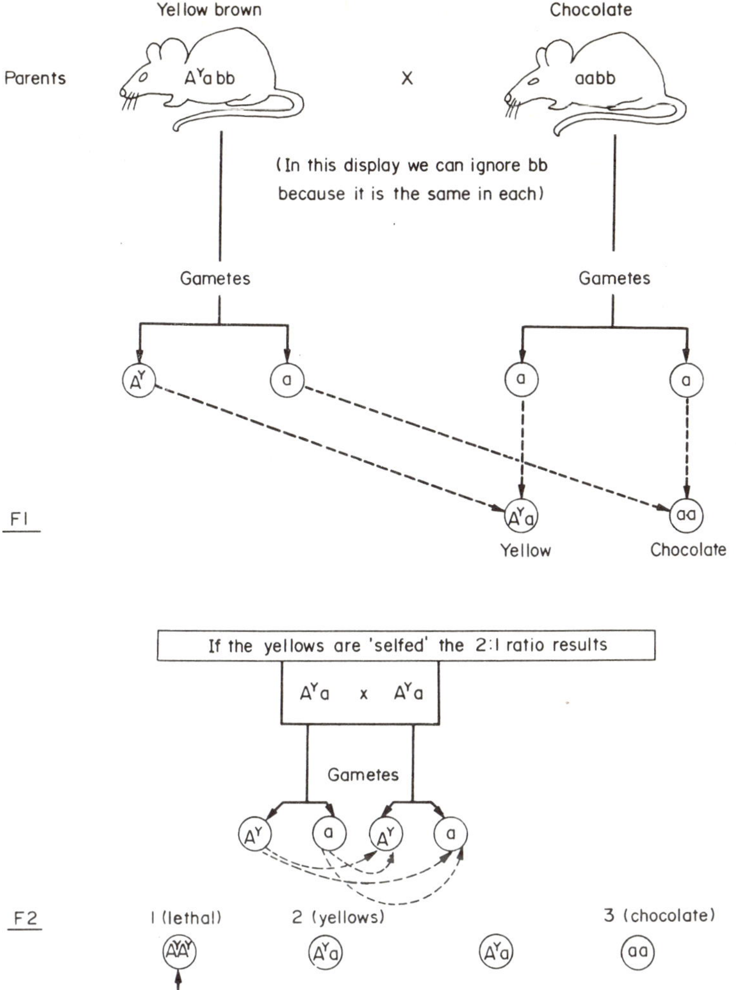

Fig. 68. Monohybrid inheritance with lethality

$$cc \text{ (albino)} \times ++ \text{ (non-albino)} = +c \text{ (all non-albinos)}.$$

These non-albino hybrids are selfed to produce the usual 3:1 ratio of a monohybrid cross.

$$\text{non-albino} \times \text{non-albino}: = 3 \text{ non-albino} + 1 \text{ albino}$$
$$(+c) \qquad (+c) \qquad\qquad (cc)$$

It is this latter albino which will be used to mate with others of its type to produce another strain of albino stock.

Inbreeding is the mating together of individuals which are related to each other through having one or more ancestors in common, (Falconer, p. 300, Bibliography).

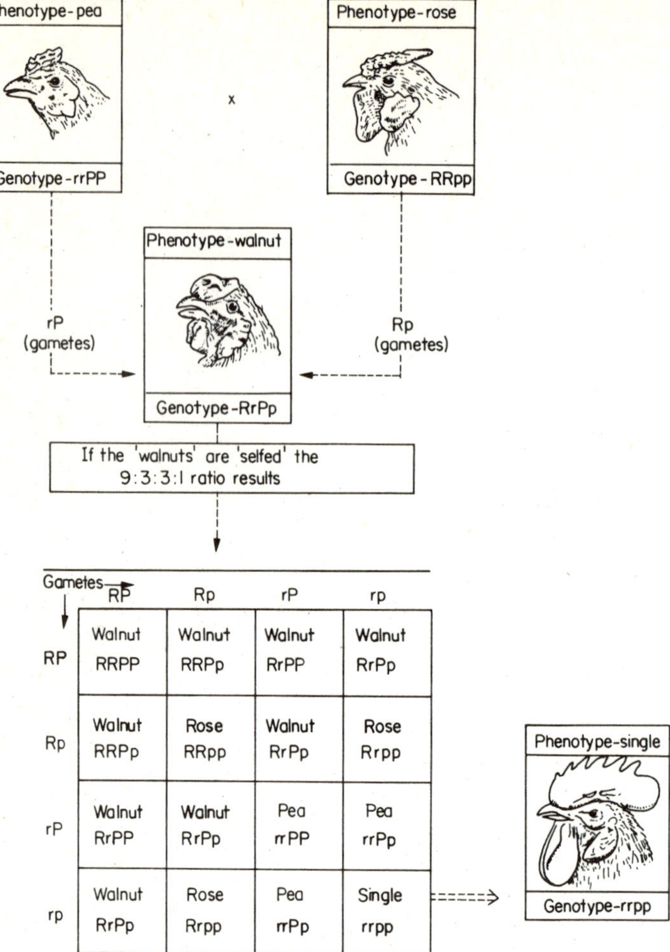

FIG. 69. Gene interaction demonstrated using different fowl comb shapes

One result of inbreeding is to reduce the number of individuals that are heterozygous for any one gene pair and to increase the number that are homozygous for one or other member of the gene pair.

Another consequence is *inbreeding depression* when the stock becomes less successful reproductively and less healthy. This is a result of the homozygous condition when more disadvantageous recessive genes may make their effects felt.

Further reading in this area may commence with the reference given previously.

5.10. Rearing non-mammals in the laboratory

The more common laboratory mammals have been discussed in some detail previously because of their importance in biomedical research. The non-mammals now considered are often used in educational science laboratories as aids to tuition or as tools in pure research.

Rearing or culturing details of a selected number of non-mammals are described in the following pages.

5.10.1. REARING THE ZEBRA FINCH IN THE LABORATORY

This little bird is an attractive addition to the animal room and is useful in behavior experiments.

The bird (*Taeniopygia castanotis*) is best bought into stock for breeding around spring time.

Culture

The birds can be bought from approved stockists and easily maintained at room temperatures and humidities in Great Britain. Adjustments may need to be made in areas where temperatures and humidities are exceptional.

Containers for the birds are available commercially, but can easily be made up if workshop facilities are on hand. A single wooden cage can be purchased with the side walls of the sliding type. If pairs, or more of single cages, are put together their intervening walls can be pulled out to produce larger living units. This is very practical when it comes to cleaning out.

A recommended cage size would be $75 \times 30 \times 45$ cm for a breeding pair or 5 or 6 non-breeders. The accessories for this cage should consist of a 1 cm diameter rough perch and a nest box area of $12 \times 12 \times 12$ cm with a 4-cm diameter opening. This box should contain some soft hay. An external lifting lid is handy to inspect the contents of the nest box. When eggs appear in the box, all excess nesting materials in the cage need to be removed to prevent overbuilding of the nest on top of the eggs.

The floor of the container should be covered with coarse sand or sawdust.

Feeding the finches involves hanging externally fitted food and water hoppers. The food will be seeds such as yellow panicum millet which can be supplemented with other seeds and some fresh plant fruits (chickweed, plantains, rye grass). These materials should be located so that they cannot be fouled by the feces. A piece of cuttlefish "bone" is to be hung up on the cage as a source of calcium.

Breeding finches show a courtship behavior and nest building. This usually happens with birds over 1 year of age.

5.10.2. REARING THE JAPANESE QUAIL IN THE LABORATORY

This bird (*Coturnix coturnix japonica*) is becoming increasingly available for laboratory workers. They have the advantage over some other birds in that they are easy to rear in the captive situation. They have been domesticated since the thirteenth century for their attractive characteristics, but have only comparatively recently come to be valued as laboratory animals.

The newly hatched quail is about 6–7 g and its growth rate to maturity is fairly rapid. By the time the bird is 1 week old its weight is increased by a multiple of 3 to 6. At around 8 weeks they weigh 120–150 g.

Culture

The quail tends to jump up into the air when it is anxious and so containing it in captivity must ensure that it is not too frequently disturbed and has adequate head room.

Containers for quail can be specially constructed of wood frame and perforated zinc (or purchased). A colony cage of 76 × 50 × 30 cm has been reported as suitable for a grouping of four males and ten females. Alternatively, caging can be improvized from two tea chests joined side by side. The intervening walls are removed and the floor scattered with straw. A double-sided mesh lid can be constructed on the top. The opaque sides of the chest reduce the incidence of disturbance. Two males and four females can be contained in this type of structure.

The temperature for maintaining the adult birds is around 18–19°C (64.4–66.2°F). Younger birds need higher temperatures in the range 27–29°C (80.6–84.2°F). The eggs are often ignored by the females and so need to be removed from the cage and incubated at 37.5°C (98.6°F) for $16\frac{1}{2}$ days. They need turning three to four times daily.

Breeding quails in the laboratory necessitates a particular attention of light (18 hours) and dark (6 hours). With this rhythm of illumination the birds come into egg-laying condition at 6–8 weeks of age. To achieve these lighting effects a time-switch mechanism is required. Two 40 watt lighting tubes mounted about a meter above the cages has been reported as satisfactory. Some interesting work has been done demonstrating that the birds kept under red light matured more rapidly and they produced more eggs. The males (100–130 g) are smaller than the females (120–160 g) and are bullied a good deal by the females. The female bird has grey tinted breast feathers with black additions whereas the male has a more chocolate brown coloration.

Feeding quails can be achieved by supplying them with turkey starter crumbs together with various brands of birdseed. This is best placed in external food hoppers to prevent the birds scattering and wasting the food.

Water needs to be supplied at ground level for the young birds but in such a manner that they are unable to stumble and drown. A few stones in a petri dish of water is suitable. The adult birds will drink from an external water dip dish.

5.10.3. INCUBATING CHICKEN EGGS

The chicken egg is a valuable tool in research. A wide range of bioactive agents may be tested by being injected into a living chick egg. The egg is a useful alternative to the employment of other laboratory animals in experimental testing procedures. It can be argued that the results have limited value in some research work and as such are not likely to replace the use of other experimental living material.

Chicken eggs under experimental conditions have had the following range of agents injected into them: viruses, bacteria, fungi, protozoans, drugs, toxins, vitamins, hormones, and living animal tissue. The experimenter would be looking for levels of toxicity or teratogenic effects.

The eggs to be incubated must be obtained from reliable dealers (in the season February–June) who can supply a high percentage of fertile eggs. A good batch will produce an 80% hatchability. The eggs should be clean, and weighing between 50 and 60 g. Before being incubated they can be stored for about 1 week (some report 2 weeks) at temperatures between 13 and 15°C (55.4 and 59°F) with a relative humidity of 75–85% (see **Bibliography**, p. 299).

Techniques

The fertile eggs will begin their embryological development at temperatures over 27°C (80.6°F) but are incubated at an optimum of 37–38°C (98.6–100.4°F) with a relative humidity of 60%.

The incubators used to maintain the eggs at constant temperature and humidity over the 21 days to hatching will vary with the demands of the experimental program.

There are two types of incubator in principle:

(a) *Still air or natural draught* types where the temperature may vary in different parts of the incubator.

(b) *Forced air or forced draught* types have mechanical devices for ventilating the incubator so that all parts are maintained at the same required temperature. Some of these types have mechanical egg-turning devices and humidifiers.

The commercial enterprise or large-scale research program will generally employ the latter type. The still-air cabinet for smaller numbers of eggs will necessitate manual egg-turning duties, at least twice daily. The water dishes must be refilled regularly and the relative humidity and temperature monitored. The box used for rearing locusts can be easily adapted for incubating eggs. Whatever system of incubation is used, it is essential to have a pre-run of at least 24 hours in order to ensure all is correct. If some of the commercially available incubators are used, it is a wise policy to check that the microswitches are functioning correctly. These switches turn the heating on and off as the observation lids are opened to turn the eggs or renew the water.

The degree of success in incubating eggs to hatching depends upon various factors, including temperature, humidity, oxygen, and carbon-dioxide concentrations, as well as the extent to which the egg is moved. Regular turning of the eggs is essential up until the 18th day but excessive vibration is likely to be fatal. The date of initial incubation can be written on the shell to act as a marker to the alternate clockwise and counter-clockwise egg rotations.

If for some reason the temperature falls to 25°C (77°F) in the first 7 days, this should not be in excess of 24 hours otherwise harmful changes are produced.

There appear to be two peaks of mortality during incubation. The first problem period is around 3–5 days when up to 25% of the eggs may die. The second peak of mortality is around 18–20 days when up to one-half of the eggs may fail to develop further.

In order to make checks upon the state of the eggs during their incubation, it is necessary to have access to, or to make, an "egg candler".

Egg candling is a method of observing the outline of the internal contents of the egg by means of transmitted light. The egg contents show up as shadowy outlines. An egg candler can be simply made for individual egg study by having a 40–100 watt bulb inside a wooden or metal box. An egg-shaped aperture, with a cushioned rim to

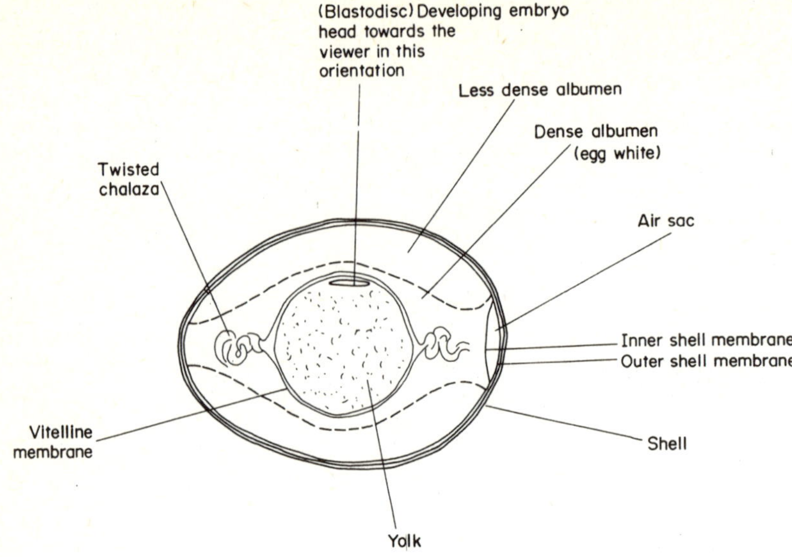

Fig. 70. Gallus egg (chicken)

support the egg is provided, so that light can only leave the box by passing through the egg. The egg is studied in a darkened room.

At $2\frac{1}{2}$–3 days an indistinct shadow with a radiating network of blood vessels will be visible. If the embryo dies between the 3rd and 7th day a "blood ring" is visible within the egg. The living embryo from day 7 onwards shows an increasing series of blood vessels close beneath the shell. As time passes, the shadowy dark area within the egg becomes larger and the outline of the air becomes more visible.

5.10.4. MAINTAINING AND BREEDING FRESHWATER FISH

Freshwater fish, which may be encountered in the laboratory, are included in the list below. These species can be used for demonstration purposes, for physiological research, or for genetics experiments.

Coldwater fish		Optimum temperature
Carassius auratus (goldfish)	Ancestorally Chinese fish that have been bred to produce a wide range of varieties. Reverts to dull color in nature	18°C (65°F)
Cyprinus carpio (carp)	A dark-brown bronze fish with mouth-edge barbels. Needs well aerated tank. Long lived. Prefer cloudy pond water. Best reared from young. Adults 60 cm or more	18°C (65°F)
Phoxinus phoxinus (minnow)	Good in large shallower tanks. Scarlet belly on breeding male. Prefers live food. Adults 7 cm to 10 cm	18°C (65°F)
Gasterosteus aculeatus (three-spined stickleback)	Pugnacious fish. Not community fish. Keep one male to three females. Scarlet belly on breeding male. Requires live food. Blue eyes, green silver sides. Adults 10 cm	River temperatures

Coldwater scavengers		Optimum temperature
Cobitis taenis (the spotted weather fish)	A scavenger loach with three pairs of barbels on upper jaw. Regular dark markings along the sides. Nocturnal in habit, burrows in sand during the day. No teeth. Bottom dwellers. Adults 10 cm	Adaptable to a range of temperatures
Ameiurus nebulosus (American catfish)	Scavenging nocturnal fish recognizable by large array of barbels. Generally kept when young. Adults 10 cm to 30 cm	

Warm water fish		Optimum temperature
Poecilia reticulata (guppy or rainbow fish)	Small tropical fish with rainbow colored males. Females paler and larger. Tolerant of foul water. "Live-bearers" with the males, anal fin used to insert sperm into female. Rarely becomes diseased	15°C–38°C (60°F–100°F)
Poecilia latipinna (black mollie)	Live-bearer with jet-black color. Male has large dorsal fin. Good community fish	24°C (75°F)
Xiphophorus helleri (Mexican swordtail)	Male has pointed swordtail. Variable colors. Males kept with several females. Easy to breed. Live-bearers. Feed on vegetable matter and live food. Adults 7.5 cm	22°C–26°C (72°F–80°F)
Xiphophorus maculatus (platy or moon fish)	All manner of color. Easy breeding (23°C–74°F). Live-bearers. Good community fish. Feed on live food and vegetable matter. Adults 5 cm	18°C–32°C (65°F–90°F)
Pterophyllum eimekei (angel-fish)	A good community fish, but tends to eat the young of other fish—not the adults. Feeds on live food. Not easy to breed. Adults 10 cm	26°C (80°F)

It is beyond the scope of this book to give detailed accounts of the aquarist skills in the breeding of fish, but an outline will be given of the breeding of one or two species.

Breeding tanks

These tanks are used exclusively for breeding young fish. The parents in general are best separated from their eggs and young because cannabalism is a possibility in most species. In order to achieve this separation a breeding tank is set up. The type of tank will depend upon the kind of eggs spawned by the female. The eggs may be adhesive or non-adhesive.

For the adhesive egg types some plants are needed for the eggs to attach themselves to. It is difficult to sterilize real plants if they are to be used as platforms for adhesive eggs so an alternative may be provided. Nylon-knitting fibers can be cut into short bundles or mops about 20 cm in length. They can be tied to a cork and floated or weighted to rest on the bottom. These "mock plants" can be boiled to sterilize, and can then be rinsed before use. Eggs will attach to these fiber lengths.

Some breeding tank suggestions are shown in Fig. 10.

Breeding fish requires that conditions are clean and the tanks must be free of pollutants. The condition of the water is very important. The pH of the water and the salts present in the water influence the hatching. There are proprietary products that can be used to produce an "instant pond" water for use when breeding fish.

Breeding goldfish (Carassius auratus)

The goldfish is an egg layer. She spawns over plants, to which the eggs attach themselves. In order to bring the adults into breeding conditions it may be necessary to carry out some preliminaries as follows.

Adult male and female are put into a breeding tank with the water at 18°C (65°F) where there are plants or plant substitutes. Separate the two by means of a glass partition for a week or so. They should be fed adequately on a varied diet, including live material. Ensure that no food is left over to pollute the tank water.

When the animals are thought to be ready for mating, the partition is removed, and the level of the water dropped, by siphoning off, to 15–20 cm. Some describe the best time to do this as on a warm evening. The spawning taking place in early morning.

Remove the plants, or plant substitutes, which have merely been floating in the spawning tank and transfer them to a hatching tank with water at 24°C (75°F). With luck the eggs will hatch out in 4 to 5 days. When the young have hatched out, the addition of algal green pond water will provide a source of microscopic food.

Rearing and feeding the fish fry can be a problem because they require minute particulate food which has to be increased in size as the fish grow.

The hatched goldfish will be able to survive a day or two on the residual egg yolk. After this time, they will need to have microorganisms in the green pond water. Drip-feed this into the breeding tank by a clip-controlled siphon. After a week or two on this type of food, they can be provided with larger particulate food such as the microscopic organisms provided by recently rotted lettuce leaves in water. This infusion (*infusoria*) of organisms can be prepared beforehand by allowing chopped clean lettuce leaf to begin the decomposition process in water. Drip feed this infusion. The water temperature over this period of time is reduced to 18°C (65°F).

After about a month the fry can be fed on pulped earthworms and water fleas. Brine shrimps (Artemia) are a good source of food for the quickly growing fry. They eat a great deal. Their feeding capacity is related to the water temperature. The warmer the water the more they eat. The brine shrimps can be purchased as dried eggs which, when added to oxygenated salt water of the correct temperature, develop into instant shrimps.

Live food is readily taken by growing fish and can be economically produced in the laboratory. It is no problem maintaining a water-flea (Daphnia and Cyclops) culture in the summer period. A culture of "microworms", small nematodes (*Anguillula silusiae*), can also be cultured at temperatures of 22°C–24°C (71.6–75.2°F) in the dark. These worms are available from traders.

The rearing of young fish is more problematic than the equivalent operation with most laboratory mammals. The control of the water purity is no small problem. No excess food should be given, it will foul the water. It is a good idea to siphon off some water, from time to time, and replace it with clean water of the same temperature. Regulated feeding of the fish as they become adult is customary. A little, often. There are automatic-feeding devices that can be used over week-ends or vacations.

The introduction of live foods can be a disadvantage in some ways because it may be a source of water contamination, in fact, a carrier of pathogens.

Breeding guppies (Poecilia reticulata)

These fish are easier to use for breeding than the goldfish, because sexing the animals is simple. They are ready for mating at about 12 weeks of age.

The males have a modified anal fin that is used as a sperm intromittent organ. They must be separated before 12 weeks to prevent unwanted matings.

These fish are live bearers unlike the goldfish which spawned eggs onto plant leaves. They can be left in a healthy tank with adequate plant cover for the young and a community will establish itself at temperatures between 20°C and 25°C (68–77°F). As live bearers the females retain the eggs internally until the young are developed enough to swim. Depending upon conditions, of temperature and food, the female may produce fifty young on an occasion. In a lifetime, there are reports of female guppies being broody between 15 and 30 times. That could be a lot of young, if they all survived.

Rearing guppies in a controlled manner requires that the females are removed from the community tank and put into a breeding tank. This tank has a perforated separator through which the young can swim to escape the cannabalistic behavior of the mother. She is put into this tank after she has been fertilized and the swollen abdomen shows a dark patch near the anal fin. The water is maintained at 24°C (75°F) to increase delivery time from a possible 3 months to 1 month or just beyond. She is fed live food (Daphnia) at this time in the breeding tank. When the young are delivered the mother is returned to the community tank. The young are kept a further 2 weeks or so in the breeder tank until they grow to a sufficient size to be safe in the community tank.

If the fry fail to swim to the surface at birth in order to fill their swim bladders, they generally fail to develop properly and must be disposed of.

5.10.5. REARING AMPHIBIA IN THE LABORATORY

The amphibia that are reared in the laboratory for purposes of experiment or demonstration, will depend upon availability. There is reason for concern in many areas because the numbers of Amphibia are being reduced in the wild, partly due to pollution, partly due to excessive collections for use in colleges and schools. Any successful method of rearing Amphibia in the laboratory is to be encouraged, if only for the purposes of conservation.

Amphibians by definition are animals that use both land and water during their life-cycle. There are some species that are adapted to live for longer periods in water or on land.

Amphibians, will for our purposes here, be put into two groups: those with tails and those without. Some examples are tabulated below:

Urodela	Amphibia with tails	*Triturus vulgaris* (*T. taeniatus*)	Common smooth newt	Europe
		Triturus helveticus (*T. palmatus*)	Palmate newt	
		Triturus cristatus	Created newt	
		Triturus pyrrhogaster (*Cynops pyrrhogaster*)	Japanese newt (red-bellied salamander)	Japan

Urodela	Amphibia with tails	*Triturus viridescens* (*Diemictylus viridescens*)	Common red-spotted newt	North America
		Triturus torosus	Giant California newt	
		Ambystoma tigrinum	Tiger salamander	
		Ambystoma maculatum	Spotted salamander	
		Ambystoma mexicanum (*Siredon mexicanum*)	Mexican salamander (axolotl)	South America
Anura	Amphibia without tails	*Rana temporaria*	Common north European frog	Europe
		Rana esculenta	Edible south European frog	
		Rana ridibunda	Marsh frog	
		Rana pipiens	Grass frog, leopard frog	North America
		Rana catesbeiana	Bullfrog	
		Hyla crucifer	Tree-frog	
		Bufo bufo	Warty toad	Europe
		Bufo regularis	Common African toad	Africa
		Xenopus laevis	Clawed toad	Africa

Rearing newts in the laboratory

These animals require specialist attention if they are to be bred in captivity. They have two phases to their life history, the aquatic and the terrestrial. Newts are generally collected from ponds when they are in their breeding season. The European newts can be kept in a suitable laboratory aquarium throughout the year. This forces them to abandon their terrestrial mode of life that normally follows on from the aquatic spawning phase. With good management and feeding they may even spawn in the following season.

The ideal set-up for the newt is a vivarium of the type described elsewhere (p. 30), a bog or woodland terrarium. In this laboratory habitat the animal is accommodated for both the aquatic and the terrestrial phase.

Aquatic accommodation for newts can be a simple freshwater aquarium (60 × 30 × 30 cm) with a water depth of about 20–30 cm. The tank should be set up in a similar manner as employed for freshwater fish. The bottom of the tank can be furnished with thoroughly washed stream or river sand to a depth of about 2–3 cm. The tank should be planted with numerous aquatic plants such as Elodea and Vallisneria. These act as places of concealment and oviposition. Keep the tank out of direct sunlight.

The surface of the water should be kept free of scum as this reduces gas exchange. An air bubbler will help to break up the water surface thus aiding oxygenation. The top of the tank needs to be covered with insect gauze netting, or similar, to prevent the newts from escaping. An overhanging lip to the tank may help to prevent escapes. Newts frequently get away from tanks by climbing up plant leaves resting against the glass sides.

The water temperatures need to be reasonably cool for the European newts; that is, 14–18°C (57.2–64.4°F). The Japanese newt requires a slightly higher water temperature 16–20°C (60.8–68°F). This latter newt is adapted to a permanent aquatic existence and only needs to leave the water for hibernation during periods of cold (6–8 weeks).

Terrestrial accommodation for newts can be provided by a bog or woodland terrarium with dimensions as large as is convenient for the circumstances

(90 × 40 × 30 cm). This tank should similarly be shielded from bright sunlight and maintained at temperatures between 15 and 18°C (57.2 and 64.4°F). This can be a problem in summer, or in the sub-tropics, so precautions need to be exercised. This also applies to centrally heated buildings in winter.

The wet vivarium to accommodate newts can also be used to hold some other amphibians, although unnatural groupings are inadvisable. For example, frogs should not generally be kept with toads.

Setting up a bog terrarium is quite straight-forward. The tank bottom needs to have a gravel covering, on top of which a loamy soil is positioned. This material is sloped to produce a land and a water portion of the tank. The junction between the two areas can be formed by a piece of natural wood. If need be, this wood can be put into position first of all and fixed by a waterproof adhesive to the tank sides and bottom to prevent water penetrating the land quarters.

The terrestrial area should be planted out with marsh plants and some mosses and ferns. The water is added after all the stones and plants are in position. This water may need to be heated thermostatically if tropical species are being housed. The air temperature can be heated by overhead lamps.

The top of the vivarium must be covered by any means that permits air circulation, but prevents escape.

Feeding newts can be on alternate days of the week, ensuring that any uneaten debris is swiftly removed before it becomes rotten and poisonous. The adult newts will take small, whole, or chopped worms. They can be fed thinly cut strips of beef-heart, or liver. This can be dropped into the tank or fed by means of blunt forceps to individuals.

Breeding newts requires that special attention is paid to the eggs and larvae which may not survive easily in artificial conditions.

Sexual activity is accompanied by bodily changes in the male. Color changes become pronounced as they enter the breeding season. There can be spots along the flanks and bright belly colorations. Crests along the back of the male make it distinctive during this mating period when the pairs swim around the tank in elaborate courtship dances.

The females of the European newts lay individual eggs on leaves over a period of a month or so in the season. They will produce 3–10 eggs each day depending upon water temperatures and species. If the female has received a male sperm pack (spermatophore) her eggs will be fertilized and with luck develop into larvae. There is no easy method of rearing larvae described in the literature, but if they do appear it is usually in water with temperatures between 15 and 25°C (59–77°F). They need to be fed on water fleas and small Tubifex worms.

Rearing salamanders in the laboratory

These animals are a larger version of the newt and their requirements are similar. There are not too many species that can be kept for long periods in the laboratory. The exceptions are those aquatic types, like the axolotl (*Ambystoma mexicanum*), which seem to survive for many years in the most basic conditions. The yellow and black terrestrial tiger salamander (*Ambyostoma tigrinum*) survives for variable periods in wet vivaria such as those described for newts. It needs to be fed with live food such as meal-worms and various insects.

Rearing axolotls in the laboratory

The axolotl, commonly kept in laboratories for research or demonstration, is a rather large "tadpole" or larva. It has all the features of the larval stage of an amphibian, yet it is sexually mature. This condition is called *neoteny*. They can be induced to metamorphose in the laboratory when they are young by special treatments, involving stimulating the thyroid gland into activity. The metamorphosed animal loses its gills and takes on the characteristics of terrestrial salamanders. In Germany there are reports of a metamorphosed axolotl that has produced offspring.

Accommodation for axolotls is not difficult to set up. The use of a large glass aquarium is suitable or a porcelain sink.

The bottom of the tank is not gravel covered which makes it easier to siphon up decaying debris. The walls tend to get somewhat slimy with algal growths which can help in the water-purification process.

The water used to keep the axolotls should be chlorine free and to a level of about 25–30 cm deep. Axolotls can jump a little so the tank needs to be covered in such a way that air is not excluded. The water temperature should not be too high. After all, they are from mountain lakes. Keep them at 14–18°C (57.2–64.4°F).

These animals are well known for their ability to regenerate damaged tissues. This was particularly well demonstrated by those in our tanks at Oxford which had some of their legs chewed off during a short stay with Xenopus whilst their accommodation was cleaned out!

Feeding axolotls is not a difficult exercise. They can be forcep fed, to reduce fouling of the water, or left to seek out food put into the water. Beef-heart or liver strips are readily accepted if they are of mouth size. They should be fed on alternate days with two or three meat strips. Some people supplement this meat diet with multiple vitamin preparations.

Breeding axolotls in the laboratory will be outlined here, but for detailed information and techniques the references in the Bibliography are recommended (p. 299).

Sexual differences are particularly noticeable in the breeding season, which is from December to June. The males are slimmer and have longer tails than the larger females. The sexually mature male has a swollen area around the urino genital opening (cloaca) at the base of the tail. Both animals are sexually mature by the time they are a year old and can be used for breeding for about 5 years.

The breeding season comes with a drop in the water temperature, and in the laboratory a drop of 10 degrees seems to trigger off mating behavior. Males and females are kept separately at 22°C (71.6°F) for a day or two before being introduced together in a tank with water at 12°C (53.6°F). The fertilized female lays large numbers of eggs (500–800). These eggs are then reared under extremely clean-water conditions at a temperature of 18°C (64.4°F). The young larvae being fed on controlled quantities of water-fleas.

Handling axolotls requires some skill because netting them out of a tank is not advisable. The fine gill structures and some digits may get entangled in the net and become damaged as the animal struggles. Two hands need to be used grasping the animal as one might pick up a log of wood from the ground.

Rearing frogs in the laboratory

There are large numbers of frogs used in laboratories for a variety of reasons, but they are by no means the easiest animal to keep in a healthy and breeding condition.

If frogs are to be kept for a short duration until used in an experiment, they should be put into a porcelain or stainless-steel sink with running water. The wastes will be flushed away and the frog's skin kept clear. The water does not need to be any deeper than 1–2 cm, and some Sphagnum moss on the bottom will give them a non-slip surface, as well as a place of concealment. Keep them cool and dark with a minimum of disturbance. They will not require feeding under these conditions.

Frog skins are particularly sensitive to chemicals and to infective agents. If any frogs in transportation are exposed to dead animals, they must be immediately washed. Always remove dead or decaying material from contact with the animals as this can be one of the main reasons for fatalities. Isolate any animals showing the inflammatory skin condition known as "red leg".

Accommodation for frogs over longer periods of time will depend upon species and the environmental conditions. The information outlined here refers to *Rana temporaria* or *Rana pipiens*, unless others are mentioned.

A wet vivarium is quite suitable for frogs where it has access to land concealment areas as well as water. The important points to consider are, keeping the temperatures down and preventing escapes. This type of terrarium is useful to accommodate grass-frogs, tree-frogs, and bullfrogs. The latter animals being in need of a good water area. It might be worthwhile repeating the caution about mixing species. They may disturb each other so that feeding may be inhibited. In the case of frogs and toads they may poison each other because of their toxic skin secretions.

A useful outdoor vivarium, constructed rather like a gardeners cold frame, has proved rather useful for rearing European frogs. This structure has the sides and tops made secure with panels of insect netting. Frogs seem to be able to escape from the smallest apertures and so the minimum number of opening lids is advisable. A food hole inlet needs to be provided so that live flies or meal-worms can be introduced. Dug into the land area of the vivarium is a self-levelling overflow porcelain sink containing plants and stones. This outdoor type of vivarium needs to have a well-planned drainage system to avoid the formation of a swamp.

Feeding caged frogs can be achieved by having dishes of fly pupae situated in the vivarium, as the flies emerge they will be taken by the frogs. Meal-worms or maggots can also be offered in this way. For a more automated approach, an insect-breeding cage can be "plumbed into" the vivarium emerging flies to enter at will. "Force feeding" with beef-heart or liver has some success in maintaining unwilling animals in good condition.

Handling and sexing of frogs is fairly straight-forward. On account of the slippery skin of the animals it is more convenient to pick up the animals by the rear legs. They can be transferred to a clean cloth for greater security.

The males of *R. temporaria* are recognizable externally by the hardened pad on the thumbs, used for gripping the females under the arms during mating. They are called the "nuptial pads".

Breeding techniques for frogs in captivity have not been well documented, but any person who has attempted to maintain frog spawn through to adult frogs will recognize the difficulties. The critical points appear to be the condition of the water and the proper food supplies.

Rearing tadpoles from frogs eggs is not a great problem, and is a popular experience for young school children. These larvae pass through different phases of feeding which must be satisfied.

Frog spawn begins to appear in pond water during springtime in the Northern Hemisphere. This can be kept in natural water in aquaria away from bright light and heat. The pond water will contain the "infusorian" food for the emerging tadpoles. As the animals grow, they change from feeding on the green algal growths to feeding on flesh. This can be given to the animals as their legs begin to appear. A fresh section of earthworm suspended on a piece of cotton will attract the tadpoles that have rasping teeth for eating meat. The flesh should be withdrawn from the water to prevent fouling.

As the tadpoles get older, they lose their tails and must now be given access to platforms above the water. This can be floating twigs or leaves. The problems of survival now begin. Most people seem to experience difficulty in rearing these froglets to maturity. This may be a problem of feeding. They do need an ample food supply of the correct type. It is best to release the froglets at this stage or to transfer them to a "near natural" outdoor vivarium. Disease in adult frogs is difficult to eradicate. The most familiar complaint being the bacterial skin disorder called "red leg". These animals are best isolated or destroyed. Some attempts at cure have been recorded, such as immersing the animal in 1–2% copper sulphate for an hour or so. Antibiotic treatment with aureomycin or teramycin may be attempted. This is achieved by administering a solution through a stomach tube, once or twice daily. This can be continued for a week during which time the animal is starved. The recommended antibiotic solution is 25 mg/per cm^3 and dosed as 0.3 cm^3 per 30 g body weight.

Rearing toads in the laboratory

These amphibia generally differ from the frogs in the sense that they have rougher, drier skins and are less liable to dehydrate on land. They could be thought of as much more alert and intelligent than frogs.

The toad more commonly found in the laboratory situation is *Xenopus laevis*, the African clawed toad. This animal is permanently aquatic and, given the correct conditions, easily bred in captivity.

Accommodation for the terrestrial toad in the laboratory can be the wet vivarium described earlier. The temperature must be kept at or above 21°C (69.8°F) if the toad is to remain active. At lower temperatures the animal will tend to burrow itself and hibernate.

Feeding toads is a little easier than feeding frogs. They require live food, such as small ground insects or flies. They may even take earthworms presented on the end of blunt forceps.

Rearing Xenopus in the laboratory

Amphibians in nature are being seriously endangered, and any attempt to rear stocks of these toads as a research alternative will make ecological sense.

Containers for the adults can be glass or polypropylene. The containers used to hold five or six pairs should be about $60 \times 40 \times 15$ cm ($24 \times 16 \times 6$ in.). They need to be covered with perforated zinc or a secure netting of 2.5 cm (1-in.) mesh to prevent the animals jumping out of the tank.

The tank in which the adults are to be kept for breeding need to have a protecting tray put into the bottom. This structure can be a sheet of perforated perspex standing on short legs. The perforations should be large enough to permit the eggs to fall through the tray onto the bottom of the container box. This prevents the adults from eating the eggs.

An alternative to a perforated perspex tray can be a suitable plastic-coated netting fitted into the bottom of the breeding box, a short distance off the base to allow the eggs to fall through.

The tadpoles and young toads may be reared in the same size containers as the adult, their numbers being thinned out as they increase in size.

Culture water needs to be chlorine free (tap-water, stand 24 hours) for both adult and tadpoles. It is not so important for the adults, but it is possible that the tadpoles will be killed by small amounts of chlorine. It is probably advisable to use rain-water or distilled water for the eggs and tadpoles (some recommend the addition of trace elements to this distilled water). The water should be to a depth of 8–10 cm (3–4 in.). In nature these amphibians live in stagnant water, but in the laboratory this water must not be permitted to become foul. The water needs to be changed about a day after any feeding operation as the tanks will become fouled with the food debris.

The temperature of the water should be held within 10–30°C (50–86°F) with a preference for the range 20–23°C (68–73.4°F) for the breeding of animals. The cultures should be kept in a more shaded area.

Feeding Xenopus adults is not a problem if the temperatures are maintained near the optimum. They will take chopped earthworms, liver, and heart slices. There are Xenopus pellets of various size-grades available. The adults can be fed individually to avoid too much debris dispersed in the water. The feeding should be twice a week.

Feeding Xenopus tadpoles requires more attention as they need to be presented with suitably sized materials. On hatching, the tadpoles need to be fed with an infusorian mixture or a green culture of Euglena. As the tadpoles develop they can be fed five or six times a day on sieved baby foods and sprinkles of nettle powder. Some protein contribution may be made by squeezing hard-boiled egg yolk through a sieve into the water. After all these feeding operations care should be taken to siphon out the debris to avoid water putrefaction.

As the tadpoles begin to metamorphose to become young toads, they change from the herbivorous to the carnivorous habit. They need to be fed well-washed Tubifex worms (to prevent introduction of infection), Drosophila (vestigial wing types), Tribolium and white worms.

Breeding Xenopus toads

Male Xenopus toads are smaller in size than the female, weighing about 90 g. The area around their cloacal region does not show any significant lips of labia.

Female Xenopus adults are larger than the males, weigh about 170 g and have significant labia in the region of the cloaca. The "stitch marks" on the sides of the abdomen are more noticeable on the females. Both sexes are mature at about 2 years. The toads do not often mate in captivity unless they are given injections of sex hormones.

Handling the toads is necessary if they are to be injected. The handling is achieved by gripping the animal over the back with the fingers between and around the rear legs. The animal is secured in this manner by using a hand held dry cloth.

Injection techniques with Xenopus toads

Injecting the toads with chorionic gonadotrophin (Pregnyl) requires disposable 2 cm³ syringes with No. 18 size needles. Both the sexes are injected twice. A "primer" brings the animals into breeding condition and a second or final injection is given to bring about the discharge of gametes.

Preparations for the injection. The following should be available:
 (i) Pairs of acclimatized mature toads.
 (ii) Tanks with "false bottoms" for the breeding pairs.
 (iii) Two 2 cm³ disposable plastic syringes with No. 18 needles.
 (iv) Ampoules of the hormone. Some 100 iu (international units) packs; some 500 iu packs. These are ampoules containing the white solid hormone (Pregnyl).
 (v) Ampoules of distilled water (1 cm³).

Giving the primer injection. The male is to be injected with 50 iu of the hormone.
The female is to be injected with 100 iu of the hormone.
Prepare a syringe for each of the sexes.
(*a*) *Injection for the male.* Break off the top of a distilled water ampoule and draw up all the water into a syringe.
Break the top off a 100 iu Pregnyl ampoule, and insert the needle of the syringe containing the distilled water. Empty all the water into this ampoule of Pregnyl, thus dissolving it.
Draw up the Pregnyl solution into the syringe. Invert the syringe, having the needle uppermost, and push the plunger in, gently expelling all the air bubbles.
For the male injection only 0.5 cm³ of this Pregnyl solution is given.
(*b*) *Injection for the female.* Carry out the same procedure, as just described for the male. All the Pregnyl solution is to be injected into the female.
(*c*) *Injecting the toads.* The hormone is injected into a dorsal lymph sac which are located either side of the mid-line of the animal near the posterior. The needle is pushed just beneath the skin on the leg side of the "stitch marks" and inserted further beneath these markings towards the mid line.
The appropriate quantity of the hormone solution is pushed into the animal. Be careful not to penetrate deeper body structures. It helps to have the animal stretched

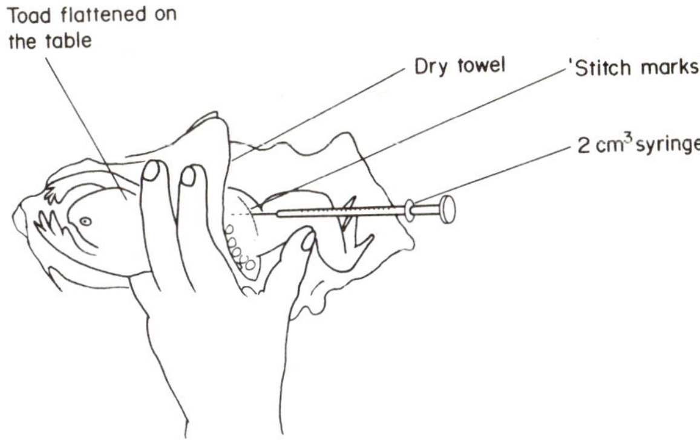

FIG. 71. Injecting the hormone into Xenopus (redrawn with permission from the Longman Group Ltd.—taken from *Nuffield Advanced Science (Biology) Laboratory Book*)

out flat on the table beneath a dry towel. Two people may need to be involved in this operation initially.

When both animals have received their primer injections, they are placed together in the breeder tanks at 23°C (73.4°F).

The second injection is administered 2 days later.

Giving the second injection. The male is to be injected with 100 iu of the hormone. The female is to be injected with 250–350 iu of the hormone.

(*a*) *Injection for the males and females.* Prepare a syringe of the dissolved hormone from a 500 iu ampoule in the manner described earlier.

Inject 0.6 cm³ of the Pregnyl solution into the female and then inject 0.2 cm³ into the male. The remaining 0.2 cm³ of solution is thrown away.

The pair of toads are returned to their breeding tanks three-quarters full of water at 23°C (73.4°F). Spawning should take place between 2 and 8 hours later and the eggs hatch after 2 or 3 days.

Hormone quantities	Prime injection	Second injection
For the male	50 iu	100 iu
For the female	100 iu	250–350 iu

Care for the tadpoles

The tadpoles begin feeding at about 36 hours. They can be started on filtered nettle powder paste. The tadpoles filter-feed at this early stage and so large particulate food is unsuitable. They can be fed daily in this manner ensuring that any excess food is siphoned off. They will be about 5 cm (2 in.) in 10–11 weeks when their legs begin to show. As the tadpoles are metamorphosing they tend to eat little or no food. The water level is dropped to about 5–6 cm (2.5 in.) as the tadpoles lose their tails. At this time the feeding habit changes, and so Tubifex worms, white worms, and Daphnia need to be fed to the young toads.

Throughout the period that the toads are growing, their tanks need to be cleaned out twice weekly as the putrifying food debris encourages disease. Avoid any detergent or disinfectant residues remaining in the cleaned out toad tanks.

Handling the tadpoles can be by means of a rubber teat pipette. As they get bigger, a small aquarium net is more suitable.

5.10.6. BREEDING LOCUSTS IN THE LABORATORY

For generations locusts have been a scourge of man and his plant foods. In the wild it is a problem to effectively eradicate locusts in those areas where they are pests. In the laboratory it is a time-consuming labor keeping them alive and breeding well.

Species of locust

The species of locust most easily reared in the laboratory is the African migratory. The desert locust is somewhat prone to disease and takes longer to mature, but is a useful dissection specimen because of its larger size.

Species	Number of pods laid per female	Number of eggs per pod	Incubations period °C 28 32	Sexual maturity of adult	Life span	Nymphs
Locusta migratoria (African migratory) Have "furry" thorax under surface and no peg-like structure between front pair of legs	6 (one every 5–6 days)	30–100	16 11 days	Yellow–dark brown 4 weeks after final molt	9–10 weeks	Hoppers 8.5 mm on hatching, 5 molts over 4 weeks at 5, 4, 4, 5, and 8 days
Schistocerca gregaria (Desert locust) Have peg-like structure between front pair of legs. Lighter colors	7	40–70	17 12 days	4–6 weeks	12–14 weeks	Hoppers 9.5 mm on hatching 5 molts over 4–5 weeks

Other species of locust are the Red locust (*Nomadacris septemfasciata*) and the Egyptian grasshopper (*Anacridium aegyptium*).

The red locust is the swarming insect that produces plagues over parts of central southern Africa.

The Egyptian species is not a swarming insect, and is found around the Mediterranean.

Accommodation and routine care

Small numbers of locusts can be maintained in the relatively cheap insect-rearing cylinders. Alternatively, a glass aquarium could be adapted for use provided there is adequate heating.

Locust cages for breeding and rearing are available commercially and their design is based upon the experience gained by large-scale breeders in research establishments.

Breeding cages. Culturing locusts in the laboratory present the following problems.

(a) They need high temperatures, illumination, and low humidity.
(b) They need perches.
(c) They need egg-laying areas.
(d) They need to be serviced for feeding, watering, and cleaning.

The cage structure illustrated (Fig. 72) conforms to all the requirements listed.

Heating and light is achieved by electric lamp bulbs. Below the removable perforated floor are one or two lamp sockets. These can be fitted with one or two 25 watt bulbs. These heating bulbs will be on continuously to keep the cage temperature above 28°C. In the cage area there is another socket to take a bulb up to 60 watt. This daylight bulb will keep the daytime temperatures up to 32–34°C. The cage room temperatures will naturally influence the cage temperatures and so experiments with various bulb sizes and positions may be required before achieving the optimum day and night time temperatures. Daylight length of 7–8 hours can be regulated by time switches on the overhead lamps, or can be turned off manually.

The relative humidity should be 45%. This water vapor being supplied by the small quantities of grass added to the cage for food. Care should be exercised if water is supplied by the "inverted test-tube" system, as it may cause unhealthy rises in humidity and the associated diseased locusts. This should only be used if no fresh vegetation is given.

Perching areas should be provided to ensure plenty of room within the cage. This can be supplied by having dried twigs standing vertically in the cage. The walls can have a perforated texture so that the insects can perch.

Egg-laying facilities are most important if one is intending to breed locusts. If a female is not provided with an area to deposit her eggs, she will put them on the floor of the cage where they will dry up quickly.

Oviposition tubes should be about 10 cm in length and 3 cm wide. There can be only two or three fitted, extra ones merely give more choice. At all times these tubes should contain a coarse heat-treated sterile sand. The sand should be damp (15% water) and packed firmly into the aluminum or glass tubes. They can be sprayed with water from time to time to replace evaporated water. They should be removed every day or two, and checked out for the presence of egg pods, in the manner described later.

Feeding locusts is fairly straight-forward. Provide fresh grass and dry bran daily. Suspend the grass in a small bundle so that it can be more easily removed at the end of a day. Decaying grass must not be too wet, nor must it have been exposed to insecticides or any bacterial infection.

When grass is not available, always leave a small vessel of dry bran and a source of water in the form of an inverted boiling tube plugged with cotton wool.

Cleaning out the locust boxes is to be done two or three times each week. All food debris needs to be removed together with feces. In order to avoid any possible bacterial infections, the cages need cleaning out with disinfectant each month. Ensure that the disinfectant is washed out of the cage before it is air dried.

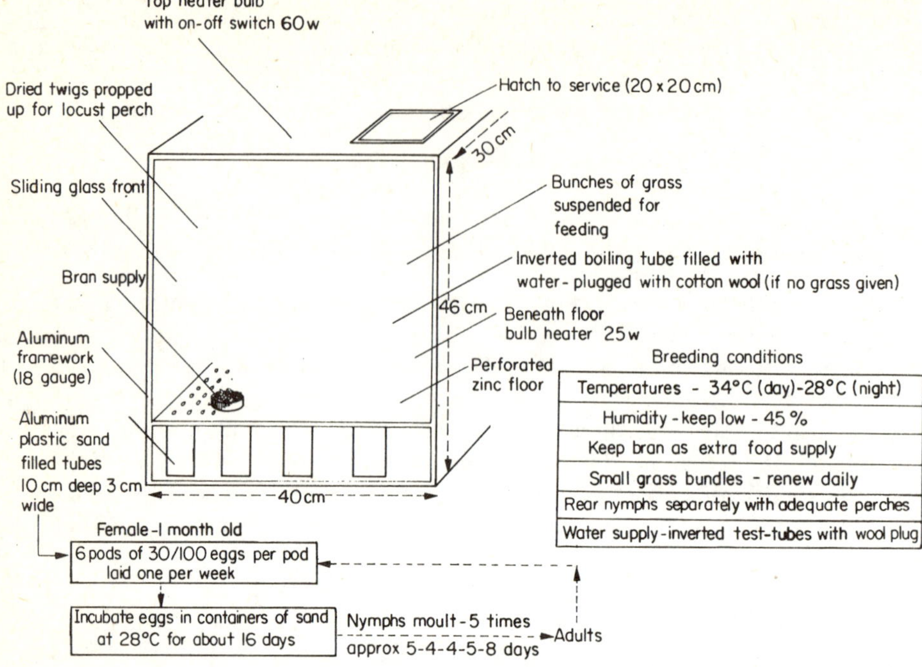

Breeding conditions		
Temperatures - 34°C (day)-28°C (night)		
Humidity - keep low - 45 %		
Keep bran as extra food supply		
Small grass bundles - renew daily		
Rear nymphs separately with adequate perches		
Water supply - inverted test-tubes with wool plug		

FIG. 72. Locust rearing

The design of cages is such that they may easily be dismantled for the purposes of cleaning and servicing.

Rearing cages. When the egg pods have been removed from the breeding cages containing 30 or 40 pairs of adults, they are incubated in the manner to be described.

A rearing cage is used to accommodate the young stages or nymphs (described as "hoppers"). There is no reason why the same type of cage that housed the adults could not be used. The only unnecessary items would be the egg-laying tubes.

In large-scale breeding operations a rearing cage is merely a large metal container of similar dimensions to the breeding cage, with heating and adequate servicing hatches. Here the insects can go through their molting operations undisturbed by the adults. It is important that adequate perching areas are provided so that the insects can suspend themselves for the molting. Deformed insects result if they cannot suspend themselves properly during molting.

Breeding and routine care

Incubating the eggs. The "egg-pod" tubes are removed, as a matter of routine, every 2 days. Fresh tubes packed with sand are kept in readiness as replacements. The tubes removed are inspected for the presence of egg pods. Brush away the surface sand and look for a crispy, white dry froth. The pods are below this froth. The young crawl through this tunnel of froth when they hatch, so it should not be destroyed or disturbed. The pods are about 100 mm in length and 8.5 mm across. They contain between 30 and 100 eggs. The female lays about six of these pods in her lifetime.

Nymph stages last about a month (called 'Hoppers')

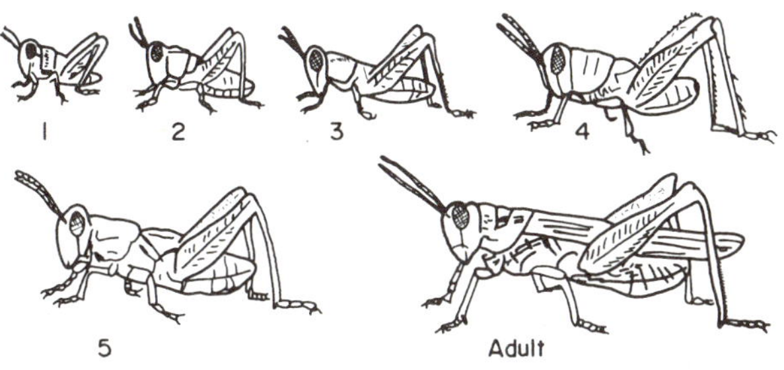

FIG. 73. Life history of the locust

The tubes containing no pods are cleaned up and refilled for further use.

The tubes containing egg pods must have the sand covering the pods and then capped to prevent drying out. The top can be covered with aluminum foil or parafilm and the tubes placed into large screw-top glass jars and incubated at 28–32°C. They will hatch between 11 and 17 days, depending upon the species. Label all tubes with expected hatching dates.

After the incubation period, remove the tubes on their expected hatching date and put them into the cage that is to be used as the rearing cage. They could be left to hatch in the glass jars if sufficient bottles are available.

The sand tubes should never be disturbed until all the hoppers have hatched out.

The young can be fed as for adults.

Caring for the nymphs. After hatching, the first stage nymph is very small (*Locusta*—8.5 mm; *Schistocerca*—9.5 mm) and an extremely good jumper. Both species of nymph molt five times over a period of a month or more. In the case of *Locusta*, they ecdyse at 5 days followed by other growth changes at intervals of 4, 4, 5, and finally 8 days. They become sexually mature 1 month after the final molt when they become adult in form. Throughout this period, the insects must be provided with ample perch positions from which they can suspend as they ecdyse.

Health and hygiene and locusts

The personnel attending locusts may develop an allergy to the insects. This is particularly true of the hoppers. A runny nose and eyes, and/or skin rashes may afflict some people who come into close contact with the insects.

The locusts do not usually develop disease within the laboratory colony unless hygiene standards are at fault. There are occasions when stock may die off with the fifth-stage hoppers and young adults turning a pink color. The corpses are soft to touch and tend to break up with much internal dark, foul fluid being released. This is said to be a bacterial infection that enters the insect's body if it is damaged during molting or handling. The bacterium (*Serratia marcescens*) can be eradicated by regular cage cleansing, cautious selection of grass for feeding, and keeping the humidity low.

Another disease sometimes found in locust colonies is attributable to parasitic nematodes, whose eggs are brought into the cage with grass. This worm (*Mermis sp.*) burrows its way into the insects body cavity. An internal examination of the dead insect may demonstrate the presence of nematodes.

5.10.7. BREEDING STICK INSECTS IN THE LABORATORY

The animals are easily bred in the laboratory.

> *Caraussius* (*Dixippus*) *morosus*—a smaller East Indian insect.
> *Extatosoma tiaratum*—a larger insect.

The first mentioned are more commonly kept because they require less attention. These insects are useful for many behavior and physiology experiments. Rearing of these insects may demonstrate one of its more interesting characteristics; that of *catalepsy.* Frequently, when attending to these animals, they appear to be dead or paralyzed in the most odd postures. These frozen postures can be broken by the proximity of heat or chemical stimuli, such as ammonia. It can be revived by a pinch near its hind end.

Accommodation and routine care

The caging of these insects can be quite simple. An adapted glass aquarium or an old glass-sided chemical balance case, with sliding doors, is quite useful. The insects need a source of plant food, such as privet. This can be placed in a pot containing water. The opening should be covered by cotton wool or aluminum foil to prevent the insect's drowning. They will also feed on ivy.

As a matter of regular routine, the feces should be swept clear from the base of the cage. To do this a larger and smaller artist's paint brush is a useful tool. At the time of cleaning up, the eggs should be separated and put into a petri dish situated on the cage floor. Dampening the brush can help the separation of eggs from the feces.

The insects are reared at 20–25°C (68–77°F) with a fairly high humidity. Keep a plastic squeeze spray bottle on hand in order to spray the leaves from time to time.

Incubating the eggs. Females lay eggs which usually hatch from a few weeks to a few months. These eggs generally hatch to become further females. The colony may probably contain no males. Stick insects exhibit parthenogenesis or "virgin birth".

The nymphs that hatch out of the eggs will be small versions of the adults. For a short time they carry the eggshell on their abdomen. It is not generally necessary to pull this off unless some difficulties are experienced in walking. Over a period of 6 months or so the nymphs molt at monthly intervals to become adults.

5.10.8. BREEDING COCKROACHES IN THE LABORATORY

Cockroaches are a pest around animal and human food stocks. They are not native to this country and so for that reason must seek out warm and damp conditions in which to live. Kitchens, bakehouses, and any warm food stores are ideal. Because these animals are prolific breeders in the correct conditions, they must be kept securely. Cockroaches are classified in the same insect order as are the locusts. They are similar in as far as they lay egg cases from which the young emerge as smaller versions of the adult. These nymphs go through a series of molts to eventually produce the adult form, the imago.

There are three types of cockroach kept for experimental or demonstration purposes:

> *Periplaneta americana*—American cockroach—large, dark golden brown. Both sexes have wings.
>
> *Blatta orientalis*—Oriental cockroach—a smaller, black. Female has vestigial wings.
>
> *Blattela germanica*—German cockroach—narrower body.

Accommodation and routine care

These insects are easy to maintain, given the correct temperature and humidity conditions. An aquarium can be fitted up to produce a suitable breeding demonstration unit for cockroaches. The plastic larval cages available commercially are adequate. One example of a cockroach breeding set-up is shown in the diagram. It is important to employ some effective "moat" to prevent escapes or invasions.

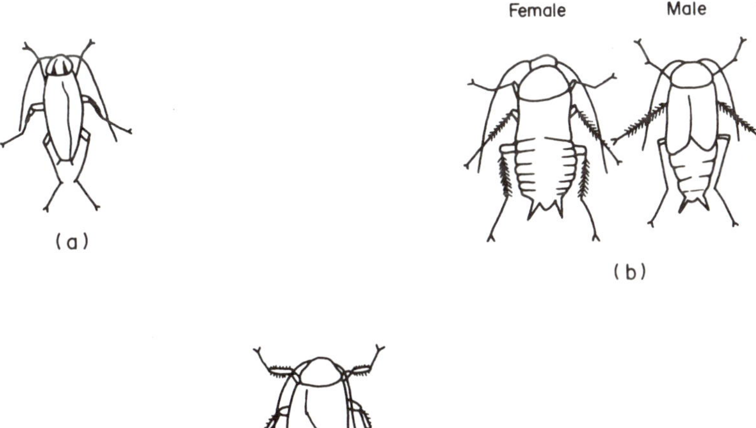

Fig. 74. Cockroaches. (*a*) German cockroach (*Blatella germanica*). (*b*) Oriental cockroach (*Blatta orientalis*). (*c*) American cockroach (*Periplaneta americana*)

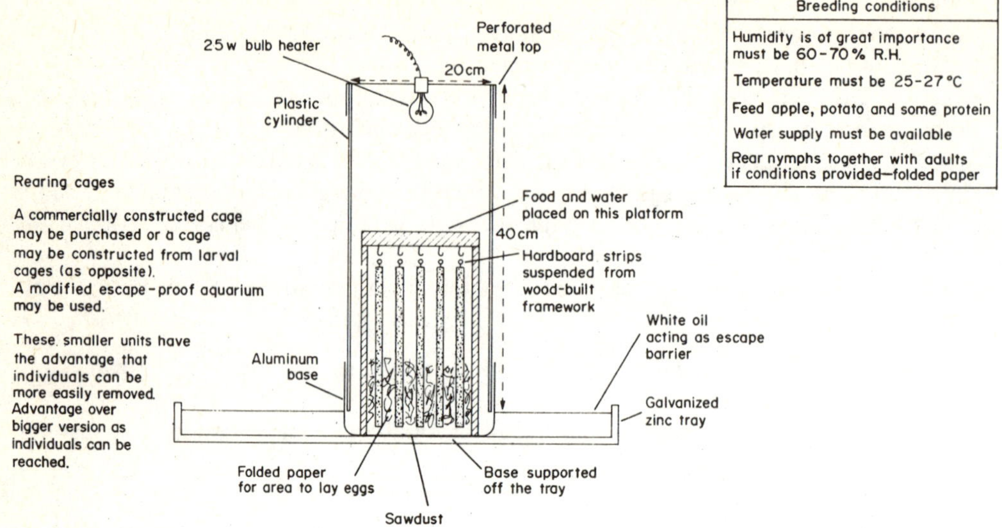

Fig. 75. Cockroach rearing

Heating and light is provided by electric lamp bulbs suitably situated on the top of the cage, ensuring that they are safely insulated. The strength of the bulbs will be determined by trial and error as a temperature range of 24°C–25°C (75.2°F–77.0°F) needs to be maintained. Reports suggest that a temperature of 28°C (82.4°F) produces the best results.

When the temperature is lowered the animals become very sluggish or inactive. At a temperature of 10°C (50°F) the animals are sufficiently inactive to make handling somewhat easier. The animals can be killed by dropping the temperature down near freezing and maintaining this for a few hours.

The relative humidity is very important for cockroaches in terms of the egg incubation, and the early days of the nymph. The adults can survive at lower humidities than is required for the success of egg hatching. A reasonable compromise must be achieved by means of trial-and-error testing or by employing an automatic humidifier to larger scale work.

If the air-moisture content is too great, then pests tend to take over in the breeding container. Mites are reported at humidities above 60% and mold growths over 75% humidity.

Perch areas are provided by the inclusion of vertical or horizontal platforms of hard board, or some such material. The presence of these shadowy areas will provide a gradient of temperatures from the upper heated to lower unheated spaces. The insect can be seen to be selective in its perching position depending upon the temperatures provided.

Cockroaches molt whilst hanging from a surface, in the manner that is clearly seen with locusts suspended from twig perches.

Egg-laying is a fairly straight-forward process for the cockroach. The female (recognizable by the more "boat-shaped" abdomen and smaller anal sensory processes) carries the egg case attached to her rear-end for varying periods of time. The pod turns a darker brown before it is deposited beneath the rubbish on the bottom of the cage. Sometimes the pods are "glued" to a suitable surface. There is no special technique required to incubate the eggs. In some circumstances, the accommodation may include an area where the egg pods are laid. The pods then being removed from time to time and put into special rearing chambers. This will control any overcrowding that may encourage cannibalism amongst the nymphs.

Cleaning out the accommodation is a temptation for the enthusiast, but it can be disastrous for the cockroach culture. The tank becomes smelly because the insects produce an offensive odor. Unless conditions are particularly foul, it is better not to attempt too many cleaning jobs. The tanks at Oxford were, at one time, all but empty because of a clean-minded technician.

Care should be exercised because the damp sawdust lining on the cage floor may contain egg pods.

| Egg numbers per case | Molts | Development temperatures | | Incubation temperatures | | | Lifecycle |
		25.5°C (77°F)	28.3°C (82°F)	24.6°C (75°F)	26.0°C (78.8°F)	29.2°C 84°F)	
Two rows totalling 16–40	9–13	520 days	195 days	58.5 days	48 days	34.5 days	4 to 6 months

Cleaning out the cockroach tank may not eradicate mites or parasitic nematodes. This may not be necessary as the insects are not apparently adversely affected. The infested cockroaches could be regarded as usefully supplying further animal cultures for laboratory study.

Food and drink is relatively easy to provide as the cockroach is fairly unselective in terms of food demands. They do also seem to be able to survive reasonable lengths of time without water.

Dry oats and fruit or vegetable, as well as proprietary mammal pellets make useful food sources.

Water can be provided by way of a wool-stoppered test tube. Do not put dishes of water in the cage as nymphs may drown in it.

Handling and sexing cockroaches

The rapid movement of these animals make them less than attractive to handle. This movement can be reduced, as mentioned elsewhere, by lowering the cage temperature. The animals are less active during the day and may be easier to deal with at this time. The fact that they regularly become more active towards dusk makes them useful in biological laboratories as demonstrations of circadian rhythm.

If a cockroach is picked up by means of long forceps, it is best that they be gripped across the mid thorax.

Sexing the insects involves examining them for the presence or absence of specific characteristics. The males of Periplaneta have two sets of projections either side of the anus. The females only have the one pair of projections (anal cerci) on the upturned, boat-shaped abdomen. The females of Blatta have vestigial wings.

Killing cockroaches in the laboratory situation may be achieved by means of chloroform vapor in a tightly sealed gas chamber, such as a chemical dessicator.

I.L.A.S.T.—H

5.10.9. BREEDING DROSOPHILA FRUIT-FLIES IN THE LABORATORY

This insect can be one of many species. The most popular type for laboratory use is described below.

The species: Drosophila melanogaster

This fly has been of great importance to the study of genetics for many years. The main reason for its importance is related to the ease with which it may be cultured in the laboratory. It is easy to handle, produces large numbers of offspring and has a short life-cycle of 10–14 days. The animal has only four pairs of diploid chromosomes which are clearly visible in the giant form in the cells of larval salivary glands. This makes the animal invaluable in the study of chromosomal inheritance.

This order of insects, the *Diptera* or flies, show a life-cycle with a larvae, pupa, and adult.

The *Drosophila* fly is found in the wild amongst decaying vegetation and fruit, from which it may be collected with the exercise of some skill and selectivity.

Stocks are normally bought in from reliable breeding sources; some of which are quoted in the *UFAW Handbook*, for the American consumer (as below).

USA stocks: (a) Biology Division, California Institute of Technology, Pasadena, California 91109.
(b) Zoology Department, University of Texas, Austin, Texas 78712.
(c) Biology Department, University of Chicago, Chicago, Illinois 60637.

UK stocks: (a) Genetics Department, University of Cambridge, Cambridge.
(b) Griffen Gerrard International, POB 14, Wembley, Middlesex HA0 1HJ.

The insects, either collected wild or bought in from breeders, are cultured on a nutrient medium that can be made up in the laboratory and put into small wide-necked bottles.

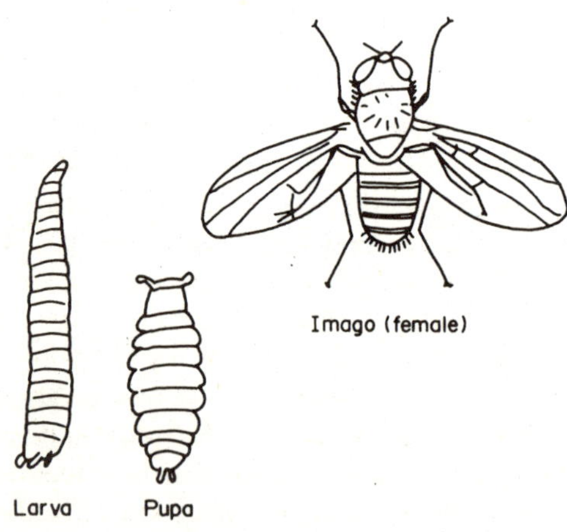

FIG. 76. Fruit-fly (*Drosophila melanogaster*)

Accommodation and routine work

The culture medium is made up as indicated below, using a cornmeal and syrup base. There are numerous formulae for making up *Drosophila* culture mediums, but it is as well to stay with one that has a reported success.

The medium (*cornmeal agar*). Mix agar with the water (750 cm^3) and stir warm to boiling-point and continue stirring to dissolve all the agar. In a separate container add the ethanol to the dried yeast. Add the syrup and the cornmeal to the yeast/ethanol mixture. Add all these ingredients to the boiling agar solution.

The cornmeal agar solution now requires boiling gently for 15 minutes and then cooling.

The major problem with these nutrient media is that they so easily become contaminated, and so protective measures need to be observed to prevent the growth of molds.

A mold inhibitor may be added in the form of propionic acid (5 cm^3), which is stirred into the cooling cornmeal mixture. Alternatively, proprietary inhibitors such as "Nipagin" or "Moldex" may be added to the medium whilst it is cooking in terms of 7–10 cm^3 to a liter.

Ingredients:

Water	1000 cm^3
Agar	15 g
Yeast (brewer's dried)	30 g
Cornmeal	100 g
Dark Treacle	100 cm^3
Ethanol (95%)	5 cm^3
"Nipagin"	(10% solution in 95% ethanol) 7–10 cm^3

If cornmeal (maize meal) is not available, a useful alternative is oatmeal (sieved) 180 g soaked in 300 cm^3 of water overnight.

Before the food medium becomes solid it needs pouring into suitable wide-necked vessels. In the United Kingdom small milk bottles can be suitable, otherwise specimen tubes will serve the purpose. It is a good idea to slope the media as it sets solid in order to give an angle down which excess fluids may run when standing in the vertical position (see Fig. 77). A piece of absorbent paper pushed into the media will help absorb some fluid as well as giving the flies a platform upon which to walk.

Leave the freshly prepared tubes lying at an angle without any cotton plug in the top in order to allow water condensation to escape. Have the tops covered with a light muslin. Plug the tops with sterile, non-absorbent cotton and then store in a cool place.

When the bottles are to be used for culturing insects they must first be autoclaved (p. 90), afterwards scatter some live yeast suspension over the surface of the food.

This will grow and act as a source of nutrient for the adult flies. The bulk of the media being food for the larvae (maggots). The pupae stages of the life history will become fixed to the dry wall of the bottle or to the paper inserted into the food.

The prepared culture tubes can now have specimens added to them, ensuring that they do not become stuck in the wet media as they are tipped into the tubes. Experience will enable a person to do this without too much difficulty.

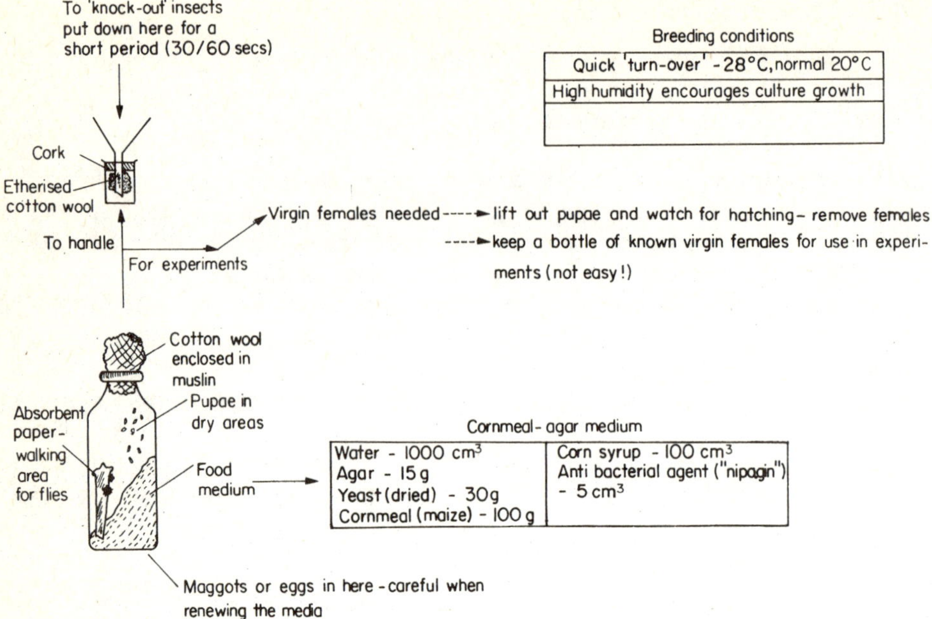

Fig. 77. Fruit-fly rearing

The insects are kept in an incubator or in a warm place at 25°C (77°F). General stock can be kept at 19–21°C (66.2–69.8°F). The relative humidity should be at about 60–70%.

Handling and sexing Drosophila

Insects will normally be bought in as virgin females and males. They will be paired up in colonies and the dates noted on the culture bottles. Each colony bottle can be left about 2 months before it is necessary to subculture. Examine weekly to see if the flies are healthy and multiplying. This means to transfer stock to new food-media culture bottles. The timing cannot be stipulated precisely because some fungus or other pollutant may necessitate earlier attention.

Breeding fruit-flies involves special techniques of handling that are outlined below.

(i) *Handling* requires that the insects are immobilized. This can be achieved by anesthetizing the flies using ether/alcohol mixture in the following manner.

(a) Apply the ether mixture to a cotton pad on the end of a suitably sized glass filter funnel. Care should be taken not to add too much ether.

(b) Stand the culture bottle on top of the empty anesthetic vessel. Tap the sides of the glass to encourage flies to move downwards. Be careful not to dislodge the food as it could also fall downwards trapping the colony. A lamp situated by the side of the ether bottle may draw some insects towards the light.

(c) When sufficient insects have moved downwards, put the ether funnel into position. Watch the flies as they are overcome by the vapor to ensure that they are not killed. A dead fly will have its wings outstretched.

(*d*) Transfer the flies to a white tile and sort them out into sexes or required genetic types. This can be done by means of a fine artist's brush, working under a high power lens or binocular microscope.

(*e*) Return the selected flies to appropriate labelled culture bottles. Caution needs to be exercised as the flies are returned as they may fall into a sticky patch of food.

(ii) *Sexing* the flies can be achieved by observing the characteristics listed below:

Males: smaller in size; blunt-ended abdomen with dark pigment; "sex" comb on the 1st tarsal joint of the forelimbs.

Females: larger in size; pointed abdomen with more stripes at the end; no "sex" combs.

Breeding Drosophila

In order to ensure that the females are fertilized by the required males then virgin females must be used. The problem being that females may be fertilized as early as 10–12 hours after they emerge from the pupa.

If all the adults are let out of a culture bottle, then those females that hatch out of the remaining pupae will be unfertilized. This emergence generally occurs at dawn or oncoming light period and so the isolation of virgin females can be done in a routine manner daily.

A separate tube of virgin females will be kept for use in experiments.

An outline of some crossing experiments with *Drosophila* is outlined elsewhere.

5.10.10. BREEDING MEAL-WORMS IN THE LABORATORY (*TENEBRIO* SPECIES)

Meal-worms are the larvae of beetles (*Coeloptera*). These animals are found in food stores because the larvae feed on flour and other stored cereal food products. Two species are mentioned here:

> *Tenebrio molitor*—yellow meal-worm.
> *Tenebrio obscura*—dark meal-worm.

These insects are cultured in the laboratory for a number of reasons, the main one being as a source of food for other laboratory animals. Meal-worms are readily taken by fish, amphibia, reptiles, birds, and others. They are expensive to buy and a laboratory stock is one way of reducing costs. They are not too difficult to culture.

The adult beetle is about 15 mm in length and the larva 25–30 mm in length. They are easily distinguished from the flour beetles which are much smaller.

Accommodation and routine maintenance

The beetles and their larvae can be kept in large metal canisters or glass jars (30 cm × 20 cm × 15 cm). If the container is transparent, some workers suggest that the outside be pointed a matt black. In any case, the tops should have ventilation and be escape proof by having an overturned inner lip. Some workers stand the culture containers on ramps in a tray of oil or paraffin (6.0 mm deep) to prevent escapes and invasions by mites. This latter point is not too easy to overcome as mites are often introduced with the cereal food.

Temperatures and humidities are important for the success of the culture. The life-cycle of these insects is somewhat lengthy and is temperature related. At 21°C (70°F) the life-cycle is about 6 months, and 4 months at 26.7°C (80°F). The relative humidity in the culture should be in the range of 60–65%. Higher humidities may encourage the population growths of mites and molds. This moisture content can be maintained by the introduction of fresh root crops or leaf vegetables. These should be changed daily so that no molds develop. Some workers devize small capillary tube water drips onto cotton wool. Cannibalism becomes evident if the culture is allowed to dry out too much.

Culture medium and food in the form of oatmeal and bran is spread across the container floor to a depth of one-third. On top of this, newspaper and soft cloths are placed for the beetles to lay their eggs on and for a place to pupate. When small larvae are seen crawling in the media, then an addition of finely chopped root and leaf vegetables is made by laying it on top of the oatmeal in a simple paper tray. Remove it daily in case of mold growth. As the larvae grow, they become more obvious in the area of the paper cover over the oatmeal. They require further supplies of vegetable food.

As pupae form on the paper and cloth, they are removed and put into a new culture container where the adult beetles emerge and are maintained. This second container will be taken through the same sequence of events as the first. In time, the first box will be cleaned out and reused. Any number of such culture containers may be set up.

A population of meal-worms may be started by putting thirty or forty adults into a 2-liter container containing the media described.

5.10.11. BREEDING FLOUR BEETLES IN THE LABORATORY (*TRIBOLIUM* SPECIES)

These beetles are universal pests of stored food products, especially milled cereals. Their use as a laboratory animal relates to experiments in genetics and the testing of insecticides. Their use for genetics experiments is somewhat limited by the compara-tively lengthy life-cycle. Two species are commonly described:

> *Tribolium confusum*—Confused flour beetle.
> *Tribolium castaneum*—rust-red flour beetle.

There are mutant forms of these two which can be used for demonstrations of genetic principles, showing eye and body color inheritance. They are smaller than the Tenebrio species. The adult beetles being 3–4 mm in length.

Accommodation and routine maintenance

The containers for these insects and their food culture can be smallish glass vessels of about 2-liter capacity. The top should be sealed by a perforated screw top or covered with muslin secured by an elastic band. For genetic experiments smaller vessels are suitable, such as glass specimen tubes (10 cm × 2.5 cm).

The temperatures and humidities are required to be regulated for successful breeding. The temperature range can be from 20°C–30°C (68°F–86°F), but shorter life-cycles are reported for cultures maintained at 25°C–27°C (77°F–80.6°F). At these latter temperatures adults are obtained in 40–42 days. At 21°C (69.8°F) the life-cycle is reported to take 56 days.

The relative humidity needs to be maintained at 60–70%. Where circumstances permit, a handy way to regulate the relative humidity is to use the technique described later (p. 290).

Culture medium and food is provided in the form of whole-meal flour (12–20 parts) mixed with powdered dry yeast (1 part). Heat sterilize. This mixture is placed into the bottom of the culture container to provide a depth of half or two-thirds the volume.

Handling and sexing these beetles is necessary if breeding experiments are to be set up. To slow the animals down, they can be chilled in a refrigerator and then the media sieved through a flour or coffee strainer to reveal the individual insects. They can be sorted out on a white tile and picked up by an insect pooter, if convenient. The sexes can be distinguished at the pupal stage by a close examination of their posterior ends beneath the binocular microscope. The pupae are about 4–5 mm in length and pale colored. The female genital projections are much longer than those on the male pupa.

For the purposes of setting up genetic crosses, the pupae are sorted into sexes and then put as pairs into appropriately labelled culture vials.

The life-cycle of these beetles can extend over 40–42 days, as mentioned earlier, at a given temperature range. Under these conditions the eggs will hatch after about 7 days. The larvae will pupate between 25 and 40 days later. After about 5–7 days the adult beetle emerges.

A population of flour beetles may be started with about a hundred adults in a 2-liter container, a third full of the food media.

5.10.12. REARING THE GRANARY WEEVIL IN THE LABORATORY

The weevil (*Sitophilus granarius*) is a pest of stored grain. It is a member of the beetle group (*Coeloptera*) of insects characterized by its beaky-head projection. The larvae live within the kernels of grains, such as wheat. The adults also live within the tunnels produced by the larvae, or they feed on the grain surface.

Culture

A glass jar containing wheat grains (free of insecticide) with a perforated metal top is all that is required.

The insects are maintained in this container at a temperature of 25°C (77°F) and a relative humidity of 60–70%.

5.10.13. REARING THE MEDITERRANEAN FLOUR MOTH IN THE LABORATORY

The larva of this insect, *Anagasta* (*Ephestia*) *kuhniella*, is a common pest of stored cereals and cereal products. It can be kept in the laboratory situation for demonstration and experimental work. The larvae are in nature injected by ichnemmon (Devorgilla) flies eggs. These eggs hatch out and the maggot stage parasitizes the moth larvae. It is a useful way of controlling this flour pest, naturally (biological control).

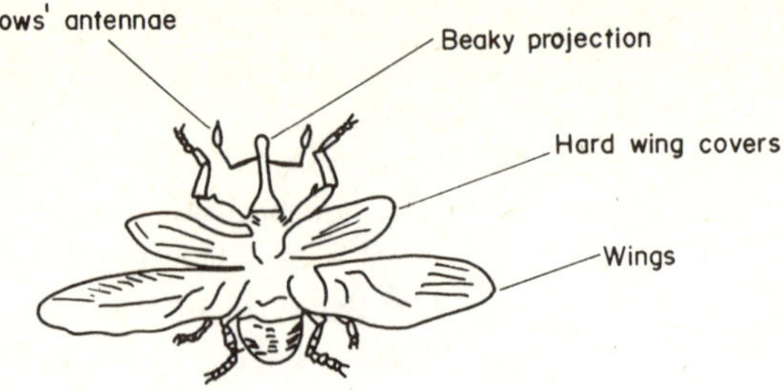

FIG. 78. Granary weevil (adult)

Culture

The easiest way to rear this insect is by having a perforated screw top glass vessel containing whole-meal flour or porridge oats, mixed in with a little powdered dried yeast (5%). This media can become infested with mites so it is better to sterilize it for 3 hours at 70–80°C (158–176°F) before use. Put 5–8 cm depth of this medium into the jar. Provide a temperature of 25°C (77°F) and a relative humidity of 60–70%. This moist atmosphere can be established by having a small tube of water plugged with open mesh muslin standing in the culture container.

After some time, the media becomes congested with the silky threads produced by the larvae. Add fresh food from time to time, but sub-culture if the media becomes heavily parasitized by mites or fungi. This may be necessary every 2 or 3 months.

If a new culture is set up, sterilize everything by autoclaving.

5.10.14. BREEDING HOUSEFLIES IN THE LABORATORY

These insects are used for the testing of insecticides in industry and as food for amphibians, reptiles, and so forth in animal laboratories. They are examples of insects that have complete metamorphosis-type life histories. The eggs and larvae are useful fish food.

The type considered here is *Musca domestica.*

Accommodation and routine maintenance

Houseflies belong to a group of diptera (flies), which includes blowflies and blue bottles. The culture of these insects in the laboratory can be a smelly operation as the maggots or "gentles" live in rotten flesh or dung.

Containers suitable for the production of houseflies should be escape proof and easy to manage through sleeve-glove openings. They should be situated away from the main laboratories on account of the offensive odors.

Cultures can be raised in smaller glass jars or larger boxes of similar dimensions to those used for locust breeding (45 × 45 × 45 cm). They can be constructed of wood or aluminum and have a glass observation window with two 13 cm diameter sleeve-glove openings beneath. Hinged and detachable panels making up the top and sides make it easier to maintain the items within the culture box.

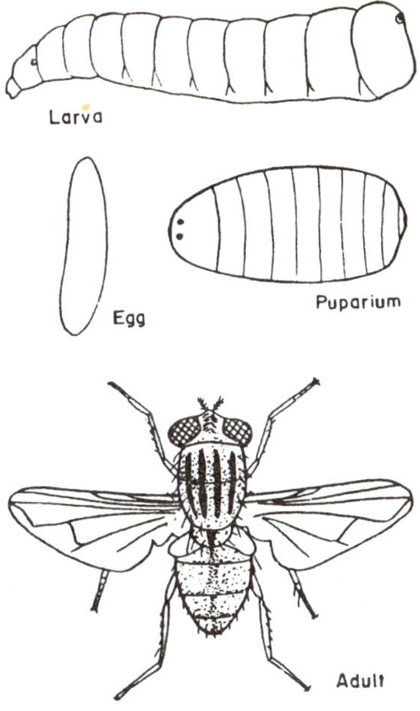

Fig. 79. Housefly (*Musca domestica*)

Stocking the containers can be achieved by placing pupae into the vessel in an open dish. These pupae can be bought in, or obtained from, a piece of rotten meat laid out in the open during the summer. This can be done by putting the meat into a small mouse box to keep other animals off it. The maggots will change to pupae and drop off the meat.

Adult flies emerging from the pupae will feed on a sugary solution provided, and lay eggs in cotton wool dampened by a milk/water mixture. This cotton wool should be removed daily and the eggs transferred to a suitable medium for the rearing of larvae.

New adults will lay eggs for about 10 days, commencing 4–5 days after emerging from the pupae.

Eggs of the fly look like tiny grains of polished rice laid one on top of another. In nature these eggs are laid in crevices of the material, such as dung, on which the maggots are to feed. The female lays between 120 and 150 eggs at one time and may deposit five or six batches in a lifetime.

If the adult flies are to be used as food, it is convenient to have them hatching in smaller bottles with muslin, or similar, sleeves. This way the insects can be introduced directly to a vivarium through the sleeve. It is better this way rather than having a "smelly" culture lying around within the tank of amphibia or reptiles.

Alternatively, pupae can be separated and put into the vivarium to hatch out.

Larvae or maggots hatch from the eggs 8–48 hours after they are laid. This depends upon the temperature. These maggots are white and legless. They burrow into their food and feed vigorously. In warm temperatures they reach full size in about $3\frac{1}{2}$ days, but in colder conditions they can take as long as 6–8 weeks.

Culture media for the larvae can be fresh horse manure, but generally this begins to dry out after a day or so and becomes less attractive as a place for flies to lay their

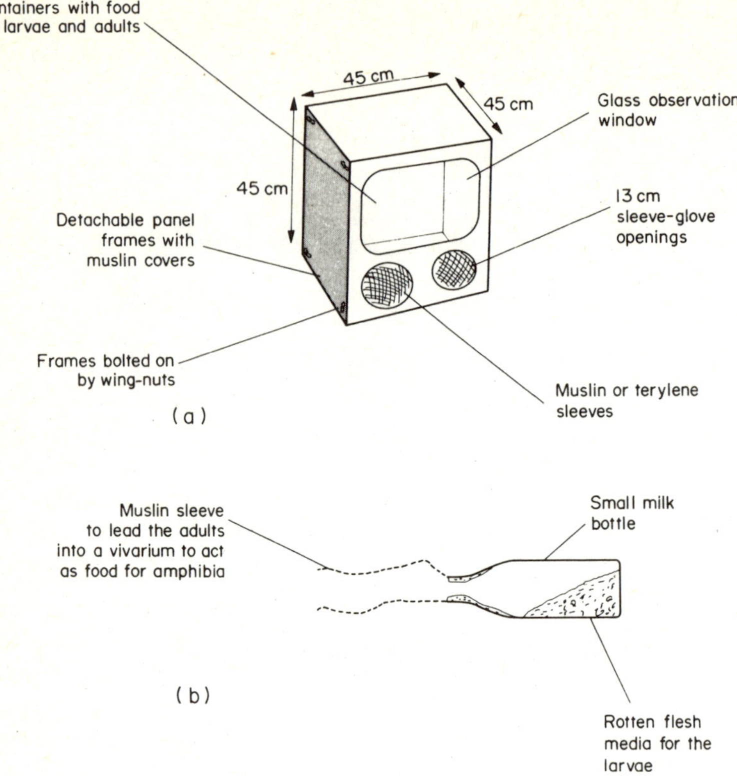

FIG. 80. Housefly rearing. (*a*) Culture box. (*b*) Larvae bottle

eggs. For this reason, an alternative has to be found if continuous breeding is to be encouraged. A well-tried culture media is tabulated below:

Bran	6 parts (by volume)
Fish-meal	2 parts (by volume)
Ground Diet 41B (rabbit pellets)	2 parts (by volume)
Powdered baker's yeast	2 parts (by volume)
Molasses, malt extract, or sugar	1 part (by volume)
Water to prepare a damp cake	

The ingredients are mixed to produce a moist cake which is then exposed to adult flies at 25°C (77°F). The cake will have pupae present on it within about a week.

Pupa or chrysalis is the motionless stage that follows on from the fully grown maggot. It remains so for $4\frac{1}{2}$ days to 3 weeks or more, depending upon environmental temperatures. It commences as a yellow puparium which changes through red, brown, and finally black.

These pupae form in the upper dry, cooler part of the culture vessel. Any pupae stuck on or within the culture cake can be gently extracted by immersing the media in warm water (25°C–77°F). The pupae will float to the top of the water with careful agitation and can be readily scooped up.

Life-cycles of these insects can take 6 days, at the quickest, but the more usual time in nature is 2 or 3 weeks from egg to adult. These adults in nature (in the UK) are capable of laying eggs after 14–18 days. All these timings are speeded up with an increase in temperature.

5.10.15. REARING WATER FLEAS IN THE LABORATORY

The term "water flea" is a generalization for microscopic freshwater crustaceans with the proper names listed below:

Daphnia pulex	*Cypris*
Cyclops	*Simocephalus*

These water fleas are useful as food for smaller inhabitants in freshwater aquaria. *Daphnia* is used as a demonstration of some basic physiological principles. They are transparent and the beating heart can be observed.

Culture

Water fleas are easily maintained in the laboratory over short periods of time. There seem to be few recorded methods of producing cultures that last the year around. It is necessary to have larger volumes of water to avoid oxygen starvation and to have adequate food. A method recommended by Harold Heath of Stanford University runs as follows. One ounce of dried sheep manure to 1 gallon of water to make the food mixture. Add this mixture to a wooden trough holding 25 gallons of water. Leave the tank for 2 or 3 days and then add the *Daphnia*. Some chopped lettuce leaves are added to the water from time to time as food. This method is rather extreme for the average laboratory requiring smaller numbers.

Containers for water-fleas need to have a large surface area in relation to the volume. This can be an aquarium or a porcelain sink with adequate lighting overhead.

The culture medium recommended is the natural pond water, which should contain a source of green algae to act as food for the water fleas. This medium needs to be aerated and should not get overcrowded with specimens. The temperature of the water should not exceed 20°C (68°F). The pond water can be oxygenated by having Canadian pond weed growing in it.

Food for the water-fleas can take the form of infusorian mixtures, or minimal sprinkles of yeast. It is better to feed a little and often. Overcrowding should be avoided by extracting some of the water-fleas and placing them into a fresh culture tank. Discard the old culture medium.

5.10.16. REARING WOODLICE IN THE LABORATORY

These crustaceans are damp dark seekers. There are four commoner types to be found under stones, leaves, tree bark, and similar places. The types referred to here are listed below. They all are known by the common names of slaters, pillbugs, woodlice, or sowbugs.

Armadillium vulgare	*Philosoia muscorum*
Oniscus asellus	*Porcellio scaber*

These animals are kept in the laboratory as a food source for other animals and for use in physiology and behavior experiments.

Culture

Woodlice prefer low illumination and moisture. These conditions can be provided by keeping the animals in opaque plastic containers with a bottom lining of moist gravel and leaf mold or soil and peat moss. This should be to a depth of about 2–3 cm. Broken pieces of ceramic plant pot can be scattered around the container to act as concealment areas.

The plastic containers referred to could be the commercially available food containers with snap on air-tight lids. When used for this purpose, the lids would need to be perforated.

The internal humidity must be high and the temperature at about 20°C (68°F). The species should be kept separately and not allowed to become overcrowded otherwise cannibalism may occur.

The humidity must be kept up by occasional sprays of water. A piece of rotting bark or pieces of cut raw potato or carrot can be left in the container as food.

5.10.17. REARING BRINE SHRIMPS IN THE LABORATORY

The brine shrimp *Artemia salina* is a ready source of food for many laboratory maintained aquatic organisms. These saltmarsh crustaceans are about 12 mm in length. Their eggs float on the surface of the concentrated salt water and dry up before they hatch. These dry eggs can survive in this condition for long periods. They are bought commercially in this condition.

Culture

The eggs purchased from a stockist must be kept dry and cool until required for use. These eggs are placed into sea water, or made up sea water of 0.1–6.0% concentration. A suitable medium can be made up by adding 10–15 g sodium chloride to 1 liter of water. Alternatively, 2 teaspoonfuls of common table salt to 1 quart of water. The containers for this salt-water culture medium can be a glass or plastic vessel with a suitable top covering to reduce evaporation. The water may need bubbler aeration. The eggs will hatch within 24–48 hours at a temperature of 20–25°C (68–77°F). The young stages may be present in many thousands and can be separated out for use as food for small aquatic inhabitants. The salt needs to be washed off them if they are to be put into a freshwater aquarium. If large populations are left in the container they will soon die off because of poor food and oxygen supplies. Rearing the shrimps to maturity for a permanent culture requires that a few of the young shrimps are transferred to a larger container of brine medium. A little yeast is added to the water each day as food. Only sufficient to cloud the water a little.

When the shrimps are mature, they will pair off and the eggs laid by the female float to the surface. They need to be scooped off and dried before being used to continue the culture.

5.10.18. REARING EARTHWORMS IN THE LABORATORY

These annelid worms are used for a variety of reasons in educational institutions. They are also a handy form of food for fish, amphibia, and reptiles. There are three types of earthworm commonly encountered in the northern hemisphere.

Lumbricus terrestris—larger earthworms.
Allolobophora species—smaller species in well-rotted soil.
Eisenia foetida—smaller "manure worm".

Collecting these worms is best achieved in springtime during or after a rainstorm. They can be put into a container with moist leaves and loamy soil. If they are to be left for any length of time, ensure that they are not overcrowded, that the temperature does not rise above 15°C (59°F), and that the moisture is retained.

Culture boxes

Earthworms can be accommodated in wooden boxes, metal biscuit tins, or trash cans. The size of container will be determined by the population of animals to be maintained. A liter of soil can accommodate about five of the larger *Lumbricus* species or twenty of the smaller *Eisenia* species of worm. A density of about twenty-five worms per cubic foot is quite suitable.

For larger quantities of worms they can be established in an outdoor "worm farm" which may be a shade patch of well-drained medium loam soil, 2 meters (6 feet) by 2 meters (or more) and 1 meter (3 feet) in depth. This "earthworm farm" area needs well-rotted manure and leafmold added to it before the worms are introduced.

Culture medium

The indoor containers need to imitate the natural outdoor conditions for good worm survival and breeding. The culture medium is made up as follows:

Mix together 3 parts of medium loam soil with 1 part of dung and 5 parts of peat. This mixture should be sprinkled with a calcium carbonate solution in order to produce an alkaline pH reading of 7.0–7.5.

The mature worms with the swollen band (the clitellum) are placed on top of the leafy top lining of the culture medium. If they are healthy, they will soon burrow beneath the top layer.

The worms need no special feeding if the soil has a surface layer of leaves and dung. Corn-meal or oats can be sprinkled on the soil surface from time to time to act as a food supplement.

The culture medium in the container must be prevented from drying out so it is best covered with a glass lid or perforated polythene. The surface should be water sprinkled from time to time and gently raked over to prevent mold growths and to encourage gas exchanges.

The culture medium needs to be renewed half yearly.

Breeding earthworms

Worms kept in the laboratory breed quicker than in nature because of the relatively higher temperatures. The eggs will hatch in 1 or 2 weeks, as compared to a month or so in nature. The worms will mature in 1 to 5 months in the laboratory, depending upon the species. In order to maintain a culture of newly hatched worms the cocoons deposited after mating need to be removed to a fresh culture medium. This can be started about 6 weeks after the culture has been set up.

The separation of the yellowy brown cocoons from the soil needs careful attention. Every 2 or 3 weeks soil has to be sieved to separate out the small spherical cocoons. The size of these cocoons varies with the species (2.5–7.0 mm). They can be "washed out" of the soil into enamel trays after the worm population has been extracted. They are then placed into new culture containers for their further developments.

5.10.19. REARING ENCHYTRAEID WORMS IN THE LABORATORY

The "white worms" (*Enchytraeus albidus*) are true worms of the oligochaete type; the group to which the earthworms belong. They are used as food for small fish and amphibia (tadpoles) and can be extracted in their thousands from compost heaps.

Culture

"White worms" can be kept in a similar way to earthworms. A container of wood or metal with a 5 cm layer of damp loam or peat on the bottom is convenient. A useful technique is to construct a container with all sides removable so that worms may be "harvested" more easily.

The worms are fed from time to time with milk-soaked bread or mashed potatoes that is placed into the peat. The culture is kept at 18–20°C (64.4–68°F).

5.10.20. MAINTAINING PLANARIANS IN THE LABORATORY

These flatworms are useful examples of free-living platyhelminthes. They are used to demonstrate their extraordinary ability to regenerate tissue if damaged.

Polycelis nigra is a common darker flatworm found in pond water in the UK. They can be collected from under stones or leaves in ponds or streams by suspending a piece of chopped earthworm on cotton in the water. The "worms" will come to feed on the meat.

Culture

Containers can be any clean vessel into which pond water is placed to a depth of 3–10 cm. This water needs to be changed on alternate days during the week. Especially after feeding. Do *not* use tap water. Maintain the animals at cooler temperatures and in shaded conditions. Temperatures should not exceed 23–24°C (73.4–75.2°F).

The animals can be fed 1–3 times per week on fresh meat, but be careful to avoid any fouling of the water. The animals will surely die.

Do not offer food to the water of any animals that are regenerating damaged tissues.

5.10.21. REARING HYDRA IN THE LABORATORY

There are two or three species of hydra found in laboratory use.

Chlorohydra viridissima—green in color; 30 mm extended body with relatively, short tentacles.

Hydra oligactis—grey brown in color; 30 mm body length with long tentacles 4–6 in number.

Hydra vulgaris (fusca)—yellowish-brown in color; 20 mm body length with tentacles longer than the body.

Culture

Containers should be sterile glass crystallizing basins covered with a glass lid. These containers become contaminated after 14–21 days and should be changed.

Media for culture can be an artificially prepared solution (see p. 228) or fresh water drawn from a pond or aquarium supporting healthy hydra. Do *not* use tap water. Stock the culture with a few sprigs of Canadian pond weed (*Elodea*) to oxygenate the water. Siphon off any debris gathering on the bottom.

Feed daily with small injected quantities of infusoria (*Daphnia*). Maintain the animals at 18–20°C (64.4–68°F).

5.10.22. REARING PARAMECIUM (AND OTHER CILIATES) IN THE LABORATORY

These are easier to culture in the laboratory than are the amebae.

The ciliates can be treated in a similar manner to that described for amebae. They are kept at about 18–20°C (64.4–68°F).

If hay infusion cultures are used, be careful that they do not turn sour. Keep the pH about 7.2 by adding a pinch of chalk.

5.10.23. REARING AMEBAE IN THE LABORATORY

This is perhaps the most common protozoan studied in biology (*Ameba proteus*). With attention to detail this animal can be cultured in the laboratory. The method described below applies to many protozoans.

Culture

Containers suitable for this work are sterile petri dishes or crystallizing basins, with lids.

Media for the rearing of amebae species can be Chalkley's solution, Brandwein's solution, or a hay infusion (details later). The pH of these culture media should be adjusted to 5.8 and 6.0. Put 2 cm of the media into the containers; add a grain of wheat (previously boiled) into the media; leave the grain standing in the media for a day or so until a growth forms around the grain. This is the food for the protozoans.

Amebae obtained from an established pure culture (or from the surface of decaying leaves in pond water) are inoculated into the culture media and kept shaded, at 18°C (64.4°F).

If pure cultures are required, regular inspections of the media should be made. Overcrowding of the media must be prevented by regular 2 monthly sub-culturing.

For alternative variations of the techniques of culturing protozoans the reader is referred to the references in the Bibliography (p. 299).

Culture solutions

Chalkley's medium. In 1 liter of distilled water dissolve 1.0 g sodium chloride, 0.04 g potassium chloride, and 0.06 g calcium chloride. This stock solution is diluted 1:10 with glass distilled water before use.

Brandwein's medium. In 950 cm³ of distilled water dissolve 1.2 g sodium chloride, 0.03 g potassium chloride, 0.04 g calcium chloride, and 0.02 g sodium hydrogen carbonate. Make up to 1 liter with a phosphate buffer. The stock solution should be diluted 1:10 with glass distilled water.

Hay infusion medium used for general protozoans. Boil 10 g of chopped hay in 1 liter of distilled water or rain water. Stand the mixture for a day in a covered flask. This stock solution should be used diluted five times with boiled water before use.

5.10.24. EXAMINING LITTER AND NESTING FOR PESTS

It may be suspected that the bedding or litter is harboring some living organism. One method of extracting these microscopic creatures from the material is outlined below.

Berlese extraction funnel method

Set up a glass filter funnel in a laboratory clamp stand. Have the funnel positioned over a container of a fluid preservative such as 5% formalin (methanal) or 70% ethanol.

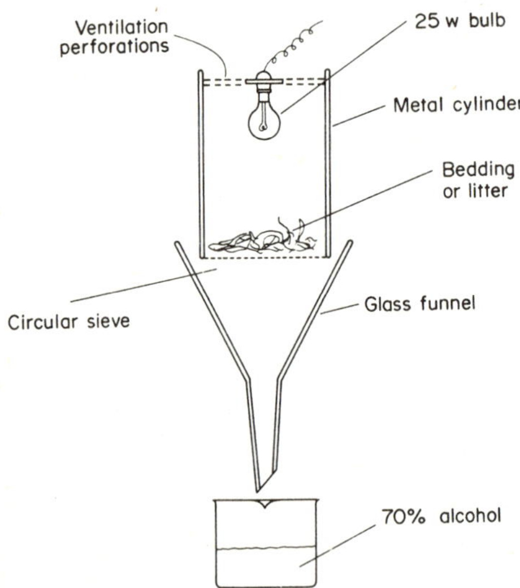

FIG. 81. Berlese extraction of pests from bedding

Construct a metal cylinder (a metal can) into one end of which an electric lamp holder is fitted with a 25 watt bulb. This end should be perforated to permit air circulation, otherwise the lamp may overheat.

The other end of the improvized metal cylinder is fitted with a detachable circle of perforated metal. Place the material to be investigated onto this sieve.

The lamp is left burning for a day or more. The heat forces the minute organisms to leave the bedding, soil, or litter and they fall through the sieve into the preservative beneath.

Examine the collected samples under the microscope and identify them.

5.11. Practical Program

TECHNIQUES IN HANDLING AND SEXING MAMMALS

The objectives of this skills based program are as follows:

(i) To introduce to student the principles and skills of handling the more common laboratory mammals.
(ii) To instruct the student in the techniques of sexing laboratory mammals.
(iii) To prepare and examine a vaginal smears taken from mammals at different times in the estrus cycle.

Handling of laboratory animals

Laboratory animals will often need to be handled; this applies particularly to mammals. Non-mammals are best left alone unless they must be handled for a specific purpose.

The correct handling of an animal is most important if the full benefits are to be gained from its presence in the laboratory. The ease with which an animal may be handled depends to an extent upon the frequency with which it has been handled since young. It is usually only necessary to handle the unweaned animals for the purposes of sexing. The weaned animals should ideally be handled daily through to adulthood. In a large animal house time will not permit technicians to be so intimate with each animal, but attempts should be made to gain the confidence of the animals by handling them when possible.

The different species need to be handled in different ways, but in general the same rules apply. The animals should be approached in a confident manner; the uncovered hands moving in from the sides of the animal in one continuous action. The animal may get frightened if unsure, jerky movements are made. Gloves or other means of defence only need to be used if the animal is diseased or fierce. The animals should not be handled by those with an infection of any kind, or if they have cuts and wounds on their hands.

The hands should be disinfected before and after handling. Handling a laboratory mammal is necessary for a number of reasons, including physical examinations and sexing. The techniques involved in these types of handling require a skill that only come with practice. The information outlined here is only the beginning. If carcases are available some previous handling practice can be achieved. There are three basic methods for lifting up an animal—lifting by the tail, lifting by the neck, or shoulders with rump support.

Species	Method description
Mouse and small rat	This is the only mammal that can be lifted by the tail. Grip the base of the tail between finger and thumb. Quickly transfer the animal to a solid surface, such as the cage top, or the fore-arm. To sex: lift the tail back over the body to examine the anogenital distance. To examine: hold the mouse by the loose skin on the back of the neck. Hold the tail between the fourth and fifth fingers. The animal's back rests in the palm of the hand.
Rat	Position the palm of one hand over the shoulders and bring the thumb around under the chin to ensure the animal cannot bite. Use the other hand to support the rump as the animal is lifted. To sex: turn the animal over to examine the anogenital area.
Guinea-pig	Lift as for rats. Be sure to support the rump and exercise care with females, it is difficult to see if they are pregnant sometimes. To sex: hold securely with abdomen outermost and examine genitalia.
Hamster and gerbil	Lift by scooping up the animal in cupped hands. Approach the animal from the head end and then to hold it firm, grasp it by the loose skin on the neck and support it as for mouse. To sex (gerbil): turn back the tail over the animal's body and examine anogenital region. To sex (hamster): hold the animal with its back in the palm of the hand.
Rabbit	Steady the animal by grasping the ears. Put the other hand beneath the abdomen. Lift the animal, taking the weight by way of the belly. Stand the animal on a rough surface. It panics if it slips on smooth surfaces. Carry the animal in front of the body supporting the weight by the rump. A hand on the head will steady the animal. To sex: support the animal by the rump and by grasping the ears, or the scruff of the neck, turn the animal so that it's legs are away from the operator.
Ferret	These animals should be approached with confidence and picked up in a manner similar to that described for rat. The rump should be supported as well. To sex: hold the animal with legs facing away from the operator and examine the anogenital distance.
Cat	This animal presents more problems than most laboratory animals. A bad-tempered cat can be very damaging, to say the least. Lifting a cat by the scruff of the neck with the arm held straight in line with the backbone, is described by experienced persons. This can be less than easy for the beginner. Moving a cat involves holding it by the scruff of the neck, tuck its head under the left arm and grip it against the body with the arm.
Dog	The nervous dog is a hazard for the handler. It is approached with confidence and not startled by rapid movements. The hand is brought along the side of the head to secure the scruff of the neck, or a collar. It is held firm with the arm behind the head in line with the backbone. Not a job for the inexperienced working with difficult animals. A dog stick may be needed.

All the instructions given above can only be regarded as indicators of skilled handling techniques. A useful movie-film cassette has been prepared by experienced technicians showing handling and sexing techniques (see Bibliography).

Sexing laboratory mammals

The importance of correct animal handling for the purposes of sexing needs no emphasis. The animal will have to be turned over, manipulated, and probed. Without skill in handling, one is likely to be bitten, or at least unsuccessful in finding out the sex of the animal.

Sexing animals requires practice and knowledge. There are useful movie films demonstrating this skill, but there is no substitute for handling the real animal. Some benefit can be gained from trials on dead animals.

A physical examination of the animal is the customary way in which to sex an animal, but observations of behavior can sometimes be useful. The general technique is to examine the under surface of the animal in order to note the following:

(a) The presence of mammary glands and nipples (caution, the male guinea-pig has a pair of nipples).

(b) The presence of the penis, which may in some mammals (guinea-pig) be extruded.

(c) The distance between and the appearance of the urinary and anal openings.

As a guide to the skill of sexing laboratory mammals, the features to be observed are summarized in tabular form below.

To use the check-list of sexual characteristics of the various laboratory species, the following key is necessary.

Key to the sexing check list of mammals.

(i) Presence of nipples and mammary glands.
(ii) Extrusion of the penis.
(iii) Body-shape difference, i.e. scrotal sacs.
(iv) Pregnant female clearly observed.
(v) Distance between urino-genital openings and the anus, different in male and female.
(vi) Distance between the openings are similar but appear different in the male and female.

	(i)	(ii)	(iii)	(iv)	(v)	(vi)	Comments
Mouse	+	−	+	+	+	−	Anogenital distance in male is twice that in female
Rat	+	−	+	+	+	−	Anogenital distance in male is greater and obvious
Hamster	+	−	+	+	+	−	Anogenital distance in male is greater and obvious
Gerbil	+	−	+	−	+	−	Pregnancies difficult to recognize
Guinea-pig	+	+	+	−	−	+	Pregnants sometimes difficult to recognize. Males have pair of nipples
Rabbit	−	+	−	−	−	+	Difficult to sex when young. Male organ the major difference
Ferret	+	−	+	+	+	−	Easy to sex by greater anogenital distance of the male
Cat	+	+	+	+	+	−	Anogenital distances not so great, but male has greater separation
Dog	+	+	+	+	+	−	Easily recognized
Primate	+	+	+	+	+		Easily seen without handling the animal

In order to achieve success in the sexing process, correct handling techniques are required. These are outlined in the previous section.

Vaginal smears and the estrus cycle

The preparation of vaginal smears is sometimes carried out in animal units in order to determine whether an animal has been mated or is ready for mating. This kind of information can be provided by the presence of spermatozoa or by the existence of a vaginal plug and so forth. A more detailed study of a stained vaginal smear can provide more precise information about the stage of the estrus cycle in which the female is at the time of examination. The sorts of changes that accompany the passage of time during the estrus cycle are summarized for mouse in the tabulation below. A stained smear taken at any time during the estrus cycle will show the presence of cells. The types of cells and their characteristics will need to be recognized in order to make the vaginal smear a useful technique.

The preparation of a vaginal smear from a mouse

Obtaining the smear involves inserting an instrument into the vagina and gently scraping out some cells from the vaginal wall. This necessitates skilled handling of the mouse. The material can be obtained in a variety of ways. Use a small glass rod with a blunted end that has been dipped into mammal saline. Rub this rod against the walls of the vagina and then transfer the material to a clean slide to make a smear.

Alternatively, a small quantity of saline may be pipetted into the vagina and then withdrawn onto a slide. A saline-dampened piece of cotton wool held in blunt-nose forceps may also be inserted into the vagina.

Examination of the smear can be carried out without staining or after staining. For a rapid assessment of the condition of the mouse unstained smears can be useful. In order to prepare a stained vaginal smear carry out the following procedure. For a more detailed reading on the staining of seminal and vaginal smears turn to the Bibliography (p. 299).

Staining procedure

(a) Fix ("pickle") the smear by covering it with equal parts of absolute alcohol and ethyl ether (1 minute). Alternatively, use equal parts *iso*-propyl alcohol and ether (10 minutes).

(b) Add alcohols of lower percentage: 90% alcohol (1 minute); 70% alcohol (1 minute).

(c) Stain the smear by covering with *Ehrlich's Haematoxylin* (10 minutes).

(d) Add some ammoniated alcohol (70%) or add alkaline tap water (to "blue" the nuclei).

(e) Add some acid alcohol to slowly remove (differentiate) the stain from the nucleus until it shows the required intensity of color when viewed under the microscope. Add more tap water to stop the stain-removal action.

Up to this point the nucleus has been stained blue. We continue to stain the cytoplasmic contents.

(f) Cover the smear with *Shorr's stain* (92–5 minutes).

(g) Add 70%, 80%, 90%, and 100% alcohols (1 minute each).

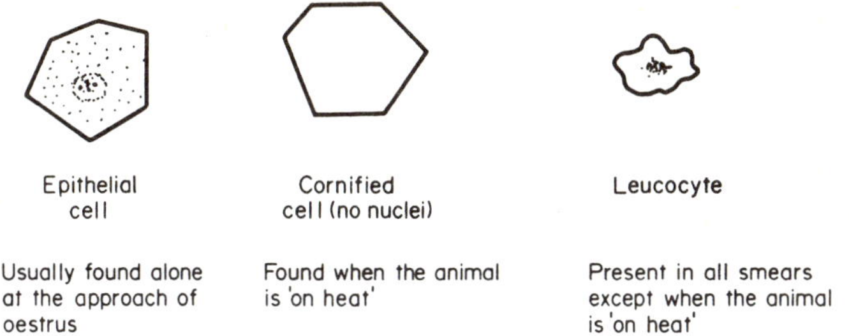

FIG. 82. Cells in a vaginal smear

(*h*) Remove the alcohol ("clear") by adding xylol or chloroform (2 minutes). If the cells are clearly stained, move on to the next stage.

(*i*) Cover the smear with a mounting agent such as DPX, Euparol, Canada Balsam, or equivalent. Add the cover glass.

Examine the completed, stained smear under the microscope.

The expected results are as follows:

> Epithelial cells—blue nuclei, green cytoplasm.
> Cornified cells—orange-pink (no nucleus).
> Leucocytes—blue nuclei, green cytoplasm.

TABLE 7.

Histological changes during estrus cycle

Stage	Duration	Vaginal epithelium	Smear
Proestrus	1 day	Epithelial layers increase in numbers and thickness. Outer layers nucleated, those beneath becoming increasingly cornified. Few leucocytes present	Epithelial cells* Cornified cells Leucocytes
Estrus	½ day	Loss of superficial nucleated layer Cornified layer now superficial Beneath this layer are nucleated cells No leucocytes	Epithelial cells† Cornified cells
Metestrus (early)	1 day	Cornified layer losing cells Leucocytes begin to appear under the surface	Cornified cells‡
Metestrus (later)	1 day	Many leucocytes in the outer epithelial layers	Cornified cells† Epithelial cells†
Diestrus	2½ days	Epithelial cells with leucocytes in the outer layers	Epithelial cells Leucocytes

* = few; † = many; ‡ = plentiful.

Reproduction, breeding, and heredity—Summary

The sperm and egg-producing structures were examined.

Androgens (testosterone) and estrogens (estradiol) were described as having important roles to play in sexual reproduction.

The changes in behavior, weight, and physiology that accompany the estrus cycle were outlined.

Successful mating of animals in captivity, it was shown, depends upon rather precise conditions.

It was suggested that some animals maintain the breeding season and rhythms of their geographical origin.

Pregnancy and pseudopregnancy changes in behavior and physiology were described, together with methods of recognizing pregnancy.

The influence of oxytocin in the birth process was mentioned.

Milk production is a hormone-related event was outlined and a review of the optimum weaning ages and weights was given.

Breeding programs were described as being in breeding or random. The former method being appropriate for producing genetic uniformity.

Different species were described as requiring different breeding methods, either in pairs, trios, or larger groups.

The units of heredity were described as being genes located on chromosomes that are distributed to offspring cells at mitosis or meiosis.

Mice and guinea-pigs were used to demonstrate the meanings of phenotype, genotype, dominance, recessiveness, and other basic genetics terminology.

A wide range of non-mammal laboratory animals commonly used in educational institutions were outlined.

Methods of sexing and handling and producing vaginal smears were described.

STUDY OBJECTIVES

UNIT 6:

Legal Requirements—Anesthesia, Laboratory Procedures, and Euthanasia

(*a*) Outlines the requirements of the 1876 "Cruelty to Animals Act" (UK).
(*b*) Gives details of the requirements for license and certificate holders.
(*c*) Outlines the requirements of the "Animal Welfare Act", 1970 (USA).
(*d*) Describes local and general anesthetics, the factors affecting their choice, and the route of their administration.
(*e*) Describes the methods whereby animals are dosed or injected with experimentally required substances.
(*f*) Describes how specimens are collected from animals.
(*g*) Describes physical and chemical methods of euthanasia.
(*h*) Describes the factors that indicate animal death.
(*i*) Suggests practical work in connection with the use of hypodermic syringes, anesthesia, and euthanasia.

UNIT 6:

Legal Requirements—Anesthesia, Laboratory Procedures, and Euthanasia

THE use of animals in the laboratory is governed by regulations stemming from government Acts. Any procedure that is likely to cause pain is covered by animal welfare Acts and demands that appropriate anesthesia and euthanasia techniques are employed. Experiments with animals usually involve injecting, dosing, or collecting body substances. All the regulations and techniques are outlined in this unit under the following headings:

6.1. The "Cruelty to Animals Act", 1876 (UK).
6.2. "Animal Welfare Act", 1970 (USA).
6.3. Anesthesia.
6.4. Dosing, injecting, and collection procedures.
6.5. Euthanasia.
6.6. Practical programme of techniques in anesthesia and euthanasia.

6.1. The "Cruelty to Animals Act", 1876 (UK)

This is the main Act of Parliament regulating the use of animals for the purposes of experiments. Since this Act was written many others have been produced to cover not only animals used for experiments but also those unconnected with scientific work. However, it is the 1876 Act which lays out the principal statements concerning the use of laboratory animals for experimental purposes. The reader is recommended to have access to the Act as well as to the *Notes on Animal Houses* produced by the Home Office (London, 1971).

The 1876 Act regulates the use of living vertebrate animals (excluding man) for the purposes of experiment calculated to cause pain. Such experiments in any event may only be carried out on premises previously registered for this purpose with the Home Office. The individual(s) performing the experiments must have a license to do so. Once a license is issued by the Home Office then a relevant certificate may be appended. It is the certificate that puts limits on the type of work the experimenter may carry out.

In this context the reader is recommended to have access to the following:

Notes for Guidance in Completing Forms of Application under the Cruelty to Animals Act, 1876 (Home Office, London, 1971).

For useful publications on the law relating to experiments on animals in Great Britain contact the Research Defence Society (p. 304).

236

6.1.1. LICENSES

Any person carrying out any experiment on a living vertebrate animal where pain is inevitable must be in possession of a license issued by the Home Secretary of the UK. The premises must also be registered. Neither the license nor the premises are transferable. A person who only holds a license and no certificate has limited scope in terms of his experimental work and he must observe the following restrictions:

(*a*) The animal must at all times be under anesthesia sufficient to prevent it feeling any pain.

(*b*) The animal must not be permitted to recover from the effects of the anesthetic if there is any chance that pain is probable, or if the animal has been injured during the experiment. The animal must be humanely killed before the effects of the anesthetic wear off.

(*c*) The animal must not be used merely to demonstrate techniques to students.

(*d*) The experiment must only be carried out if it advances knowledge in the fields of physiology or medicine.

The above is only an outline of the main features of the responsibilities of the license holder.

In order to obtain a license one has to submit a complete dossier on the experiments intended together with full details of personnel qualifications.

The emphasis throughout is prevention of pain, sometimes described as "the pain clauses or conditions".

6.1.2. CERTIFICATES

In order for a license holder to be permitted release from some of the restrictions just mentioned he must be certificated. These certificates are issued by authorized signatories such as university professors of relevant academic departments such as Medicine or Physiology. The Home Secretary oversees the whole process of certificate issue. The restrictions imposed by the license are modified by certificates in the following way:

Certificate A

This permits the holder to carry out simple experimental procedures on animals without anesthesia being administered. The operations are confined to injections, withdrawal of body fluids, dietary variations, exposures to radiations, and sensory stimuli. Details of these procedures must be written into the conditions of the certificate.

Certificate B

This permits the experimenter to allow the animal to recover from the anesthesia if the object of the experiment can only be achieved in this way. The animal must be killed as soon as the object of the experiment has been achieved. This work includes such as gland-tissue removal or destruction, transplanting of tissues including tumour material.

Certificate C

This permits animals to be used for lecture demonstrations. Notification of the intended event must be given with full details of audience, time, place, and intentions. The animal must be anesthetized and not permitted to recover from the anesthetic before being killed.

Certificate E

This allows work to be carried out on cats and dogs. It is used in conjunction with certificate A.

Certificate EE

This allows work on cats and dogs when experiments under certificate B are being carried out.

Certificate F

This is required if work is to be carried out using members of the horse family. It is held in conjunction with certificates already mentioned above.

6.1.3. INSPECTIONS

In an attempt to enforce the conditions implied in the Act, the Secretary of State has an inspectorate of veterinary or medical persons who visit registered premises. This they do without warning. These inspectors can instruct that any "unsatisfactory experiment" be terminated and the animal humanely destroyed.

6.1.4. ANIMAL RETURN OF EXPERIMENTS

Another way in which the Home Office keeps in touch with the activities of licensees is by receipt of completed forms describing all experiments undertaken within the meaning of the Act.

6.1.5. LITTLEWOOD COMMITTEE RECOMMENDATIONS, 1965

The 1876 "Cruelty to Animals Act" appears to be rather open to abuse and not wide enough in its reference. Careful reading of the 1876 Act calls forward much criticism and it was in this respect that the British government set up a committee to look into the Act in detail and then to make recommendations. The committee, chaired by Sir Sydney Littlewood, put forward many powerful recommendations for changes in the law. These recommendations have not to this time been put into effect.

6.1.6. OTHER "ANIMAL ACTS"

Numerous "anti-cruelty" to animal Acts have been passed, notably the Protection of Animals Act, 1911. All the other Acts deal with farm animals, domesticated animals, etc., their breeding, treatment, and slaughter. They put in detail with respect to any animal kept by man for his purposes. These Acts together with the 1876 Act will be studied in greater depth in a later study program.

6.2. "Animal Welfare Act", 1970 (USA)

Not infrequently laboratory animal personnel will carry out work on both sides of the Atlantic. In this context it is relevant to give space to a brief survey of the "legal considerations" in the USA.

For starters, it is of interest to note that the technician's manual in the UK commences with a chapter on "the law" and the obligations to cause no pain. The USA manual has no such reference throughout. Its main stress is on efficient management. The concern with "feelings" implied in the UK Act might be thought of as sentimentality by some!

The Animal Welfare Act, 1970 is an amendment of an earlier 1966 Act. It is an amalgam of provisions relating to the commercial use of animals as well as to the use of animals in "research facilities".

Section 13. "The Secretary shall promulgate standards to govern the humane handling, care, treatment, and transportation of animals by dealers, research facilities, and exhibitors. Such standards shall include minimum requirements with respect to handling, housing, feeding, watering, sanitation, ventilation, shelter from extremes of weather and temperatures, adequate veterinary care, including the appropriate use of anesthetic, analgesic or tranquilizing drugs, when such use would be proper in the opinion of the attending veterinarian of such research facilities, and separation by species when the Secretary finds such separation necessary for the humane handling, care or treatment of animals.

"That the Secretary shall require, at least annually, every research facility to show that professionally acceptable standards governing the care, treatment, and use of animals, including appropriate use of anesthetic, analgesic, and tranquilizing drugs, during experimentation are being followed by the research facility during actual research or experimentation."

Section 16. "The Secretary shall promulgate such rules and regulations as he deems necessary to permit inspectors to confiscate or destroy in a humane manner any animal found to be suffering as a result of a failure to comply with any provision of this Act or any regulation or standard issued thereunder...."

From the foregoing it can be expected that animal welfare agencies are concerned that both the UK and USA Acts do not go far enough nor can they correctly be implemented with so few inspectors. They imply that research facilities know when an inspector is arriving and that government facilities are not covered by the Acts! A very readable article on this topic is found in the *Biologist* (Vol. 26, No. 1).

6.3. Anesthesia

The animal welfare Acts considered previously make it law that we should treat animals in a humane manner. Animals that are to undergo experimentation procedures need to be anesthetized in order that they feel no pain. Anesthetics may be categorized as follows:

> *Local anesthetics*—eliminate sensation in a small area by blocking nerve activity.
> Examples: lignocaine, procaine.
> *General anesthetics*—eliminate sensation over the whole body by depressing central nervous system activity resulting in a temporary loss of consciousness.
> Example: inhaled gases—ether, nitrous oxide, intravenous injections—thiopentone.

The use of general anesthetics can produce unpleasant and unwanted side effects and it is for this reason that pre-anesthetics or premedicants are administered. This is not the place to go into further detail of premedication, it is sufficient to say why they are given.

> *Pre-anesthetics*—Atropine sulphate reduces saliva and mucus flow.
> Pethidine is a pain reliever (analgesic).
> Phenothiazine drugs are sedatives (tranquilizers).

The use of pre-anesthetics is determined by the type of anesthetic to be administered. Some anesthetics may produce excessive salivation, others muscular rigidity or vomiting. The premedication must be matched to the anesthetic and to the experimental procedures.

6.3.1. FACTORS AFFECTING THE CHOICE OF ANESTHETIC

The anesthetic to be used will depend upon the following factors:

(a) *Species of animal.* The anesthetic to be used must not have undesirable side effects, if possible. It must also have the desired anesthetic effects. Different species respond in different ways to different anesthetics.

(b) *Locality of surgery.* The site of the operative procedures will determine whether an inhalation gas or an injected drug is used. Clearly the use of gaseous anesthesia would be unsuitable for an animal involved in thoracic-cage surgery.

(c) *Time needed in anesthesia.* An operation of long duration will necessitate different anesthesia procedures from the animal undergoing simple surgical manipulations.

(d) *Post-surgery fate.* An animal that is to be destroyed painlessly at the end of surgery before it recovers consciousness can have an anesthetic that would be inappropriate to an animal that is to recover.

(e) *Health of the animal.* If the animal in experiment is old or has ill health it may react differently to the anesthetic. This must be taken into consideration.

6.3.2. ROUTE OF ANESTHETIC ADMINISTRATION

The mode of administering the anesthetic will also depend upon the factors outlined above to a large extent. The ways in which the anesthetic may be given are as below:

(a) *Topical anesthesia*—skin-surface application. This can be a volatile agent that evaporates rapidly "freezing" the site. There is a danger of tissue death using this method. *Ethyl chloride* can be used in this manner. Perhaps more reliable is the surface application of an agent in jelly or solution form. This is a local anesthetic technique.

(*b*) *Infiltration anesthesia*—local injection technique where the active agent, such as *procaine*, is injected directly at the site of surgery. This is similar to the dental method.

(*c*) *Regional anesthesia*—a regional injection technique whereby regional nerve blockage is produced. This *extradural* or *epidural* method puts the active agent into the spinal canal or major nerves leaving the spinal cord, thereby blocking the spinal nerves in that region. The chemical agent can be *lignocaine*.

(*d*) *Inhalational anesthesia*—respiratory technique whereby gases or volatile liquid vapours are inhaled to induce anesthesia. These are general anesthetics such as: (volatile liquids), *diethyl ether*, (ether) *halothane*; (gas), *nitrous oxide*.

TABLE 8.
Anesthesia and anesthetics

Type of anesthesia	Mode of anesthesia	Anesthetic	Comments
Local	Topical	Ethyl chloride	Skin spray or skin application
	Infiltration	Procaine	"On site" injection
	Regional	Lignocaine	Quick onset, longer action
General	Inhalation	Diethyl ether	Explosive
		Halothane	Non-explosive
		Nitrous oxide	("Laughing gas")
	Intravenous	Thiopentone sodium	"Sagatal"
		Pentobarbitone sodium	"Nembutal"
	Intraperitoneal	Pentobarbitone sodium	

In order to maintain the required level of anesthesia these agents must be administered continuously in a controlled manner. This is achieved by one of the systems outlined below.

(i) *Chamber system*—generally used for small laboratory animals. The vessel can be a glass jar resembling a chemical dessicator. The animal is placed on top of a metal grid beneath which the volatile agent is soaked into cotton or muslin. The liquid should not be permitted to contact the animal. This system has no regulatory device.

(ii) *Open system*—a mask cone is applied to the nose of the restrained animal. A measured number of drops of the anesthetic liquid are put onto the lint cloth mask. The vapor induces anesthesia as the animal breathes in the air/vapor mixture.

This system is regulated only to the extent that the drops of anesthetic administered are counted.

The above two systems have obvious disadvantages for any form of lengthy surgery.

(iii) *Semi-closed system*—consists of a cylinder of oxygen to supply the animal's needs through a mask or by way of a tube inserted into the trachea (endotracheal tube). This oxygen passes through an anesthetic vaporizer on its way to the animal in order to supply a controlled oxygen/anesthetic mixture.

(iv) *Closed system*—consists of an oxygen cylinder and an anesthetic vaporizer as for the "semi-closed" set-up, but expired gases are cleared of carbon dioxide by passing through a soda lime cannister into a rebreathing bag. This system is also able to regulate the oxygen/anesthetic mixture.

(e) *Intravenous, intraperitoneal, drug injection general anesthesia.* The non-volatile anesthetics can be introduced to the animal by way of injections directly into a vein (intravenous) or directly into the abdominal cavity (intraperitoneal). This is done because the active agent may be destroyed if given by mouth, or irritant if injected intramuscular. This route speeds up the action of anesthesia.

The drugs administered by these routes include the following:

Thiopentone sodium is a commonly used intravenous general anesthetic which requires skill upon administration and dosage calculation.

Pentobarbitone sodium ("Nembutal") is widely used but because of the more lengthy recovery time has to some extent been superceded by the above mentioned. May be administered intraperitoneal or intravenous.

The use of anesthetics is a specialist area and requires that technicians are license holders or are acting under the close supervision of a license holder whilst administering anesthesia.

Animals in both the pre- and post-operative period need special attention backed up by a knowledge of the problems confronting the animals at this time.

Pre-operative care involves withholding food for a recommended period before the anesthetic is administered. Water must not be withheld. If need be premedicants are administered.

Post-operative care involves keeping the animal warm and devoid of too much stimulation. Offer normal diet unless specifically instructed to the contrary.

6.4. Dosing, injection, and collection procedures

If an animal is to be dosed with antibiotics, food extras, or physiologically active drugs for the purposes of an experiment, the operative must be a license holder. If body fluids are to be collected from the live animal the same restrictions apply.

This section is concerned with a brief description of the terminology and techniques employed in the areas of dosing laboratory animals and collecting substances from the same.

6.4.1. DOSING AND INJECTING ANIMALS

When a drug or similar is to be administered to an animal there are several routes possible. For convenience these routes may be defined as follows:

(a) *By way of the alimentary canal*—entry by the mouth or by the rectum.

(b) *By way of skin, blood vessels, muscles, etc. (parenteral).*

As a generalization, it might be said that injections are given in preference to administration by the mouth in the following circumstances:

(i) If the drug or active agent is likely to be altered or destroyed on its passage through the alimentary canal.

(ii) If the administered substance is irritant to the alimentary canal or unable to be absorbed by the gut lining.

(iii) If the substance is required "on site", as in the case of some local or regional anesthetics.

(iv) If the alimentary canal route is too slow in producing the required effects, i.e. anesthesia.

It is true that some drugs are irritant to the tissues in general and are therefore restricted to administration directly into the bloodstream, i.e. intravenous thiopentone injections.

These techniques generally employ *disposable* sterile-packed hypodermic syringes. This offsets the need for extra duties in terms of keeping a sterile supply of syringes. The sizes of needles used for the various parenteral injections tabulated below vary from 6.4 mm 24 gauge to a 3.2 cm 19 gauge needle. The needles must be the correct size for the job, sharp and sterile—i.e. cardiac puncture bleeding: mouse, 23 gauge; guinea-pig, 19 gauge.

Route	Terminology	Descriptions	Comments
Alimentary canal	Oral	Capsules or tablets are placed into the mouth or throat. A stomach tube may be needed	Useful for antibiotics and food additives. Rodents, rabbits, guinea-pigs, cats, dogs
	Per rectum intrarectal	Taken by way of the rectum	
Percutaneous	Topical	The active agent is taken through the skins surface	
	Intravenous	Injection into a vein that is made more clearly visible by warming. Located in the margin of an ear or at the base of the tail or near the tip	Rodents, rabbits, cats, dogs, primates
	Intraperitoneal	Injection into the abdominal cavity. Care necessary unless viscera such as intestines or bladder are punctured in error. Injection site just above the bladder in the mid-line	Rodents, rabbits, guinea-pigs, cats, dogs, primates
Parenteral	Intramuscular	Injection into a muscle. Usually into the muscle on the hind area of the upper thigh. Care not to make the injection intravenous, not to touch nerves	Rodents, rabbits, guinea-pigs, cats, dogs, primates
	Subcutaneous, intracutaneous, and intradermal	Injections beneath the skin. Usually put into a fold of skin behind the head or on the abdomen	Rodents, rabbits, guinea-pigs, cats, dogs

Route	Terminology	Descriptions	Comments
Parenteral	Intrathoracic	Injections into the thoracic cavity. Care not to penetrate diaphragm	Primates
	Intracardiac	Injections into the heart	
	Intranasal and intratracheal	Injections into the nose or trachea	
	Intracerebral	Injections into the brain whilst the animal is anesthetized	Rodents, rabbits
	Epidural, extradural intraspinal	Injections into or near the spinal cord	Useful for anesthetics for regional use

6.4.2. COLLECTING BODY SUBSTANCES

For the purposes of monitoring the progress of experimental work it may be necessary to take specimens from the laboratory animal. Specimens are also taken in order to aid in the diagnosis of disease or to monitor the degree to which an animal is host to various parasites. The techniques here summarized must be practised on dead animals with an experienced instructor before any attempt is made on a living animal.

Collection area	Mode of collection	Reasons for collection
Oral	A mouth or throat specimen smear prepared	Investigations for bacteria
Integument	A skin, fur, or feather sample is taken	Examined for ringworm fungi, surface parasites, eggs of parasites, and other disease agents
Bladder	Urine sample taken by inserted catheter or delivered in metabolism cage sample vessel	Examination of the physical and the chemical properties of the urine are diagnostic aids to disorders of the kidneys, pancreas (diabetes) and clues to metabolic disorders, bacterial infections, and tumors
Alimentary canal	Feces samples taken direct from the rectum or collected in a vessel beneath a metabolism cage	Examinations for the eggs or life-history stages of gut parasites. Investigation of digestive processes. Examination for evidence of microbiological agents or any disease
Vagina	A smear taken from the walls of the vagina	Examination of the cells in the stained smear indicate the stage of estrus in a breeding female animal
Spinal cord	A sample of cerebro spinal fluid is extracted by syringe whilst the animal is in some cases anesthetized (lumbar puncture)	Examination of the CSF for disease-causing agents and unusual chemical components
Heart	Intracardiac withdrawals of blood by heart puncture whilst the animal is anesthetized. Used with rats, guinea-pigs, rabbits, hamsters, and kittens	Examination of the physical and chemical properties of the blood as diagnostic aids to disease

Collection area	Mode of collection	Reasons for collection
Veins	Intravenous withdrawal of blood by *venepuncture*—insertion of a suitable hypodermic needle. The vein can be in the ear margin, at the base or tip of the tail, in a limb *Venesection* can be the severing of a small vein in the ear margin by means of a sharp blade	Blood cell counts Oxygen transportation ability Coagulation ability
Orbitalsinus	A needle into the eye socket; mice, hamsters	

Specimens that have been collected for laboratory examination must be presented for investigation as soon as possible or preserved in such a manner as to prevent their deterioration. The bottling of specimens demands attention to detail as later tests on the material will be invalidated if the containers are already unclean before the blood or urine was added.

If investigations on the collected material are to be delayed then refrigeration will be the first avenue of preservation but this has a limited value for any length of time with blood. If tissues and organ structures are to be preserved then the preserving agent must be carefully chosen in case the chemical agent interferes with the materials to be investigated. This type of information is available in most laboratory texts on histology.

Whatever the technique employed to preserve or contain the collected specimens they must be labelled fully with date, nature of specimen and origin, preservative, experimenter, coded information (if appropriate), investigations requested, deadline dates, etc.

6.5. Euthanasia

This is the killing of animals in a humane, painless manner. The methods of euthanasia described below are normally to be put into operation only by experienced personnel. Trainees should practice on dead animals and not be permitted to exercise this skill on living animals if any doubt exists concerning their ability to produce rapid unconsciousness and subsequent death. A swift termination of the animal must be done in conditions that do not induce panic or any distress before the event.

The method of euthanasia chosen will depend upon the following:

(*a*) *Species of animal*—the size of the animal and any problems relating to restraint.
(*b*) *Fate of the carcasses*—if the dead animals are for disposal then clearly damaged tissues are not going to present problems. However. if certain tissues are going to be the subject of laboratory examination then methods of euthanasia that create tissue change are inadvisable.
(*c*) *Numbers of animals to be killed*—if large colonies of animals are to be destroyed then an appropriate method must be chosen to be efficient and at the same

time being humane. This method may differ from that employed for the single animal.

The reasons for killing animals in the laboratory situation are various, including the following:

Culling or reducing stock numbers that are surplus to requirements.
Preventing an animal from suffering with a disease or injury.
Examining of internal organs after an experimental period.

6.5.1. METHODS OF EUTHANASIA

The recommended methods of killing laboratory animals fall into two groups:

Physical methods—stunning, cervical dislocation (neck dislocation), decapitation, shooting, electrocution.
Chemical methods—drug injections, gas or vapour inhalation.

Each of the above-mentioned methods is outlined below. Before any animal is killed it must be removed from the animal room and not killed in the presence of other animals or in front of students.

6.5.2. PHYSICAL METHODS OF EUTHANASIA

(a) *Stunning.* This method renders the animal unconscious in a swift painless manner. For smaller animals with relatively thin skulls death is instantaneous. For an animal such as guinea-pig or some birds the stunning may be followed by the throat being cut or decapitation. The instruments employed for stunning the animal all have the same characteristic, they deliver a heavy blow to the animals cranium thereby irreversibly damaging the brain.
The method is usually only employed with animals that present few problems of restraint.
The method is clearly inappropriate for animals whose brains are required for scientific investigation.
Small animals can be stunned by holding the body firmly and striking the cranium swiftly against a hard surface. One blow should be enough. This can be suitable to stun rats, guinea-pigs, and mice. Fish, amphibia and reptiles, and some birds can be despatched in this manner also.
Larger animals can be stunned by striking the back of the cranium with a wood or metal instrument. Fowls, guinea-pigs, and rabbits can be killed in this manner.
Larger animals can be stunned by means of mechanical devices that resemble pistols with stunning-bolts that are fired against the beast's cranium. By this method the animal is generally rendered unconscious and is despatched by bleeding as the result of a cut jugular vein in the throat.
Electricity is used to stun some larger animals such as pigs, sheep, and cattle.

(b) *Cervical dislocation.* This method dislocates the neck vertebrae or in more familiar terms "breaks the neck". The spinal cord at the base of the skull is damaged or severed in this killing method.

Smaller animals can have their necks dislocated by comparatively little physical action. The mouse, for instance, is placed on a non-slip surface and held firmly by the base of the tail. A pencil is pressed across the back of the skull and a sharp pull exerted on the tail. The neck is broken in this manner. The fowl and small birds can be killed by swift neck dislocations.

Large animals such as the guinea-pig are sometimes killed by a neck wrench upwards as the animal stands on a table top. A rabbit can be killed by the "rabbit chop" as the animal is supported by the rear legs. All these methods require skill that only comes with correct training and should never be attempted by the inexperienced.

(c) *Decapitation.* This method removes the animal's head in a quick and clean fashion. In common with the "throat-cut" technique of bleeding after stunning much blood loss will occur. Decapitation is usually only employed with smaller animals such as some birds, mice, guinea-pigs (less common), fish, amphibia, and reptiles.

(d) *Shooting.* This method involves penetrating the brain of an animal employing a pistol-like device that fires a bolt through the cranial bones. This bolt may fire free from the gun as a bullet, or it may be the type where the bolt only fires from the end of the muzzle and is retained in the device (captive bolt).

Both the described devices employ blank cartridges and as such these humane killers require a police permit in the UK.

When these devices are used the animals need to be restrained and the technician using the instrument must know the line of fire to be employed with each animal. Even though the brain suffers extensive damage the animal's heart may continue to beat for a while after the lethal shot has been delivered. To hasten the cessation of this heart activity a cut across the neck blood vessels may be necessary.

These humane killers are used with cats, dogs, goats, pigs, cattle, horses, and sheep.

(e) *Electrocution.* This method employs A.C. electrical energy which is passed through the brain and heart of the animal. This "stepped up" current produces instantaneous stunning and death in a matter of seconds.

This is a skilled technique and requires special apparatus. It is unlikely to be encountered in the majority of laboratories. The use of electricity as a humane killer for laboratory animals is a subject of discussion.

The physical methods outlined above appear to be less pleasant than the chemical methods to be reviewed now, but there is a case for suggesting that they are more rapid and thereby more humane.

6.5.3. CHEMICAL METHODS OF EUTHANASIA

(a) *Drug injections.* This method involves putting the lethal dose into the animal by the most effective route possible for the species concerned. In most cases the animal will need to be restrained. This can involve some skilled operations on behalf of the technician if aggressive animals are concerned. Specially designed cages need to be employed for the primates for instance, so that the animal can be pulled towards the cage bars for the injection.

(i) *Barbiturates* such as pentobarbitone sodium ("Nembutal") may be injected intravenous or intracardiac to produce a swift death. This is not always so easy with a restless animal and so the intraperitoneal route may be used with a more concentrated dose of the drug. The dosage given will be relative to the animal's size. In any event the animal should be dead within 15 minutes. Checks on death characteristics should be made in case the animal is merely in deep anesthesia. This applies particularly to animals administered "Nembutal" by the oral route in the form of capsules as may be used for kittens. The "Characteristics of Death" are dealt with later in this unit.

(ii) *Magnesium sulphate* as a saturated solution administered intravenously is another rapid way in which to destroy an animal. The dosage again is relative to the animal's size. Since this drug is comparatively cheap, "Nembutal" or some similar sedative dose may be given to quieten an animal before being vein injected with the magnesium sulphate solution.

(b) *Gas or vapor inhalation.* These methods employ toxic gases or vapors that may also be flammable and consequently skilled techniques and common wisdom combine to prevent danger to both animals and personnel.

(i) *Toxic gas inhalation* necessitates the use of a lethal chamber so that the gases employed do not leak out into the atmosphere. The gases most commonly used for the purposes of euthanasia are carbon-monoxide and carbon-dioxide.

Carbon-monoxide gas is the toxic gas present in the exhaust fumes of the combustion engine. Some organizations use filtered exhaust fumes to destroy small animals especially when the numbers to be despatched are large. This gas is also present in smaller quantities in coal gas.

Carbon-dioxide gas is the gas expired after respiration. It is heavier, non-flammable gas, unlike the previously mentioned carbon-monoxide. It is also of lesser toxicity than the monoxide and thereby a safer gas for use. The gas is colorless, odorless, and non-irritant and can be supplied under pressure in cylinders.

In order to use the gas for euthanasia purposes it has to be conducted from the supply cylinder to a lethal chamber. This can be a clear plastic bag into which carbon-dioxide is slowly pumped, or it could be an animal cage which is placed inside a large clear plastic bag. As the gas begins to fill up the bag from the base upwards the animal slumps into unconsciousness. The animal is left in this state for as long as 30 minutes to ensure death.

This method can be used for most small laboratory animals (except dogs) and is recommended by most animal welfare organizations.

A euthanasia cabinet made of wood with a perspex viewing wall on the top is recommended by UFAW and construction details may be referred to in the appropriate manual mentioned in the Bibliography.

The comparatively quiet manner in which the animal is overcome by the gas recommends this method for those creatures likely to become excited, by any other method, and so damage themselves. In the case of birds the beating of wings can be a problem overcome by this carbon-dioxide lethal chamber method.

(ii) *Volatile vapor inhalation* also necessitates the use of a lethal chamber. This can be a smallish glass dessicator as described earlier for anesthetizing small rodents. In this case chloroform or ether vapor will be used. The disadvantage of chloroform is that it is a liver poison and a skin irritant. It is not inflammable as is ether which makes ether less than useful for regular euthanasia. Any room where ether is used should have no spark or flame sources and all electrical apparatus must be spark free. These gases should not be used in the vicinity of any animals that are not to be disposed of.

Bigger animals can be destroyed in large specially constructed lethal chambers made of metal with a glass viewing window and an aperture into which a pad soaked with the anesthetic is inserted. There is a vent through which air can pass and the animal takes in an air/anesthetic mixture. When the animal drops unconscious this vent is closed so that the animal now respires pure anesthetic.

The use of vapors in the manner described has several disadvantages and may well be superceded in most cases by the use of the carbon-dioxide gas chamber.

TABLE 9.
Euthanasia techniques

Type of euthanasia	Method of euthanasia	Comments	Suitable species
Physical methods	Stunning	Of limited value for laboratory animals if required undamaged. Mechanical and electrical stunners for larger animals	Rat, guinea-pigs, rabbits, some birds, larger fish, amphibia, reptiles. Pigs, sheep, cattle, horses
	Neck dislocation	A skilled technique only useful on smaller restrained animals	Mice, rats, guinea-pigs (occasionally), rabbits, some birds
	Decapitation	Much blood loss. Requires the correct knowledge and instrument	Mice, rats, and guinea-pigs (occasionally), some birds, fish, amphibia, reptiles
	Shooting	Captive and free-bolt methods produce brain damage that may be followed by throat bleeding to ensure death	Cats, dogs, goats, pigs, cattle, horses, sheep
Chemical methods	Barbiturate injections	Pentobarbitone sodium ("Euthetal") injected intravenous or intracardiac or intraperitoneal	Rabbits, cats, dogs, primates, hamsters
	Magnesium sulphate injections	Intravenous injections of the saturated solution	Rabbits, dogs, primates
	Gas inhalation	Carbon-dioxide inhalation in a lethal chamber	Mice, rats, hamsters, guinea-pigs, rabbits, cats, birds, reptiles
	Vapour inhalation	Ether, chloroform, halothane toxic volatile vapors in a lethal chamber. Not highly recommended if other methods available	Mice, rats, guinea-pigs, rabbits, cats, hamsters, amphibia, reptiles
	In solution	*MS 222 (tricaine methane sulphonate) (1:1000–1:3000 dilutions) *(F. J. Bové, Sandoz, Basle/Switzerland), p. 299	Aquatic species can be narcotized and then transferred to chloroform/water mixture

TABLE 10.

Euthanasia techniques for non-mammals

Type of animal	
Sponges (Porifera)	Add 70% alcohol to the water
Sea anemones	Float menthol crystals on the water of extended individuals—leave for 12 hours or more and then add 10% formalin to the water
	Alternatively, drip clove oil into the sea water containing an extended individual
Jellyfish (Coelenterata)	Add magnesium sulphate crystals to the water
	Alternatively, float menthol crystals on the water
	Both can be preserved in 2% formalin solution
Flatworms, tapeworms (Platyhelminthes)	Put the parasitic flukes or free-living Planarians into a small quantity of water in a watch glass
	Spray hot 2% formalin onto the animals
Roundworms (Nematoda)	Plunge the worms into hot 2% formalin solution, or hot 70% alcohol
Earthworms, leeches, aquatic worms (Annelida)	Immerse the animals in water and add one of the following: 30% alcohol, menthol crystals, or magnesium sulphate
Land snails, freshwater snails	Asphyxiate by immersing the snails in water that has been previously boiled to remove the air. Leave overnight. Alternatively, leave them in water containing magnesium sulphate
Sea water specimens	Add menthol crystals to the water or drip alcohol or 1% formalin into the water
Starfish, sea-urchins (Echinodermata)	When the animals are expanded add magnesium sulphate to the water. Leave them overnight until unresponsive
Crab, crayfish (Crustacea)	Immerse in fresh water or air-free water to asphyxiate
Insects	Land insects may be killed by placing them in a killing bottle with ethyl acetate
	Aquatic insects can be killed by dripping chloroform into the water with a few drops of detergent
Spiders	Drop into 70% alcohol
Fish	Add 2% formalin to the water or alternatively place the animal into 70% alcohol
	Marine fish will suffocate if removed from water
	Fish may be narcotized by the proprietary drug MS-222 Sandoz
Amphibia	Immerse in 0.05% MS-222 solution
Reptilia	Put the animal into a plastic bag and place it in the deep freeze
Birds	Use a carbon-dioxide killing-box

6.5.4. POST-EUTHANASIA—INDICATORS OF DEATH

It is important to know whether the animal has died as a result of the method of euthanasia employed. An animal that is thought to be dead must be examined in order to check on the vital systems such as the central nervous system, heart and respiratory system. The following checks may be carried out:

(*a*) Check that there is no heart-beat.
(*b*) Check that there is no respiratory action.
(*c*) Check for signs of muscular rigidity (rigor mortis).
(*d*) Check for pupillary reflex by touching the eyeball gently with a piece of cotton or bringing a bright light in front of the eyeball. The pupil will respond by closing up if nervous function is intact.

It is better to be on the safe side rather than dispose of a deeply anesthetized animal which may recover, so it is advisable to wait until all signs of death are present.

6.6. Practical Program

ELEMENTARY TECHNIQUES IN ANESTHESIA AND EUTHANASIA

The objectives of this skills based practical work are as follows:

(i) To instruct the students in the range of hypodermic syringes and needles, their correct use and maintenance.

(ii) To introduce the student to the techniques and skills involved in intravenous injections and blood collection using dead animal specimens.

(iii) To instruct the student in some physical and chemical methods of euthanasia using dead laboratory animals.

(i) *Using a hypodermic syringe.* A large variety of syringes are available and students are advised to examine as many different types as possible. There are the glass metal variety which can be sterilized before re-use and the plastic disposable types. The size of the barrel can be 1 cm^3 through 20 cm^3 and even larger. Onto the nozzle of the barrel the needle of chosen size is pushed with a twisting action. The smaller the needle, the better, as it can cause less damage when it is pushed through tissues for the injection.

When a fluid of a known volume is required in the barrel of the syringe it is drawn into the hypodermic from a preparation supplied in a rubber-topped vial. Withdrawing fluid from a vial requires a special technique to avoid air being drawn into the syringe barrel. Similarly, drawing up fluid from an open vessel must be achieved without air being trapped. To ensure that all air is expelled before the injection some of the fluid in the barrel is ejected from the needle into the air.

Any syringes that are to be used again must be correctly washed out with sterile, distilled water and then sterilized before being used again. If it is the practice to use needles more than once they must also be sterilized and their points previously checked and adjusted for sharpness. Syringes and needles that are disposable must be got rid of in an approved manner so that there is no risk of their being used again.

(ii) *Intravenous injections and simple blood collections.* In order to acquire the skills involved in administering substances by the venous route, students are advised to observe the technique being demonstrated on dead animals. For the purposes of demonstration the mouse, rat, and rabbit may be used.

In all cases it is appropriate to clean the injection site with an alcohol wipe. The mouse in life could be restrained in a special holder with the tail only being exposed. The vein area of the tail is warmed and the injection made using a small needle gauge 24. The adult rat is thought to be less suitable for this injection route but younger animals can be used for the practice procedures.

The tail blood vessels can be used to supply blood specimens either by a venesection or venepuncture. This technique might well be observed at this point in the practical program.

A common practice with mouse and rat is to warm the tail tip, cleanse it with alcohol, and then to cut off the tip of the tail when it is clean and dry. The sample is

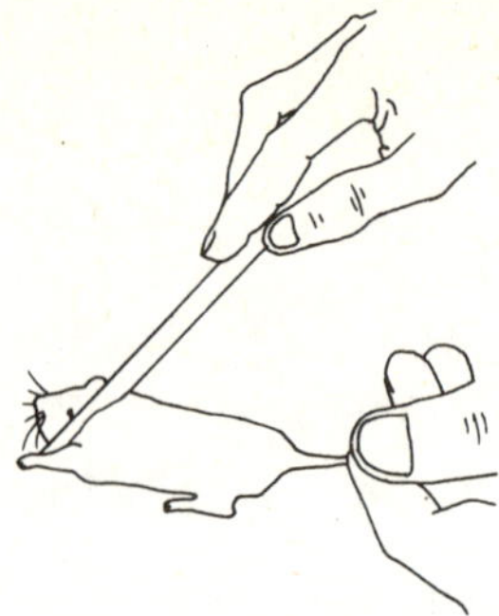

FIG. 83. Cervical dislocation of the mouse (drawn from a photograph, *UFAW Handbook*, Churchill-Livingstone, 4th edn., 1973)

small and may be picked upon a slide or in a pipette. Alternatively, a small gauge needle may be inserted into the tail tip and blood withdrawn in this manner.

To inject or bleed a rabbit, the ear needs to be warmed and a needle is inserted into a marginal blood vein. The area is previously cleansed. Alternatively, venesection bleeding may be used to produce small samples of blood from the rabbit ear.

(iii) *Physical and chemical methods of euthanasia.* The physical methods of euthanasia require more skill than the chemical methods to be demonstrated later. Humane killing methods which require that the technician use his own hands in the process take some time to perfect and time for the individual to overcome the natural disinclination to destroy animal life.

The cervical dislocation method can be practised using a dead mouse. With a pencil pressed hard behind the skull and the tail held firmly near the base, the animal is pulled quickly towards the operator.

The stunning method can be tried out using a dead mouse or rat. The animal is gripped firmly around the abdomen and the head struck heavily against a hard surface such as a sink edge or table top. Do not hold the tail as the sheath may come away.

Chemical methods of euthanasia can be demonstrated for laboratory mice and rats using a lethal chamber constructed for the purposes of using ether and/or chloroform. Such a chamber may be a modified glass chemical dessicator or a commercially available metal chamber. In the first instance, the animal is given the volatile liquid together with air. When the animal is unconscious the air supply is cut off. A carbon-dioxide lethal chamber set-up should be demonstrated for use with larger animals such as the rabbit. This involves having a carbon-dioxide cylinder leading gas into a secure clear plastic bag into which the animal is introduced for the euthanasia procedure.

Legal requirements, anesthesia, laboratory procedures, and euthanasia—summary

The major implications of the "Animal Welfare Acts" in both UK and USA were outlined. In each case the importance of the "pain clauses" was indicated even though somewhat difficult to monitor in all experimental situations.

The registration of premises and personnel was discussed and the fact that premises and licenses are not transferable was stressed.

The reasons for choosing a certain anesthetic were outlined and the manner of administration was described in terms of speed and compatibility with the species of animal.

The modes of dosing animals with experimental chemical agents was outlined in terms of the efficiency of getting the agent to the required tissue area with the intravenous mode being a popular choice, where appropriate.

The collection of specimens from laboratory animals was outlined with emphasis being laid on keeping the specimens uncontaminated and undamaged for investigation purposes.

The methods of humane killing were outlined as being physical and chemical. The physical methods appeared more "brutal", but may be considered more humane in terms of speed.

The indicators of animal death were considered as those features associated with brain death and cardiovascular cessation.

Some practical work connected with introductory training techniques in anesthesia and euthanasia was suggested.

7: *Data Capsules*

THIS appendix includes information in tabular form for quick reference. Every effort has been made to ensure that the numerical data are correct. The author would value any enlightenment that readers can provide if in their experience some of the data is at variance with their understanding. In other words, if anything appears to be wrong, please write to the author.

The encapsulated data provided covers the following:

7.1. Animal data

Mouse	Fowl	Locusts
Rat	Fowl eggs	Cockroaches
Rabbit	Canary	Fruit-flies
Guinea-pig	Budgerigar	Meal-worms
Hamster	Quail	Flour beetles
Gerbil	Frogs	Houseflies
Ferret	Xenopus	Stick insects
Cat	Guppy	Earthworm
Dog		
Monkey		
Man		

7.2. Outline classification of the animal kingdom

7.3. Numerical data

(i) Constants and conversions.
(ii) Calculations.

7.4. Chemical data

(i) Strengths of solutions.
(ii) Isotonic saline solutions.
(iii) Gases in cylinders.
(iv) Humidity regulating solutions.

7.5. Safety data

(i) General laboratory safety checklist.
(ii) Safe practice checklist.

254

(iii) First-aid—outline procedures.

(iv) Plants poisonous to animals.

Animal data

Mouse—Mus musculus

Adult weight (g)—male	20.0–40.0
Adult weight (g)—female	25.0–90.0
Average life-span (years)	
Male	2–3
Female	1–2
Minimum cage dimensions	
Height (cm)	12.5
Area (cm^2)	500.0
Number of adults per cage	Pair or trio and a litter
(size as above)	
Environmental temperature (°C) (°F)	24.0–25.0 (72.2–77.0)
Environmental relative humidity (%)	45–55
Approximate daily diet consumption per adult (g)	3.0–4.0
Approximate daily water consumption per adult (cm^3)	4.0–7.0
Approximate daily urine release (cm^3)	1.0–2.0
Approximate daily fecal mass (g)	1.0–1.5
Normal body temperature (°C) (°F)	(37.1) (98.8)
Heart rate	600
Average number of beats per minute	(480–740)
Ventilation rate	163
Average number of breaths per minute	(84–230)
Tidal volume (cm^3)	0.15 (0.09–0.23)
Oxygen consumption (cm^3/g/hour)	2.50 (1.50–3.50)
Basal metabolic rate for animals of	
mass (g)	30.0
surface area (m^2)	0.007
joules/m^2/day	3.094, 045.0
kcal/m^2/day	739.0
Arterial blood pressure (kilo–pascals)	Diastolic: 2.67–12.00 kPa
	Systolic: 12.67–18.67 kPa
Haemoglobin (g per 100 cm^3 of blood)	
at birth	10.5
at 3 weeks	11.0
at 2 months	16.0
Erythrocyte count (million per 100 cm^3 of blood)	
at birth	3.7
adult	4.7–12.5
Haematocrit (%)	41
Average weight at maturity (g)	28.0 (male) 20.0–25.0 (female)
Age of maturation of male (weeks)	6
Age of maturation of female (weeks)	6
Recommended minimum breeding age (weeks)	10 (6–10)
Recommended maximum breeding age (months)	12
Estrus cycle duration (days)	4–5 (influenced by presence of male)
Estrus (heat) duration (hours)	9–20
Time of ovulation (hours after estrus)	2.3
Type of ovulation	Spontaneous
Mating methods	Monogamous pairs or trios
	for genetics work
	Harems for larger numbers

Mouse—Mus musculus (*continued*)

Breeding life (months)	
Male	6–18
Female	6–12
Fertilization (hours after ovulation)	5
Numbers of eggs shed	6+
Viability of eggs (hours)	10–12
Gestation period (days)	19–21 (longer if post-partum mated)
Usual litter numbers	8–11 (inbreds smaller)
Litter frequency number (per year)	8–12
Weight at birth (g)	1.0–1.5
First estrus cycle after birth (hours)	20–24 post-partum then following after lactation
Optimum weaning age (days)	18–21
Optimum weaning weight (g)	10–12
Chromosome number (diploid)	40

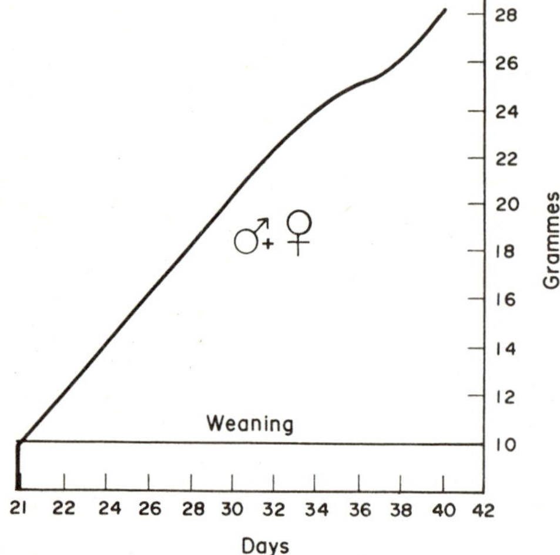

FIG. 84. Growth curve—mouse (with permission from Bantin & Kingman Ltd.)

Rat—Rattus norvegicus

Adult weight (g)—male	200–500
Adult weight (g)—female	250–350
Average life span (years)	
Male	2–3
Female	
Minimum cage dimensions	
Height (cm)	25.0
Area (cm²)	1000.0
Number of adults per cage (size as above)	Pair
Environmental temperature (°C) (°F)	21–24 (70–76)
Environmental relative humidity (%)	45–55 (55–65 avoids "ringtail")
Approximate daily diet consumption per adult (g)	15.0–20.0
Approximate daily water consumption per adult (cm³)	24.0–35.0
Approximate daily urine release (cm³)	11.0–15.0
Approximate daily fecal mass (g)	9.0–15.0
Normal body temperature (°C) (°F)	(37.3) (99.2)

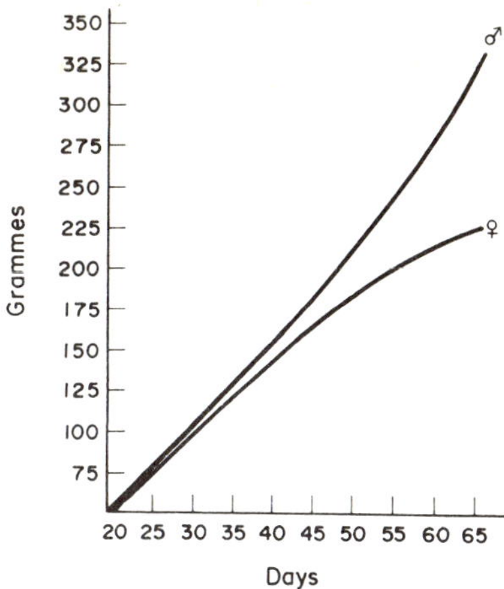

FIG. 85. Growth curves—rat (with permission from Bantin & Kingman Ltd.)

Rat—Rattus norvegicus (*continued*)

Heart rate	375
Average number of beats per minute	(260–600)
Ventilation rate	100
Average number of breaths per minute	(66–210)
Tidal volume (cm^3)	0.86 (0.60–1.25)
Oxygen consumption (cm^3/g/hour)	2.00
Basal metabolic rate for animals of	
mass (g)	300.0
surface area (m^2)	0.04
joules/m^2/day	3,357, 813.0
kcal/m^2/day	802.0
Arterial blood pressure (kilo–pascals)	Diastolic: 12.13 kPa
	Systolic: 12.67 kPa
Haemoglobin (g per $100\,cm^3$ of blood)	
at birth	—
at 3 weeks	—
at 2 months	14.8
Erythrocyte count (million per $100\,cm^3$ of blood)	
at birth	—
adult	8.9
Haematocrit (%)	46
Average weight at maturity (g)	170–210 (male) 150–170 (female)
Age of maturation of male (weeks)	4.0–5.0 (4.0–8.0)
Age of maturation of female (weeks)	4.0–5.0 (4.0–8.0)
Recommended minimum breeding age (weeks)	12 (9–14)
Recommended maximum breeding age (months)	(12–15) (9–11)
Estrus cycle duration (days)	4–5
Estrus (heat) duration (hours)	9–20 (12 average)
Time of ovulation (hours after estrus)	8.0–11.0
Type of ovulation	Spontaneous
Mating methods	Monogamous or trios
	Polygamous
Breeding life (months)	
Male	—
Female	9–11
Fertilization (hours after ovulation)	7–10
Numbers of eggs shed	10+
Viability of eggs (hours)	10–12
Gestation period (days)	21–22

Rat—Rattus norvegicus (*continued*)

Usual litter numbers	9–11 (6–15)
Litter frequency number (per year)	7–9
Weight at birth (g)	4–5
First estrus cycle after birth (hours)	20–24 post-partum then following on after lactation
Optimum weaning age (days)	21 (18–23)
Optimum weaning weight (g)	35–40
Chromosome number (diploid)	42

Rabbit—Oryctolagus cuniculus

Adult weight (g)—male (buck)	1500.0 (variable) 5000.0
Adult weight (g)—female (doe)	1500.0 (variable) 6000.0
Average life span (years)	
Male	4.0–6.0 (variable)
Female	—
Minimum cage dimensions	
Height (cm)	45.0
Area (cm²)	5500.0
Number of adults per cage (size as above)	One
Environmental temperature (°C) (°F)	10–20 (50–67)
Environmental relative humidity (%)	40–50
Approximate daily diet consumption per adult (g)	30.0–300.0
Approximate daily water consumption per adult (cm³)	60.0–300.0
Approximate daily urine release (cm³)	20.0–200.0
Approximate daily fecal mass (g)	30.0–45.0
Normal body temperature (°C) (°F)	(38.5) (101.5)
Heart rate	180
Average number of beats per minute	(130–300)
Ventilation rate	53
Average number of breaths per minute	(38–65)
Tidal volume (cm³)	21.00 (19.30–24.60)
Oxygen consumption (cm³/g/hour)	0.60–0.85
Basal metabolic rate for animals of	
mass (g)	3,500.0
surface area (m²)	0.20
joules/m²/day	3,387, 121.0
kcal/m²/day	809.0
Arterial blood pressure (kilo–pascals)	—
Haemoglobin (g per 100 cm³ of blood)	
at birth	—
at 3 weeks	—
at 2 months	12.8
Erythrocyte count (million per 100 cm³ of blood)	
at birth	—
adult	6.3
Haematocrit (%)	39.8
Average weight at maturity (g)	2,500–3,000
Age of maturation-of male (weeks)	16.0–24.0
Age of maturation of female (weeks)	16.0–24.0
Recommended minimum breeding age (weeks)	24 (24–35)
Recommended maximum breeding age (months)	24
Estrus cycle duration (days)	Possibly 15.0–16.0
Estrus (heat) duration (hours)	Prolonged in absence of buck
Time of ovulation (hours after estrus)	10.0–12.0
Type of ovulation	Induced
Mating methods	Introduce doe to buck in his quarters

Rabbit—Oryctolagus cuniculus (*continued*)

Breeding life (years)	
Male	—
Female	4–5 years (3 years)
Fertilization (hours after ovulation)	—
Numbers of eggs shed	10.0
Viability of eggs (hours)	8.0
Gestation period (days)	31–32
Usual litter numbers	6–8 (5–10)
Litter frequency number (per year)	7–9
Weight at birth (g)	30–70
First estrus cycle after birth (hours)	35
Optimum weaning age (days)	42–55
Optimum weaning weight (g)	1000–1500
Chromosome number (diploid)	44

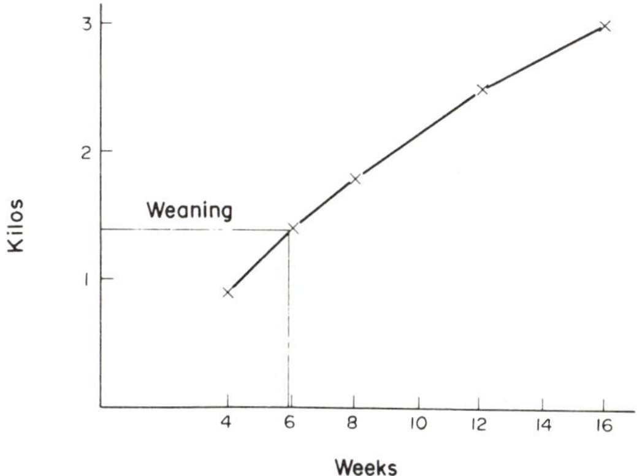

FIG. 86. Growth curve—rabbit (with permission from Bantin & Kingman Ltd.)

Guinea-pig—Cavia porcellus

Adult weight (g)—male (boar)	1000 (900–1500)
Adult weight (g)—female (sow)	850 (700–1300)
Average life span (years)	
Male	Up to 8 years
Female	(2.0–3.0)
Minimum cage dimensions	
Height (cm)	30
Area (cm^2)	2500
Number of adults per cage (size as above)	Pair
Environmental temperature (°C) (°F)	(15–18) (59–64.4)
Environmental relative humidity (%)	45–55
Approximate daily diet consumption per adult (g)	20–30
Approximate daily water consumption per adult (cm^3)	80–150
Approximate daily urine release (cm^3)	20–25
Approximate daily fecal mass (g)	10–15
Normal body temperature (°C) (°F)	(38.3) (101)
Heart rate	250
Average number of beats per minute	(150–300)
Ventilation rate	85
Average number of breaths per minute	(70–150)
Tidal volume (cm^3)	1.80 (1.00–3.90)
Oxygen consumption (cm^3/g/hour)	0.80

Guinea-pig—Cavia porcellus (*continued*)

Basal metabolic rate for animals of	
mass (g)	—
surface area (m^2)	—
joules/m^2/day	—
kcal/m^2/day	—
Arterial blood pressure (kilo–pascals)	Diastolic: 5.87 kPa
	Systolic: 13.20 kPa
Haemoglobin (g per 100 cm^3 of blood)	
at birth	—
at 3 weeks	—
at 2 months	12–15
Erythrocyte count (million per 100 cm^3 of blood)	
at birth	
adult	4.5–7.0
Haematocrit (%)	45
Average weight at maturity (g)	500 (male) 450 (female)
Age of maturation of male (weeks)	10.0
Age of maturation of female (weeks)	4.0–5.0
Recommended minimum breeding age (weeks)	12–16
Recommended maximum breeding age (months)	18–24
Estrus cycle duration (days)	14–16
Estrus (heat) duration (hours)	15 (9–20)
Time of ovulation (hours after estrus)	10.0
Type of ovulation	Spontaneous
Mating methods	Harems
Breeding life (months)	
Male	
Female	18–24
Fertilization (hours after ovulation)	7–10
Numbers of eggs shed	2–4
Viability of eggs (hours)	20.0
Gestation period (days)	63 (59–72)
Usual litter numbers	3–5
Litter frequency number (per year)	3
Weight at birth (g)	90 (male) 85 (female)
First estrus cycle after birth (hours)	20–24 post partum
Optimum weaning age (days)	14 (4–28)
Optimum weaning weight (g)	180 (170–180)
Chromosome number (diploid)	64

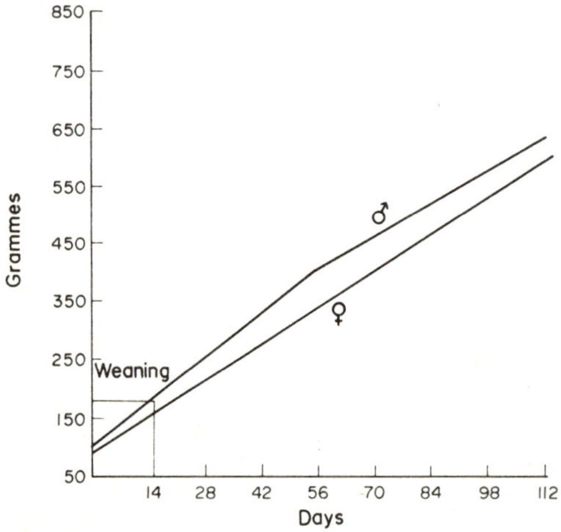

FIG. 87. Growth curves—guinea-pig (with permission from Bantin & Kingman Ltd.)

Hamster (Syrian)—Mesocricetus auratus

Adult weight (g)—male	90–120 (100–150)
Adult weight (g)—female	95–140 (90–140)
Average life span (years)	
Male	1–1.5 (1.0–3.0)
Female	
Minimum cage dimensions	
Height (cm)	20
Area (cm²)	1000.0 (960)
Number of adults per cage (size as above)	Pair or single female and litter
Environmental temperature (°C) (°F)	(21–24) (69–74)
Environmental relative humidity (%)	40–60
Approximate daily diet consumption per adult (g)	10–15
Approximate daily water consumption per adult (cm³)	8.0–12.0+
Approximate daily urine release (cm³)	7.0
Approximate daily fecal mass (g)	2.0–2.5
Normal body temperature (°C) (°F)	(37.8) (99.5)
Heart rate	400
Average number of beats per minute	(350–500)
Ventilation rate	82
Average number of breaths per minute	(60–110)
Tidal volume (cm³)	0.80
Oxygen consumption (cm³/g/hour)	2.90
Basal metabolic rate for animals of	
mass (g)	—
surface area (m²)	—
joules/m²/day	—
kcal/m²/day	—
Arterial blood pressure (kilo–pascals)	Diastolic
	Systolic: 9.33–14.67 kPa
Haemoglobin (g per 100 cm³ of blood)	
at birth	
at 3 weeks	
at 2 months	12.0–16.0
Erythrocyte count (million per 100 cm³ of blood)	
at birth	—
adult	4.0–10.0

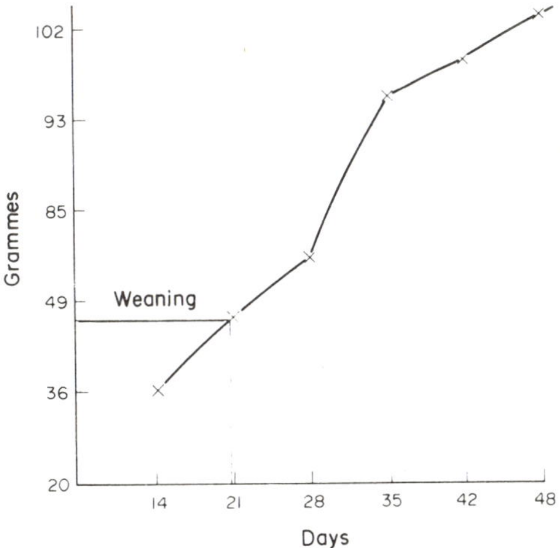

FIG. 88. Growth curve—hamster (with permission from Bantin & Kingman Ltd.)

Hamster (*Syrian*)—Mesocricetus auratus (*continued*)

Haematocrit (%)	49–54.8
Average weight at maturity (g)	80 (male) 75 (female)
Age of maturation of male (weeks)	8.0
Age of maturation of female (weeks)	6.0
Recommended minimum breeding age (weeks)	10 (6–12)
Recommended maximum breeding age (months)	12
Estrus cycle duration (days)	4.0–5.0
Estrus (heat) duration (hours)	4.0–23.0
Time of ovulation (hours after estrus)	Early in estrus
Type of ovulation	Spontaneous
Mating methods	Female to male (colonies)
Breeding life (months)	
Male	
Female	12
Fertilization (hours after ovulation)	12
Numbers of eggs shed	1–2
Viability of eggs (hours)	10
Gestation period (days)	16 (15–18)
Usual litter numbers	5–7 (7–9)
Litter frequency number (per year)	3–4
Weight at birth (g)	1.5–2.5
First estrus cycle after birth (hours)	1.0–8.0
Optimum weaning age (days)	20–21
Optimum weaning weight (g)	30 (male) 25 (female) (40)
Chromosome number (diploid)	44

Mongolian gerbil—Meriones unguiculatus

Adult weight (g)—male	46.0–131.0
Adult weight (g)—female	53.0–133.0
Average life span (years)	
Male	2–3 (2.5–6.0)
Female	
Minimum cage dimensions	
Height (cm)	20.0
Area (cm²)	1000.0
Number of adults per cage (size as above)	Pair
Environmental temperature (°C) (°F)	(18.0–22.0) (64.4–71.6)
Environmental relative humidity (%)	45–55
Approximate daily diet consumption per adult (g)	5.0–7.0
Approximate daily water consumption per adult (cm³)	Small quantities—mainly from fresh vegetables
Approximate daily urine release (cm³)	Drops only
Approximate daily fecal mass (g)	1.5–2.5
Normal body temperature (°C) (°F)	(37.4) (99.3)
Heart rate	360
Average number of beats per minute	(260–600)
Ventilation rate	90
Average number of breaths per minute	(70–120)
Tidal volume (cm³)	—
Oxygen consumption (cm³/g/hour)	—
Basal metabolic rate for animals of	
mass (g)	—
surface area (m²)	—
joules/m²/day	—
kcal/m²/day	—
Arterial blood pressure (kilo–pascals)	—

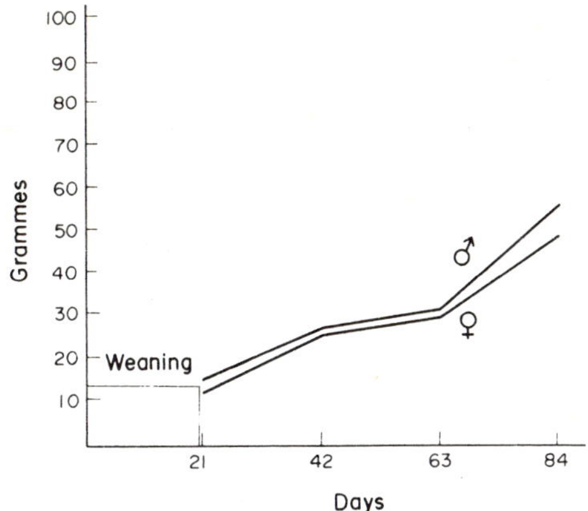

FIG. 89. Growth curves—mongolian gerbil (with permission from Bantin & Kingman Ltd.)

Mongolian gerbil—Meriones unguiculatus (*continued*)

Haemoglobin (g per 100 cm³ of blood)	
at birth	—
at 3 weeks	—
at 2 months	13.9–14.4
Erythrocyte count (million per 100 cm³ of blood)	
at birth	7.7–8.7
adult	
Haematocrit (%)	45–47
Average weight at maturity (g)	61–70
Age of maturation of male (weeks)	9–18
Age of maturation of female (weeks)	9–12
Recommended minimum breeding age (weeks)	9–12
Recommended maximum breeding age (months)	7–20 (3–10 litters)
Estrus cycle duration (days)	4–6
Estrus (heat) duration (hours)	12–18
Time of ovulation (hours after estrus)	—
Type of ovulation	Spontaneous
Mating methods	Monogamous permanent pairing
Breeding life (months)	
Male	24 +
Female	7–20
Fertilization (hours after ovulation)	—
Numbers of eggs shed	4–9
Viability of eggs (hours)	—
Gestation period (days)	24–26
Usual litter numbers	4–6
Litter frequency number (per year)	8–12
Weight at birth (g)	2.5–3.5 (male)
	2.5–3.5 (female)
First estrus cycle after birth (hours)	Immediately post-partum
Optimum weaning age (days)	21–28
Optimum weaning weight (g)	12–18
Chromosome number (diploid)	44

Ferret—Mustela furo

Adult weight (g)—male (hob)	400–3500 (males heavier)
Adult weight (g)—female (gill)	400–3500

Ferret—Mustela furo (*continued*)

Average life span (years)	
Male	
Female	5–6
Minimum cage dimensions	
Height (cm)	30.0
Area (cm^2)	5500.0
Number of adults per cage (size as above)	Three females or two males
Environmental temperature (°C) (°F)	(7–25) (44.6–77)
Environmental relative humidity (%)	45–55
Approximate daily diet consumption per adult (g)	
Approximate daily water consumption per adult (cm^3)	
Approximate daily urine release (cm^3)	
Approximate daily fecal mass (g)	
Normal body temperature (°C) (°F)	(38.8) (101.9)
Heart rate	—
Average number of beats per minute	—
Ventilation rate	
Average number of breaths per minute	33–36
Tidal volume (cm^3)	—
Oxygen consumption (cm^3/g/hour)	—
Basal metabolic rate for animals of	
mass (g)	—
surface area (m^2)	—
joules/m^2/day	—
kcal/m^2/day	—
Arterial blood pressure (kilo–pascals)	—
Haemoglobin (g per 100 cm^3 of blood)	
at birth	—
at 3 weeks	—
at 2 months	—
Erythrocyte count (million per 100 cm^3 of blood)	
at birth	—
adult	8.5
Haematocrit (%)	—
Average weight at maturity (g)	700–800
Age of maturation of male (weeks)	—
Age of maturation of female (weeks)	—
Recommended minimum breeding age (weeks)	32–52
Recommended maximum breeding age (months)	36
Estrus cycle duration (days)	April through July
Estrus (heat) duration (hours)	Prolonged in absence of male
Time of ovulation (hours after estrus)	30
Type of ovulation	Induced by mating
Mating methods	Monogamous
Breeding life (seasons)	
Male	
Female	3 seasons
Fertilization (hours after ovulation)	12
Numbers of eggs shed	—
Viability of eggs (hours)	—
Gestation period (days)	41–43
Usual litter numbers	3–15 (6–10)
Litter frequency number (per year)	2
Weight at birth (g)	6–12
First estrus cycle after birth (days)	End of lactation (16 days later) or the next season can occur during lactation
Optimum weaning age (days)	49–56
Optimum weaning weight (g)	300–450
Chromosome number (diploid)	—

Cat—Felis species

Adult weight (kg) male (toms)	3.5–5.9
Adult weight (kg) female (queens)	2.25–3.0
Average life span (years)	
Male	variable 8 +
Female	
Minimum cage dimensions	
Height (cm)	45–50
Area (cm^2)	5500
Number of adults per cage (size as above)	Queen and litter
Environmental temperature (°C) (°F)	(15–22) (59–71.6)
Environmental relative humidity (%)	45–60
Approximate daily diet consumption per adult (g)	Variable
Approximate daily water consumption per adult (cm^3)	30 (on wet diet)
Approximate daily urine release (cm^3)	68.6
Approximate daily fecal mass (g)	
Normal body temperature (°C) (°F)	(38.7) (101.6)
Heart rate	
Average number of beats per minute	110–130
Ventilation rate	
Average number of breaths per minute	20–30
Tidal volume (cm^3)	—
Oxygen consumption (cm^3/g/hour)	—
Basal metabolic rate for animals of	
mass (g)	—
surface area (m^2)	—
joules/m^2/day	—
kcal/m^2/day	—
Arterial blood pressure (kilo—pascals)	—
Haemoglobin (g per 100 cm^3 of blood) at birth	
at 3 weeks	14.0 (male)
at 2 months	12.7 (female)
Erythrocyte count (million per 100 cm^3 of blood)	
at birth	
adult	7.14

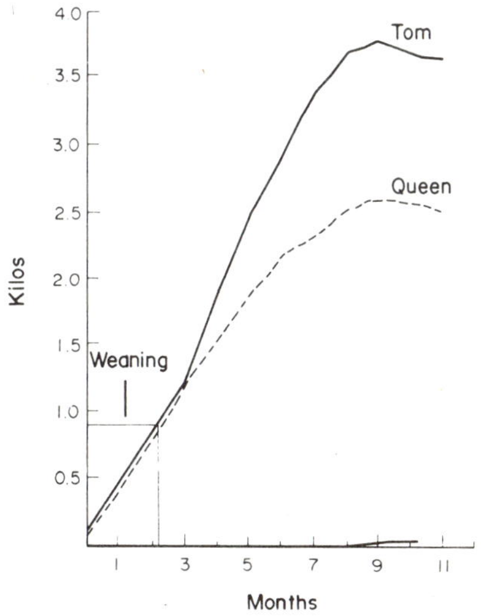

FIG. 90. Growth curves—cat (from *UFAW Handbook*, Churchill-Livingstone, 4th edn., 1973)

Cat—Felis species (*continued*)

Haematocrit (%)	37
Average weight at maturity (kg)	3.5 (male) 2.5 (female)
Age of maturation of male (months)	12
Age of maturation of female (months)	5–8
Recommended minimum breeding age (months)	7–9 (females)
Recommended maximum breeding age (years)	6 years (males) 8 years (females)
Estrus cycle duration (days)	14
Estrus (heat) duration (days)	3–6 (3–10)
Time of ovulation (hours after estrus)	25–27
Type of ovulation	Induced by mating, Sometimes spontaneous
Mating methods	Monogamous—queen taken to tom's territory (uneconomic) Polygamous—one tom to 5–15 queens
Breeding life (years)	
Male	8 months—8
Female	12 months—6
Fertilization (hours after ovulation)	24–36
Numbers of eggs shed	—
Viability of eggs (hours)	—
Gestation period (days)	63 (64–66)
Usual litter numbers	3–6
Litter frequency number (per year)	2
Weight at birth (g)	90–140
First estrus cycle after birth	4th week of lactation
Optimum weaning age (weeks)	4–5
Optimum weaning weight (g)	800–1000 (variable)
Chromosome number (diploid)	38

Dog—Canis familiaris

Adult weight (Kg) male (dog)	Variable, e.g. Beagle, 10–12 Kg
Adult weight (Kg) female (bitch)	Variable
Average life span (years)	
Male	
Female	Variable 10–15
Minimum kennel dimensions inside area (m²/Kg body wt)	Variable depending upon weight 1 × 0.75 m (14 Kg⁺ animals) 1.3 × 1.0 m (14–23 Kg animals)
Number of adults per cage (size as above)	One
Environmental temperature (°C) (°F)	(10–22) (50–71.6)
Environmental relative humidity (%)	45–55
Approximate daily diet consumption per adult (g)	Late afternoon feeding 400 +
Approximate daily water consumption per adult (cm³)	1200 +
Approximate daily urine release (cm³)	24–41 per Kg body wt. per day
Approximate daily fecal mass (g)	—
Normal body temperature (°C) (°F)	(38.0) (100.4)
Heart rate	62–80 (larger breeds)
Average number of beats per minute	90–130 (smaller breeds)
Ventilation rate	
Average number of breaths per minute	16–30
Tidal volume (cm³)	—
Oxygen consumption (cm³/g/hour)	—
Basal metabolic rate for animals of	
mass (g)	—
surface area (m²)	—
joules/m²/day	—
kcal/m²/day	—

Dog—Canis familiaris (*continued*)

Arterial blood pressure (kilo–pascals)	—
Haemoglobin (g per 100 cm^3 of blood)	
at birth	—
at 3 weeks	—
at 2 months	12.0–18.0
Erythrocyte count (million per 100 cm^3 of blood)	
at birth	
adult	5.5–8.5
Haematocrit (%)	37.0–55.0
Average weight at maturity (g)	Variable
Age of maturation of male (weeks)	Variable
Age of maturation of female (months)	Variable 8–14
Recommended minimum breeding age (months)	10–18
Recommended maximum breeding age (years)	Variable
Estrus cycle duration (days)	Variable
Estrus (heat) duration (days)	7–13
Time of ovulation (hours after estrus)	At beginning of estrus
Type of ovulation	Spontaneous
Mating methods	Mate in packs. Mate 10–16 days after beginning of estrus
Breeding life (years)	
Male	
Female	Variable—6
Fertilization (hours after ovulation)	—
Numbers of eggs shed	—
Viability of eggs (hours)	—
Gestation period (days)	60–66 (63 average)
Usual litter numbers	2–3 (small species)
	4–12 (larger species)
Litter frequency number (per year)	2
Weight at birth (g)	Variable
First estrus cycle after birth	Next breeding season
Optimum weaning age (weeks)	7–8
Optimum weaning weight (Kg)	Variable (1.0–1.5)
Chromosome number (diploid)	78

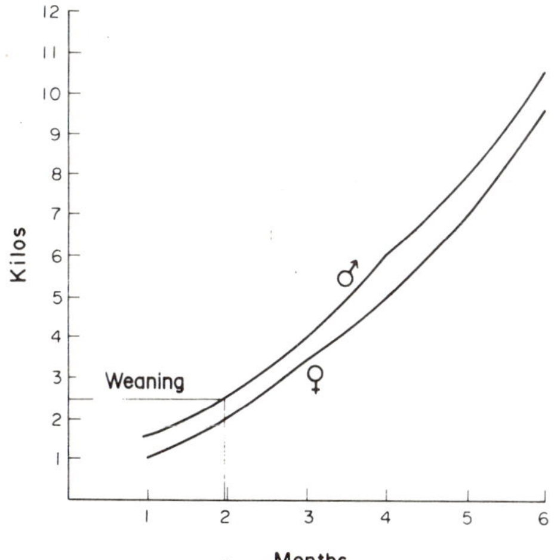

FIG. 91. Growth curves—dog (with permission from Bantin & Kingman Ltd.)

Rhesus monkey (Macaca mulatta)

Adult weight (Kg)—male	8–12
Adult weight (Kg)—female	6–10
Average life span (years)	
Male	
Female	Up to 30
Minimum cage dimensions	
Height (cm)	78 cm × 78 cm by 91.0 cm high
Living Space (m³)	1.3 m³
Number of adults per cage (size as above)	One
Environmental temperature (°C) (°F)	22–24 (71.6–75.2)
Environmental relative humidity (%)	40–60
Approximate daily diet consumption per adult (g)	Approx 4–5% body weight
Approximate daily water consumption per adult (cm³)	
Approximate daily urine release (cm³)	
Approximate daily fecal mass (g)	
Normal body temperature (°C) (°F)	(38.5) (101.5)
Heart rate	193
Average number of beats per minute	(165–240)
Ventilation rate	40
Average number of breaths per minute	(31–52)
Tidal volume (cm³)	—
Oxygen consumption (cm³/g/hour)	—
Basal metabolic rate for animals of	
mass (g)	—
surface area (m²)	—
joules/m²/day	—
kcal/m²/day	—
Arterial blood pressure (kilo–pascals)	—
Haemoglobin (g per 100 cm³ of blood)	
at birth	—
at 3 weeks	—
at 2 months	12.6
Erythrocyte count (million per 100 cm³ of blood)	
at birth	
adult	6.5
Haematocrit (%)	—
Average weight at maturity (Kg)	4.5–9.0
Age of maturation of male (years)	—
Age of maturation of female (years)	—
Recommended minimum breeding age (years)	4½ female
	5½ male
Recommended maximum breeding age (years)	—
Estrus cycle duration (days) (menstrual cycle)	28
Estrus (heat) duration (hours)	—
Time of ovulation (hours after estrus)	Generally 12–13 day of the cycle
Type of ovulation	Spontaneous
Mating methods	Monogamous
Breeding life (years)	5½–6
Male	4½–
Female	Up to 15
Fertilization (hours after ovulation)	—
Numbers of eggs shed	—
Viability of eggs (hours)	—
Gestation period (days)	168 (150–174)
Usual litter numbers	One
Litter frequency number (per year)	Once
Weight at birth (g)	490 ± 60 (male)
	465 ± 70 (female)
First estrus cycle after birth (months)	3

Rhesus monkey (Macaca mulatta) (*continued*)

Optimum weaning age (months)	3–6
Optimum weaning weight (g)	700–1000
Chromosome number (diploid)	42

Man—Homo sapiens

	Physiological data
Average birth weight	3.48 kg (7 lbs 11 oz)
Average birth number	One
Body temperature	37.8°C (98.4°F)
Blood volume	8.5–9.0% body weight
Bleeding time	3–8 minutes
Clotting time	5–15 minutes
Erythrocyte count	4.6–6.2 million/mm^3 (male)
	4.2–5.4 million/mm^3 (female)
Leukocyte count	5.000–10.000/mm^3
Platelets	150.000–350.000/mm^3
Haematocrit	40–54% (male)
	37–47% (female)
Haemoglobin	14–18 g/100 cm^3 (male)
	12–16 g/100 cm^3 (female)
Sedimentation rate (Wintrobe)	0.50 mm in 1 hour (male)
	0.15 mm in 1 hour (female)
Blood pH	7.35–7.45
Normal heart rate	72–78 beats per minute

Blood pressure—arterial (normal range) mm of Hg (kPa)

systolic	*diastolic*	
105–135 (14–18)	60–86 (8–11.47)	(males 16–19)
105–140 (14–18.67)	62–88 (8.27–11.73)	(males 20–24)
110–150 (14.67–20.00)	70–94 (9.33–12.53)	(males 40–44)
115–160 (15.33–21.33)	70–98 (9.33–13.07)	(males 50–54)

(female readings are slightly less)

Blood groups—white races (regional variations)	AB—5%
	A—40%
	B—10%
	O—45%
	Rhesus positive—86%
	Rhesus negative—14%
Normal respiratory rate (at rest)	15–20 breaths per minute
Urine volume	600–2.000 cm^3 in 24 hours (average 1200 cm^3)
Urine pH	4.7–8.0 (usually acid pH 6.0)
Urine specific gravity	1003–1030
Blood chemistry:	
Calcium	9–10 mg per 100 cm^3
Chlorides	580–620 mg per 100 cm^3
Cholesterol	150–240 mg per 100 cm^3
Glucose	80–120 mg per 100 cm^3
Potassium	14–20 mg per 100 cm^3
Protein (total)	6–8 g per 100 cm^3
Sodium	310–340 mg per 100 cm^3
Urea	20–40 mg per 100 cm^3
Uric acid	3–6 mg per 100 cm^3
Chromosome numbers	46 (xx female sex chromosomes)
	(xy male sex chromosomes)
Chromosome abnormalities:	
Turner's syndrome	45 (xo female)
Klinefelter's syndrome	47 (xxy male)
Down's syndrome (mongolism)	47 (additional chromosome on pair 21)

Homo sapiens: *Infectious Diseases* (*Zoonoses and others*)

Diseases	Incubation period (days)	Quarantine period (days)	Infectivity
Anthrax	3–7 (maybe a few hours)	None	Spores up to 15 years in soil
Botulism	12–36 hours	None	Not applicable
Brucellosis	14–30 hours	None	Rarely from man to man
Hepatitis (infective)	10–40	None	Up to 7 days from onset
Influenza	1–3	None	5 days after onset
Leptospirosis	6–12	None	Rarely from man to man
Measles	8–15	14	Four days before; 5 days after the rash
Mumps	12–26	28	Two days before and 9 days after the swelling
Paratyphoid	1–10	Until repeated swabs are negative	As long as fecal swabs are positive
Psittacosis	7–14	None	During acute illness
Rabies	14–63	None	Rarely from man to man
Rubella	14–21	10–21	Four days after catarrhal symptoms
Salmonella (food poisoning)	12–48 hours	Until repeated negative swabs	Three days to 3 weeks—very variable
Scarlet fever	3–5	10	Ten days to 3 weeks in untreated
Smallpox	10–14	14	Two to 3 weeks. Until all scabs disappear
Syphilis	14–112	None	Two to 4 years
Tetanus	4–21	None	Not from man to man
Typhoid	7–21	Until repeated swabs are negative	Variable, may be 3 months or longer
Typhus	6–14	14	Infective for lice, 3 days after temperature normality

Fowl—Gallus domesticus

Introduction to the species	Evolved from red jungle fowl (*Gallus gallus*) of South-east Asia. Extensive hybridization
Life cycle	
Sexual maturity	6–8 months
Eggs laid	280–300 per year
Incubation period	21 days
Breeding	Run 1 cock to 9–10 hens (larger breeds)
	Run 1 cock to 14–15 hens (smaller breeds)
Sexing	Red combs on cockerels. Characteristic call
Rearing temperatures	
Laying fowls	13–18°C (55.4–64.4°F)
	(Not below 15°C (59°F) for good egg production)
Day-old chicks	32°C (89.6°F) reduced over 5 weeks to 18°C (64.4°F)
Relative humidity	A maximum of 75%
Feeding requirements	A wide range of feeding and water equipment available. Feed proprietary mashes, crumbs, and pellets. Water ad lib
Containers	Movable runs or deep litter housing with automated services. A floor litter of white wood shavings
Lighting:	
Constant day	10-hours darkness
	14-hours light
Increasing day	6–8-hours light from day-old to 18 weeks then an increase of 20 minutes per week until maximum of 16 hours.
	Essential to good egg production and maximum growth
Chromosome number (2n)	78

Gallus *egg incubation*

Weights of chicken eggs	50–60 g
Egg season	February–June
Storage period before incubation	7–14 days (variable)
Storage temperatures	13–15°C (55.4–59°F)
Storage humidities	75–85%
Incubation temperatures	37–38°C (98.6–100.4°F)
Incubation humidity	60%
Incubation time	21 days
Turning the eggs	Twice daily (or more if automated) until day 18
Critical factors for survival	Oxygen concentration. Carbon-dioxide concentration. Temperature and humidity movement
Candling the eggs	$2\frac{1}{2}$–3 days indistinct dark areas. 7 days blood network shows. Remove any "clears".

Canary—Serinus canarius

Introduction to the species	Wild bird in green from Canary Islands. New colors produced by hybridization. Bred as a pet by Spanish since 1478
Life-cycle	
Sexual maturity	12 months
Eggs laid	1–6 per clutch at 24-hour intervals
Incubation period	14–18 days
Breeding	Separate the sexes in winter—keep cocks singly. Pair in February or March in UK when they are 1 year-old
Sexing	Difficult. Male heavier and more active with good song
Handling	As for budgerigar
Rearing temperatures	10–16°C (50–60.8°F)
Body temperature (cloacal)	Body temperature (cloacal) 41–42°C (105.8–107.6°F)
Respiratory rate	96–144 per minute
Weight ranges	12–29 g
Erythrocyte count (millions per cm^3 of blood)	4.5–5.0
Haemoglobin (g per 100 cm^3 of blood)	10.4
Heart rate (beats per minute)	560–1000
Average life span (in years)	5–6 (reports of up to 14 years)
Feeding requirements	Canary seed, cracked hemp for adults. Soft egg yolk and biscuit mix until 6 weeks of age
Containers	As for budgerigar
Chromosome number (2n)	80

Budgerigar—Melopsittacus undulatus

Introduction to the species	From the aboriginal (Betcheroygah) meaning "good food". Australian wild small green parakeet
Life-cycle	
Sexual maturity	10–12 months
Eggs laid	4–8 per clutch
Incubation period	18 days
Breeding	Separate the sexes in the winter. Pair in February or March in UK

Budgerigar—Melopsittacus undulatus (*continued*)

Sexing	
Male	Upper part of beak blue
Female	Upper part of beak brown (at 3 months plus)
Handling	Hold neck between 1st and 2nd fingers with the back in the palm. The thumb and 1st finger enclosing the body. Have the wings folded against the back of the body
Rearing temperatures	15°C (59°F) for the young (minimum). 10°C (50°F) for the adults (minimum). They prefer warmth
Feeding requirements (use externally fitted hoppers)	Millet seed, grass seeds. Supplement with vegetables and some protein foods (i.e. egg yolk)
Containers	Breeding cage or aviary
Length	65–100 cm (25–40 in.)
Height	45 cm (18 in.)
Depth	30 cm (12 in.)
Perches	15 mm diameter oval and rough
Litter	Sand tray
Nesting	Box 15 × 15 cm (6 × 6 in.) entrance 4 cm diameter (1.5 in.)
Chromosome number (2n)	58

Japanese quail—Coturnix coturnix japonica

Rearing temperatures	
Adults	18–19°C (64.4–66.2°F)
Chicks	27–29°C (80.6–84.2°F)
Illumination rhythms for breeding	18 hours daylight 6 hours night
Sex characteristics	
Males	100–130 g—browner breast feathers
Females	120–160 g—grey, black breast feathers
Sexual maturity	6–8 weeks begin egg production (first 2 weeks not always fertile)
Eggs	
Incubation temperature	37.5°C (98.6°F)
Incubation time	$16\frac{1}{2}$ days
Weight	9–10 g
Numbers	300 recorded in first year
Chick weights	6–7 g—day old 18–36 g—7 days old
Containers: colony cage (4 males, 10 females)	76 × 50 × 30 cm (30 × 19.5 × 12 in.)
Food requirements	High protein diet—turkey starter crumbs, supplement with bird seed. Water ad lib.

Laboratory Frogs

Rearing temperatures (active)	
Rana temporaria	15–20°C (59–68°F) February–October
Rana esculenta	22–27°C (71.6–80.6°F) April–October
Rana ridibunda	22–27°C (71.6–80.6°F) April–October
Rana p. pipiens	20–25°C (68–77°F) April–June
Rana p. berlanderi	22–27°C (71.6–80.6°F) Year around
Rana catesbeiana	22–27°C (71.6–80.6°F) February–October

Laboratory Frogs (continued)

Hibernating conditions (inactive)	3–4°C (37–39°F)
	Keep in refrigerated box 45 × 30 × 15 cm with water depth 10 cm plus.
R. temporaria	Avoid overcrowding and rapid temperature changes. September to May
R. esculenta *R. ridibunda*	As above. Short-term forced hibernation possible. Natural period October to May
R. p. pipiens	As above. Short-term forced hibernation possible. Natural period October to May
R. p. berlanderi	Do not attempt to hibernate this species
R. catesbeiana	Do not attempt to hibernate bullfrogs imported from southern US
Feeding requirements	Live food recommended. Adult insects, larvae, worms. Larger frogs accept new born mice.
	R. temporaria and *R. catesbeiana* reluctant feeders if crowded
Accommodation	A large wet vivarium with two-thirds land area with concealment crevices. Regular changes of water to prevent fouling and skin infections. Avoid overcrowding and "heaping". 45 cm upright measurement minimum to permit jumping without damage
Sex characteristics	Males have nuptial pads on the thumbs in *R. temporaria*
Transport	In damp sphagnum moss or in well-ventilated container containing a little water

Xenopus laevis

Rearing temperatures adults	Tolerate a range 10–30°C (50–86°F)
	Optimum 23°C (73.4°F)
Breeding temperatures Tadpoles	20–23°C (68–73.4°F)
Containers Adults	Commercially available stacking polypropylene boxes are ideal: 40 × 30 × 15 cm; 60 × 40 × 15 cm (24 × 16 × 6 in.)
	Chlorine-free water, depth 8–10 cm (3–4 in.) Box cover—perspex, perforated zinc or similar
Tadpoles	Breeding tanks three-quarters full of chlorine free water at 23°C (73.4°F). 30 × 20 × 20 cm
	Special false (escape) floors needed for breeding (see text)
	Clean out regularly
Food requirements Adults	Feed proprietary diet pellets in late afternoon, alternate days:
	40 mm adults—Xenopus pellets (No. 1)
	60 mm adults—Xenopus pellets (No. 2)
	Full grown—Xenopus pellets (No. 3)
	Feed mixed heart, liver, meal-worms, Tubifex worms as a natural alternative. Feed daily for young toads
Tadpoles	Feed proprietary "nettle soup" (small quantities, frequently). Alternatively, use "infusorians". Siphon off debris
Sex characteristics Males	Smaller in size. 90 g adults. No significant lips around the urino-genital cloaca
Females	Larger than the male. 170 g adults. Significant lips around the cloacal opening
Breeding details	Sexually mature at 2 years of age. Special gonadotrophin hormone injections used to induce mating (see text, p. 204).
	Necessary to keep eggs and tadpoles away from the parents by means of perforated floor (see text, p. 203)
	Syringes for injection 2 cm³ with No. 18 needles. Eggs hatch 2–3 days after spawning. Tadpoles begin feeding 36 hours later. Metamorphosis to toads 10–12 days later (or longer)

LABORATORY FRESHWATER FISH

Poecilia reticulate (*Lebistes reticulatus*) (guppy)

Rearing temperatures	20–25°C (68–77°F)
	15–32°C (59–89.6°F) range
Life history	Viviparous (live bearers)
Sexual maturity	2–3 months
Fully grown	6 months
Gestation period	4–6 weeks or longer (depends upon conditions); 3 weeks at 25°C (77°F)
Brood size	20–100
Brood frequency	15–30 times
Sex characteristics	
Males	Smaller than female (2.5 cm); brightly colored; sharply pointed modified anal fin
Females	Larger than males (5–6 cm); shows darker patch near anal fin when fertilized accompanied by abdominal swelling
Accommodation	Community aquarium with aeration and heater
	Breeding tank with perforated separator on the bottom to permit young to escape the mother (24°C–75°F). Return mother to community tank
Food requirements	Omniverous—dried and live food such as Daphnia, Artermia, Tubifex to adults. Use "infusoria" for the young fry
Chromosome number (2n)	46

LABORATORY INSECTS

Drosophila melanogaster

Breeding temperatures	25°C (77°F)
Storage temperature range	19–21°C (66.2–69.8°F)
Life-cycle	10–14 day cycle
Eggs laid	
First instar hatches	1 day (22 hours)
First molt (2nd instar)	2 days (47 hours)
Second molt (3rd instar)	3 days (70 hours)
Pupa forms	5 days (11.8 hours)
"Pre-pupal" molt (4th instar)	5 days (122 hours)
Pupal head, wings and legs become visible	$5\frac{1}{2}$ days (130 hours)
Pupal eyes pigmented	7 days (167 hours)
Adult emerges	9 days (215 hours)
Females mate after emerging	10–12 hours
Sex characteristics	
Males	Smaller size; blunt, darker end to the abdomen; "sex" comb on the first tarsal joint, of the forelegs
Females	Larger size; pointed, striped abdomen; no "sex" combs
Virgin females	Obtain direct from pupae; paler in color; leave alone 1–2 days to check virginity (if necessary)
Culture medium requirements	Agar, cornmeal, dried yeast, syrup, anti-fungal agent (nipagin) water (many alternatives—see text)
Relative humidity in culture vessels	90% plus most successful
Culture vessels	2.5 cm × 10.0 cm vials; milk bottles—$\frac{1}{3}$ and $\frac{1}{2}$ pint with foam or sterile cotton plugs
Sterilization procedures	Media and containers must be sterile
Hot oven	1–2 hours at 150°C (302°F)
Steam	30 minutes
Autoclave	100 kPa (15 lb in.$^{-2}$) for 15 minutes
Chromosome number (2n)	8

Locusta migratoria

Breeding temperatures	32°C (89.6°F) daytime
	28°C (82.4°F) night time
Relative humidity in the cage	Low 45%
Life-cycle	9–10 weeks life span
Eggs laid	In moist sand (15% water)
Number of eggs	Each pod contains 30–100
Number of pods	6 (one every 5–6 days)
Eggs hatch	16 days incubated at 28°C (82.4°F); 11 days incubated at 32°C (89.6°F)
Nymph stages	4 weeks (approx.) at 32°C (89.6°F)
	5 molts at 5, 4, 4, 5, and 8 days
Sexual maturity	4 weeks after the final molt
Sex characteristics	
Males	Yellow; upturned end to the abdomen; longer projections (cerci) at rear of abdomen
Females	Darker brown; abdomen with "beak like" structure at rear; shorter projections either side of the anus. (The young are paler grey)
Caging	
Aluminum, plywood, hardboard, etc.	Variable dimensions
	Average type 50 cm × 40 cm × 40 cm
Egg tubes—glass, aluminum	3 cm wide, 10 cm length
Oviposition—sand	Sand 100 parts to 15 parts water
Food requirements	Daily grass. Dry wheat bran with powdered dried yeast (5 to 1). Some water provided
Summer	
Winter	Dry wheat bran (10 parts by volume)
	Dried milk (10 parts by volume)
	Dried grass (10 parts by volume)
	Powdered dried yeast (1 part by volume)
	Fresh daily water provided
Chromosome numbers (2n)	Male, 23
	Female, 24

Periplaneta americana (*American cockroach*)

Rearing temperatures	25.5°C (77°F) development 520 days
	28.3°C (82°F) development 195 days
Relative humidity in the cage	70% for most purposes
	High humidity important, but mite and mold invasions between 60 and 75% a problem
Life-cycle	
Preoviposition period	14–20 days (approx.)
Incubation times for eggs	58.5 days at 24.6°C (75°F); 48.0 days at 29.2°C (84°F)
Period between egg cases	6–7 days
Egg cases per female	12–59
Eggs per case	16–40
Number of molts	9–13
Duration of life cycle	4–6 months
Life span of female	225–706
Life span of male	82–362
Sex characteristics	
Males	Abdominal anal styles
Females	Upturned, boat-shaped abdomen with one pair smaller anal cerci
Caging	
Metal, glass, or plastic	Require perches or platforms. Security surround of oil needed to prevent escapes. Bottom lining of damp sawdust
Food requirements	Fresh fruit and vegetables with supplements of balanced diet pellets. Constant water required

Tenebrio *species* (*mealworms*)

Rearing temperatures	21°C (70°F) 6 months life-cycle; 26°C (80°F) 4 months life-cycle
Relative humidity of the culture	60–65% Caution—mites and molds at higher moisture levels. Cannabalism at low humidities
Containers	Keep in secure metal or glass containers (30 cm × 20 cm × 15 cm) for larger colonies. Two liter containers suitable 30–40 adults
Food requirements	Oatmeal and bran to a depth of one-third volume of the container. Supplement with root and leaf vegetables for the larvae and to keep up humidity

Tribolium species (*flour beetles*)

Rearing temperatures	21°C (69.8°F) 56-day life-cycle 25–27°C (77–80.6°F) 40–42 day life-cycle
Relative humidity	60–70%
Life-cycle	
Eggs	Hatch in 7 days (approx.) at 25°C (77°F)
Larvae	Pupate between 25 and 40 days later. Pale colored 4–5 mm in length. Adult emerges 5–7 days later
Pupa	
Sex characteristics	
Males	Pupae with short genital projections
Females	Pupae with longer genital projections
Containers	Keep in secure 2-liter glass vessel, suitable for up to 100 beetles. Smaller 10 cm × 2.5 cm vials for experimental work. Half fill with the culture media
Food requirements (culture medium)	Wholemeal flour (12–20 parts) mixed with powdered dry yeast (1 part)

Musca domestica—*housefly*

Rearing temperatures	25–27°C (77–80.6°F)
Relative humidity	55–58%
Life-cycle	Varies with temperature—6 days to 3 weeks
Eggs laid	In moist dung, rotting flesh, or culture cake
Number of eggs	120–150 per batch
Frequency of laying	5–6 batches per female life-time
Eggs hatch	8–48 hours (depends upon temperature)
Larvae	$3\frac{1}{2}$ days feeding to become full grown; 6–8 weeks feeding at lower temperatures
Pupae	$4\frac{1}{2}$ days to 3 weeks plus (depending upon temperature). Adults emerge
Adults	Lay eggs 4–5 days later over a period of 10 days
Sex characteristics	
Males	Eyes larger and closer together
Females	Eyes have wider separation area
Containers	Variable dimensions. Larger wood- or metal-framed cubes with detachable side panels covered in insect netting. Smaller cultures in 15 × 20 cm glass vessels
Handling	Cooling or the use of carbon dioxide as an anesthetic
Food requirements (culture media)	
Larvae	Damp "cake", made of bran, fish-meal, yeast, sugar, food pellets
Adult	Milk, sugar, water mixture

Carausius morosus (*East Indian Stick Insect*)

Rearing temperatures	20–25°C (68–77°F)
Relative humidity in the cage	High. Keep the leaves sprayed with water
Life-cycle	
Eggs	Oval, dry shelled
	Remove and place in a separate rearing cage. Hatch in a few weeks or months.
	Females (generally) only as these insects exhibit parthenogenesis (virgin birth).
Nymphs	Smaller editions of the adult. Six molts once a month (generally)
Food requirements	Privet leaves or some species of ivy

Lumbricus terrestris and related worms

Rearing temperatures	10–13°C (50–55.4°F). Fatal over 15°C (59°F)
Containers	46 × 38 × 220 cm (holding 200 worms)
Density of populations	5 *Lumbricus* per liter of culture medium. 20 *Eisenia* per liter of culture medium (25 worms per cubic foot of medium)
Culture medium	3 parts loam soil, 1 part dung (horse manure), 5 parts peat (made alkaline pH 7.0–7.5 with chalk)
Food requirements	Leaf-mold. Horse manure. Cornmeal or oatmeal
Routine duties	Water spraying. Gentle raking. Examine for cocoons every 2 or 3 weeks
Egg incubation periods	1–2 weeks
Maturation times	1–5 months (varies with species)
Renewing culture medium	6-month intervals
Cocoon culturing	Sieve and separate spherical, yellow brown cocoons by water washing off the compost. Sizes 2.5–7.0 mm in length

Outline classification of the animal kingdom

Animals are grouped into two main divisions

> *Invertebrates*—non-backboned animals.
> *Vertebrates*—backboned animals.

Within these divisions there are families (phyla) of animals. Individual animals are given at least two names. The first name (genus) is written with a capital letter; the second name (species) is written with a small letter, e.g. *Mus musculus* (mouse).

A system of classification (Taxonomic system) puts the animals into groupings that are as far as possible natural groupings, with related animals together.

Invertebrate Phyla	Examples
Protozoa	*Amoeba proteus. Paramecium caudatum. Euglena viridis*
Coelenterata	*Chlorohydra viridissima. Actinia equinum* (sea anemone). *Aurelia aureata* (jelly fish)
Platyhelminthes	*Polycelis nigra* (planarian). *Fasciola hepatica* (liver fluke). *Taenia solium* (tapeworm)
Nematoda	*Ascaris lumbricoides* (roundworm of pig). *Toxocara canis* (puppy roundworm)
Mollusca	*Planorbis planorbis* (ramshorn pond snail). *Helix pomatia* (Roman or Apple land snail). *Mytilus edulis* (edible mussel). *Sepia* (cuttle fish). *Limax, Arion* (slugs). *Loligo* (squids)

I.L.A.S.T.—K

Invertebrate Phyla	Examples
Annelida	*Tubifex* (mud worms). *Lumbricus terrestris* (earthworm). *Enchytraeus albidus* (white worms). *Nereis virens* (ragworm). *Arenicola marina* (lugworm). *Hirudo medicinalis* (leech)
Arthropoda	*Daphnia pulex* (water flea). *Gammarus* (freshwater shrimp). *Carcinus* (crab). *Asellus* (water louse). *Astacus* (cray fish). *Artemia* (brine shrimp). *Oniscus* (woodlouse). *Musca domestica* (house fly). *Drosophila melanogaster* (fruit or vinegar fly). *Periplaneta americana* (cockroach). *Locusta migratoria* (locust). *Tenebrio molitor* (meal-worms). *Tribolium confusium* (flour beetles). *Caransius morosus* (stick insect). *Anagaster kühniella* (Mediterranean flour moth).
Echinodermata	*Asterias rubens* (starfish). *Echinus esculentus* (sea urchin).

Phylum chordata contains all animals with backbones. Those animals with the backbone made up of bony vertebrae are called *vertebrates*.

Vertebrate classes	Examples
Pisces	*Scyliorhinus caniculus* (dogfish). *Clupea* (herring). *Cyprinus* (carp). *Anguilla* (eel).
Amphibia	*Triturus* (newt). *Ambystoma* (axolotl). *Rana* (frog). *Bufo* (toad). *Xenopus* (clawed toad)
Reptilia	*Lacerta* (lizard). *Anguis* (slow worm). *Natrix* (grass snake). *Chelonia* (green turtle). *Testudo* (tortoise)
Aves	*Struthio* (ostrich). *Coturnix* (Japanese quail). *Gallus* (fowl). *Taeniopygia castanotis* (Zebra finch)
Mammalia	*Mus musculus* (mouse). *Rattus norvegicus* (rat). *Oryctolagus cuniculus* (rabbit). *Meriones unguiculatus* (gerbil). *Mesocricetus auratus* (hamster). *Cavia porcellus* (guinea-pig). *Felis catis* (cat). *Canis familiaris* (dog). *Mustela furo* (ferret). *Macaca mulatta* (rhesus monkey)

Numerical data

(i) CONSTANTS AND CONVERSIONS

In scientific work the units that are used are internationally recognized and known as the "Systeme International d'Unites" (*S.I. unit system*). The basic SI units that are likely to be met are as follows:

Mass—kilogram (kg)
Length—metre (m)
Time—second (s)
Electric current—ampere (A)
Temperature—Kelvin degree ($^\circ$K)
Luminous intensity—candela (cd)
Unit of force—newton (N)
Unit of work—joule (J)
Unit of pressure—newton per metre^2 (Nm^{-2}) also known as pascal (Pa)

The *calorie* so commonly used in biology as a measure of energy is not included in the SI system. The joule is used for both energy and heat

1 Calorie = 1 kilocalorie = 1000 calories = 4185.5 joules

Conversions

  Weight or mass

$$1 \text{ kilogram (kg)} \begin{cases} 2.2046 \text{ pounds (lb)} \\ 35.274 \text{ ounces (oz)} \\ 15{,}432.000 \text{ grains (gr)} \end{cases}$$

$$1 \text{ gram (g)} \begin{cases} 15.432 \text{ grains (gr)} \\ 0.03527 \text{ ounce (oz)} \end{cases}$$

1 milligram (mg)—0.015432 grains (gr)

$$1 \text{ pound} \begin{cases} 0.4536 \text{ kg} \\ 453.59 \text{ g} \end{cases}$$

$$1 \text{ ounce} \begin{cases} 28.35 \text{ g} \\ 437.5 \text{ gr} \end{cases}$$

1 grain—64.799 mg

  Fluid volume or capacity

1 gallon (UK)—4.546 liters (8 pints)

1 gallon (USA)—3.651 liters

1 pint (UK)—0.568 liters

1 pint (USA)—0.4544 liters

$$1 \text{ fluid ounce (fl. oz)} \begin{cases} 28.412 \text{ cm}^3 \\ 480 \text{ minims} \end{cases}$$

1 minim (min)—0.05919 cm^3

$$1 \text{ liter (1)} \begin{cases} 0.22 \text{ gallon} \\ 1{,}7598 \text{ pints} \end{cases}$$

$$\left. \begin{array}{l} 1 \text{ milliliter (ml) or} \\ \quad \text{cubic centimeter (cm}^3) \end{array} \right\} 16.894 \text{ minims}$$

  Lengths

$$1 \text{ metre (m)} \begin{cases} 1.0936 \text{ yards} \\ 39.370 \text{ inches} \end{cases}$$

1 centimetre (cm)—0.3937 inches

1 millimetre (mm)—0.03937 inches

1 inch—25.400 mm

1 foot—30.48 cm

1 yard—0.914 m

Microscopic and sub-microscopic

Old system	SI system
1 μ	1 μm
1 mμ	1 nm
1 Å	0.1 nm
(Angstrom)	(nanometer)

viruses $\rightarrow \frac{1}{10}\,\mu$m $-\frac{1}{100}\,\mu$m

---

Areas

$$1 \text{ square meter (m}^2) \begin{cases} 10.7639 \text{ feet}^2 \\ 1.196 \text{ yards}^2 \end{cases}$$

1 cubic meter (m³)—1.308 yards³
1 square inch—6.4516 cm²
1 cubic inch—16.387 cm³

Pressures and temperatures

1 lb f/in²—6.9 kN/m²
1 mm mercury—0.1333 kilo-pascals (133 N/m²)
120 mm mercury—16.00 kPa
760 mm mercury—101.3 kPa

The centigrade (Celsius) scale is normally used in scientific work. In order to convert from the Fahrenheit scale to Centigrade and vice-versa, one can employ the simple calculation as shown below. A rough estimate of conversions may be determined by using the Fig. 69.

Fahrenheit to Centigrade	*Centigrade to Fahrenheit*
Subtract 32	Multiply by 9
Multiply by 5	Divide by 5
Divide by 9	Add 32

British–U.S. capacity conversions

British or U.S. unit	British Capacity Fluid ounces	British Capacity Metric	British Volume cu. in.	British Wt of water at 62 °F	American Capacity Fluid ounces	American Capacity Metric	American Volume cu. in.	American Wt of water at 62 °F
Gallon	160	4.546 liters	277.42	10 lb.	128	3.651 liters	231.00	8.326 lb.
Quart	40	1.136 liters	69.35	2½ lb.	32	0.946 liters	57.75	2.08 lb.
Pint	20	0.568 liters	34.67	1¼ lb.	16	0.454 liters	28.87	1.04 lb.
Gill	5	142.06 cm³	8.66	5 oz.	4	118.29 cm³	7.218	4.16 oz.
Fluid ounce	1	28.41 cm³	1.733	1 oz.	1	29.57 cm³	1.80	1.04 oz.

British or U.S. unit	British Capacity Metric	British Capacity British	British Wt of water at 62 °F	American Capacity Metric	American Capacity U.S.	American Wt of water at 62 °F
1 cu. ft.	28.316 liters	6.2288 gal.	62.28 lb.	28.316 liters	7.48 gal.	62.28 lb.
1 cu. in.	16.387 cm³	0.57674 fl. oz.	0.576 oz.	16.387 cm³	0.544 fl. oz.	0.576 oz.

(ii) CALCULATIONS (*Laboratory*)

Technicians are from time to time called upon to make up solutions, calculate diets, and so forth. The arithmetic operations covered here should be worked through with many examples. An incorrect dose of chemical injected into an animal can be fatal and so if any doubts exist *never* use the solution that has been made up.

Percentage solutions

Percentage refers to the strength of a solution, which may be a weight in volume (w/v) or a volume in volume (v/v) solution.

A w/v solution means that the *solute is a solid* which must be measured in weight, whereas the solvent is a liquid and must be measured in volume units.

A v/v solution means that *both the solute and the solvent* are liquids and, therefore, both are measured in volume units.

A volume % may also be a measurement of the amount of gas dissolved in a liquid. For example, $10\,cm^3$ of gas dissolved in $100\,cm^3$ of fluid—the concentration can be expressed as 10 vols. %.

A 1% solution (metric) is $1\,g$ in $100\,cm^3$; a 1% solution (apothecaries) is 1 grain in 110 minims.

For ease of reference:

$1\,cm^3$ of water weighs $1\,g$
$100\,cm^3$ of water weighs $100\,g$

so,

a 5% salt solution contains $5\,g$ salt in every $100\,cm^3$
a 50% salt solution contains $50\,g$ salt in every $100\,cm^3$
a 0.9% salt solution ("normal saline") contains $0.9\,g$ sodium chloride in every $100\,cm^3$

If the "active ingredient" is a liquid then

a 1% solution means $1\,cm^3$ liquid dissolved in every $100\,cm^3$

Dilution of solutions

How to dilute solutions from one strength to another

(*a*) If using solutions labelled in terms of percentage, the volume of stock solution needed may be calculated as follows:

$$\frac{\text{strength (\%) required}}{\text{strength (\%) of stock solution}} \times \text{volume required}$$

i.e. $\dfrac{\text{want}}{\text{have}} \times \text{amount}$

Examples

1. Prepare 500 cm³ of a 1% solution from a 5% stock solution.
   $\frac{1}{5}$ × 500 = 100 cm³ of 5% solution
   add 400 cm³ of water (i.e. to make up to 500 cm³)
   = 500 cm³ of 1% solution

2. Prepare 1 liter of a 0.1% solution in spirit from a 5% stock solution.
   $$\frac{0.1}{5} \times 1000 = \frac{1}{50} \times 1000 = 20 \text{ cm}^3 \text{ of 5\% solution}$$

   add 980 cm³ spirit (make up to a liter)
   = 1000 cm³ of 0.1% solution

(*b*) If using solutions with strengths expressed in "parts" (e.g. 1 in 10), the amount to be taken of the stock solution is calculated by using formula

$$\frac{\text{strength required}}{\text{strength of stock solution}} \times \text{volume required}$$

Examples

1. Prepare 600 cm³ of 1 in 30 solution from a stock solution of 1 in 10

   $$\frac{\frac{1}{30}}{\frac{1}{10}} \times 600.$$

   (N.B. To divide by a fraction—turn it "upside down" and multiply—$\frac{10}{1}$.)
   $\frac{1}{30} \times \frac{10}{1} \times 600$ cm³ stock solution

   = 200 cm³ stock solution (1 in 10)

   As 600 cm³ of 1 in 30 solution is required, 400 cm³ of water must be added to the 200 cm³ stock solution to produce 600 cm³ at the solution of 1 in 30.

2. Prepare a liter of a 1 in 80 solution from a 1 in 20 solution

   $$\frac{\frac{1}{80}}{\frac{1}{20}} \times 1000 = \frac{1}{80} \times \frac{20}{1} \times 1000 = 250 \text{ cm}^3 \text{ stock solution (1 in 20).}$$

   As 1000 cm³ of 1 in 80 solution is required, 750 cm³ of water must be added to the 250 cm³ stock solution to produce 1000 cm³ at the dilution of 1 in 80.

(*c*) If using solutions with strengths expressed in both ratio and percentage, convert one to the other.

Example

Prepare 400 cm³ of a 0.02% solution from a stock solution of 1 in 200 solution.

$$\frac{\text{strength required \%}}{\text{strength stock solution \%}} \times \text{volume required}.$$

$$\frac{0.02}{\frac{1}{200} \times 100} \times 400 \text{ cm}^3$$

$$\frac{0.02}{0.5} \times 400 = 16\,\text{cm}^3 \text{ stock solution.}$$

$$400 - 16 = 384\,\text{cm}^3$$

$16\,\text{cm}^3$ stock solution $= 384\,\text{cm}^3$ water $= 400\,\text{cm}^3\ 0.02\%$ solution.

Percentage in small volumes

Reference is made here to small ampoules (e.g. $1\,\text{cm}^3$ to $20\,\text{cm}^3$) with contents expressed in percentage, e.g. 1% is $10\,\text{mg}$ in $1\,\text{cm}^3$; $100\,\text{mg}$ in $10\,\text{cm}^3$.

Examples

1. How much X is there in $2\,\text{cm}^3$ of a 2.5% solution?
   1% is $10\,\text{mg}$ in $1\,\text{cm}^3$

   2.5% is $10 \times \dfrac{2.5}{1} = 25\,\text{mg}$ in $1\,\text{cm}^3$, or $50\,\text{mg}$ in $2\,\text{cm}^3$.

2. How much X is there in $10\,\text{cm}^3$ of a 15% solution?
   1% is $10\,\text{mg}$ in $1\,\text{cm}^3$
   15% is $150\,\text{mg}$ in $1\,\text{cm}^3$
   $= 1500\,\text{mg}$ in $10\,\text{cm}^3$, or $1.5\,\text{g}$ of X.

Injection doses

Reference is made here to calculations of doses when the amount required is smaller or larger than that contained in an ampoule.

Examples

1. How would you give $6\,\text{mg}$ of a drug from an ampoule containing $10\,\text{mg}$ in $1\,\text{cm}^3$?
   $10\,\text{mg}$ in $1\,\text{cm}^3$
   $6\,\text{mg}$ in $1 \times \frac{6}{10} = \frac{3}{5}\,\text{cm}^3$, or $0.6\,\text{cm}^3$.
2. How would you give $75\,\text{mg}$ of a drug from an ampoule containing $100\,\text{mg}$ in $2\,\text{cm}^3$?
   $100\,\text{mg}$ in $2\,\text{cm}^3$
   $75\,\text{mg}$ in $2 \times \frac{75}{100} = \frac{150}{100} = 1\frac{1}{2}\,\text{cm}^3$ (or $1.5\,\text{cm}^3$).

3. How would you give $0.6\,\text{mg}$ of a drug, using ampoules containing $0.4\,\text{mg}$ in $1\,\text{cm}^3$?
   400 micrograms ($0.4\,\text{mg}$) in $1\,\text{cm}^3$
   600 micrograms in $1 \times \frac{600}{400} = 1.5\,\text{cm}^3$ (using two ampoules).

4. How would you give $15\,\text{mg}$ of a drug from an ampoule containing $20\,\text{mg}$ in $1\,\text{cm}^3$?
   If there are $20\,\text{mg}$ in $1\,\text{cm}^3$ there are $15\,\text{mg}$ in $1 \times \frac{15}{20} = \frac{3}{4}\,\text{cm}^3$ (or $0.75\,\text{cm}^3$).

Converting ratios and percentages

(i) To convert a ratio to a percentage, multiple by 100.
   For example, $1:25 = \frac{1}{25} \times 100 = 4\%$
(ii) To convert a percentage to a ratio, divide by 100.
   For example, $20\% = \frac{100}{20} = 5 = 1:5$.

(iii) BREEDING PROGRAM CALCULATIONS (supplied by James Davys FIAT., OLAC 1976, Oxford.)

Introduction

Breeding program calculations involve the use of two skills: firstly, the use of known breeding data; secondly, some simple equations. Assumptions should always be stated, as this ensures that errors can easily be recognized.

When deciding on the base line data to use in a program, only figures for each particular strain of animals may be used, which should be brought up to date regularly. If answering an examination question, it is advisable to use whole numbers if possible as the strain of animal is not usually specified and, therefore, a hypothetical strain may be used.

The basic equation for any breeding program is the calculation of numbers of young produced by a breeding female, of the species concerned, each week. Once this part of the program has been calculated, the rest is simple multiplication. Therefore, it is advisable, if at all possible, to deal in weeks rather than days.

The calculations used in this chapter concern rats, mice, and guinea-pigs but one can simply change the breeding data to any species required as the formula remains the same; therefore, if one wished to devise a program to produce rabbits, for instance, by assuming that a rabbit produces an average of 6 young with a litter interval of 6 weeks—i.e. 4.5 weeks gestation and 1.5 weeks to remate, allowing for induced ovulations affected by lactation—the number of young produced per female per week = 1.

When calculating any breeding program, the requirements of the researcher must be taken into consideration from the first instance, particular attention must be paid to the strain of animal selected. Breeding data should be obtained for each individual strain one wishes to breed from, because this can vary considerably in any species, particularly if inbred strains are required. The researcher may wish to choose the breeding system to be used. One must, in an ongoing breeding program, allow for regular weekly replacement of an equal proportion of breeding stock, in this way no fluctuations occur as the proportional age and breeding status of the colony remains the same.

When considering a breeding program, one must allow for breeding fluctuations which occur in a colony from time to time. The smaller the colony, the greater percentage of extra production must be allowed. As a general rule, if production of a species is below 500 animals per week, 15% should be allowed; between 500 and 1000 per week, 10%. For production of a species over 1000 animals per week, 5% extra production would be programmed for. Greater percentages should be allowed for inbred strains and larger laboratory species.

By making all these allowances, one may obtain a regular production requirement of any laboratory animal.

Examples of a breeding program calculations

Question: Devise a continuous breeding program to supply 1000 random-bred female mice with a body weight of 18/20 g including space allocated for production of these animals. Assuming unlimited number of breeding animals available?

Calculation in 5 stages:

1. Breeding data:

Gestation	3 weeks
Length of estrus cycle	4–5 days
Age at weaning	3 weeks
Average litter size at weaning	8

Assuming good productivity, the litter interval should be 4 weeks—i.e. 3 weeks gestation + 1 week to remate, due to delayed implantation caused by lactation and allowing for a high proportion of post partum mating.

Therefore: $\dfrac{\text{average young weaned}}{\text{average litter interval}}\ \dfrac{8}{4} = 2$ young per female, per week.

Since this is a relatively small number they would be housed in Monogamous pairs, 1000 breeding ♀ and 1000 breeding ♂ would be required. The method of production is selected by strain preference for pair trio or harem mating.

2. 5–10% should be allowed for breeding fluctuations.
   $\therefore$ 5% for this program is 50 pairs.

3. RBS replacement (Replacement Breeding Stock) assuming a 5-litter economic breeding life × 4 week litter interval = 20 weeks. Therefore, all breeding stock must be replaced $\frac{1050}{20}$ approx. 53 pairs each week.
   An extra 27 breeding pairs would be allowed to produce these animals.

4. Cage space is needed for:

   (*a*) 1077 breeding pairs.
   (*b*) RBS from weaning to mating: 53 × 3 = 159.
   (*c*) Space for stock until despatch (assuming it takes 3 weeks for a good random bred strain of mouse to reach 18/20 g), 3000 stock mice would be culled if no outlet could be found for them.

5. To start breeding program—set up over 4-week period $\frac{1077}{4}$ = 269 pairs mated per week approx.
   Issue point: 3 weeks gestation + 3 weeks to weaning + 3 weeks to grow on = 9 weeks.

Question: Devise a breeding program to produce 570 breeding pairs for 4 weeks from 21 *breeding pairs of random bred* mice?

Assuming 3 weeks gestation period and average litter size of 5 males and 5 females

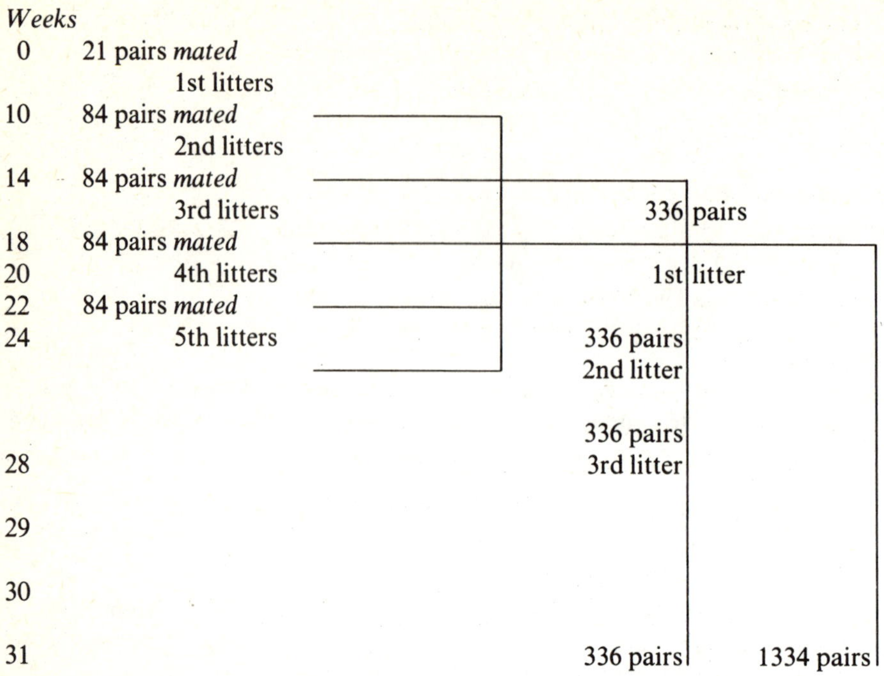

Weeks

```
 0    21 pairs mated
                  1st litters
10    84 pairs mated
                  2nd litters
14    84 pairs mated
                  3rd litters                          336 pairs
18    84 pairs mated
20                4th litters                           1st litter
22    84 pairs mated
24                5th litters              336 pairs
                                           2nd litter

                                           336 pairs
28                                         3rd litter

29

30

31                                         336 pairs    1334 pairs
```

Therefore animals could be mated from week 28 to 31 inclusive and animals would be weaned on week 31.

Question: Devise a continuous breeding program to produce 500 random bred male/female rats per week at 100/50 gm, including space allocated for these animals. Assuming unlimited numbers of breeding animals available?

Calculation in 5 stages:

1. Breeding data: Gestation period   3 weeks
                     Length of estrus cycle   4–5 days
                     Age of weaning   3 weeks
                     Average litter size at weaning   10

Assuming high fecundity, using a harem-breeding system of 5 females to 1 male and separation of gravid female rats. Separate cages would be used for parturition and suckling, therefore a 7-week interval would be expected between litters—i.e. 4 weeks for mating and gestation and 3 weeks to weaning.

Therefore: $\dfrac{\text{average litter size}}{\text{litter interval}} \dfrac{10}{7} = 1.4$ young ♀ per week.

Therefore 352 breeding females would be required. These would be mated in harems of 5—i.e. all females were boxed out when pregnant, males would be used with more than one harem, and only 37 breeding males would be required as 185 males would be suckling and 185 females would be in harems with breeding males. Breeding females would be returned to male once litter is weaned.

2. 10% would be allowed for breeding fluctuations.
   $\therefore$ an extra 35 breeding females and 5 breeding males would be used.

3. RBS replacement assuming a 5-litter economic breeding life $\times$ 7 weeks litter interval = 35 weeks. Therefore, all breeding stock must be replaced every 35 weeks.
   $\frac{377}{35}$ = 11 breeding females and 2 breeding males would be replaced each week.

To produce these, an extra 7 breeding females would be required + 2 breeding males.

4. Space is needed for:
   (*a*) 57 breeding males and 394 breeding females.
   (*b*) RBS from weaning to mating—i.e. 6 weeks $\times$ 9 = 54 boxed females and 12 males.
   (*c*) Space for stock until despatch—i.e. 2 weeks, therefore, 1000 female rats. Surplus $\male$ and $\female$ would be culled unless an outlet could be found for them.

5. A breeding program set up over a 7-week period:
   $\frac{394}{7}$ = 55 approx. breeding females + 11 males would be mated over a 7-week period approx.
   One would examine all females after they had been mated for 3 weeks and boxed out all pregnant animals. These would be checked every 2 days after this.

Issue point: 3 weeks gestation + 3 weeks to weaning + 2 weeks to grow on
$\qquad\qquad$ = 8 weeks.

Question: Devise a continuous breeding program to produce 150 male/female random bred guinea-pigs per week at 190/210 g, including space allocated for these animals? Assuming unlimited numbers of breeding animals available.

Calculation in 5 stages:

1. Breeding data:

Gestation	10 weeks
Length of estrus cycle	14 days
Age at weaning	2 weeks
Average litter size	4

Summary of breeding performance of some laboratory animals

Species	Cat	Dog	Guinea-pig	Syrian hamster	Mouse	Rabbit	Rat	Rhesus monkey
Estrus cycle	15–21 days	Biannual	16 days	4 days	4–5 days	..	5 days	Menstrual cycle 28 days
Gestation range	52–60 (63) days	60 days	65–72 days	16 days	19–21 days	30–32 days	22 days	150–180 (165) days
Litter size range	3–6	3–6	3–5	7–9	6–10	5–10	6–12	1
Weaning age	42 days	56 days	14 days	21 days	18–21 days	42 days	21 days	3 months
Weaning weight	800 gms	Variable	180 gms	40 gms	10–12 gms	Depending on species	40–50 gms	800 gms
Breeding age	6–9 months	1 year (2nd season)	12 weeks	6 weeks	42 days	6–9 months	63 days	4 years
Remate female	4th week lactation	Next season	Post-partum	End of lactation	Post-partum	4th week lactation	End of lactation or post-partum	After weaning
Economic breeding life	4 years	6 years	18–24 months	12 months	6–10 litters	36 months	9 months or 6 litters	12 years

Source—James Davys, OLAC 1976. Oxford.

Assuming good productivity, the litter interval would be 12 weeks—i.e. 10 weeks gestation + 2 weeks to remate, due to delayed implantation caused by lactation and allowing for a high proportion of post partum mating.

Therefore: 450 breeding females would be required. These are mated in harems of 5, therefore, 90 breeding males are required.

2. Breeding fluctuation should be allowed for by addition of 10% of 450 = 45 breeding females and 9 breeding males would be allowed for this.

3. RBS assuming economic breeding life of 2 years. Therefore, 5 females and 1 male would be replaced each week. 30 extra breeding females and 6 males would be allowed to produce these animals.

4. Space is needed for:

   (*a*) 525 breeding females and 105 males in harems.

   (*b*) RBS from weaning to mating 10 weeks—i.e. 50 females and 10 males.

   (*c*) Space for stock until issue—i.e. 1 week, assuming good random-bred strain of guinea-pig would reach 190/210 g 7 days after weaning.

5. One would mate the harems of 1 male + 5 females per week over a 12-week period. Therefore, 9 harems per week approx.

Issue point: 12 weeks gestation + 3 weeks to grow on = 15 weeks.

Chemical data

(i) STRENGTHS OF SOLUTIONS

Molar solutions (M) contain 1 gram molecular weight (1 mole) per litre of solution.

For example, hydrochloric acid (HCl) is
$$
\begin{aligned}
H &= 1.0 \\
Cl &= 35.5 \\
\hline
&36.5 \\
\hline
\end{aligned}
$$

A molar solution of HCl is therefore 36.5 g in a litre of water.

Normal solutions (N) contain 1 gram equivalent per litre.

For example, the gram molecular weight of HCl is the same as its gram equivalent weight

A normal solution of HCl is therefore 36.5 g in a litre. However, sulphuric acid (H_2SO_4) has molecular weight $\frac{98}{2}$ and therefore its equivalent weight is 49.

A normal solution of H_2SO_4 is 49 g per litre of water.

Millimolar solutions (milli-moles mM) contain 1/1000 of a gram molecular weight per litre and is often used to describe biological fluids.

Milli-equivalents per litre solutions contain 1/1000 of a gram equivalent weight in a litre. This is used to express the concentration of molecules and ions in blood plasma.

(ii) ISOTONIC SALINE SOLUTIONS

In many physiological experiments it is necessary to have the animal tissues bathed in solutions which are isotonic (equal salt water concentration) with the body fluids of the animal. These salines are often referred to as Ringer's solutions, named after the author of original formulae for salines.

Invertebrate saline—0.75% sodium chloride solution.
Amphibian saline—0.64% sodium chloride solution.
Mammalian tissue saline—0.9% sodium chloride solution.
Mammalian blood saline—0.6% sodium chloride solution.

All these salines are made up in deionized water. (For the exact recipes, see chemical handbooks.)

(iii) GASES IN CYLINDERS

Gases for use in scientific work are generally supplied under pressure in metal cylinders. There are standard color codes for these gases for ease of reference.

British standard color code on cylinders	Gas (US codes)
Black with white top (white—international code)	Oxygen (green)
Grey	Carbon dioxide (gray)
Black bottom with grey and white top	Oxygen and carbon-dioxide mixture (green and gray)
Grey with black top	Nitrogen
Blue	Nitrous oxide (blue)
Grey bottom with black and white top	Air (yellow—breathing air)
Brown	Helium (brown)

(iv) HUMIDITY REGULATING SOLUTIONS

It is possible to provide a required environmental humidity, for the purposes of small animal culturing, by placing into the breeding container the appropriate chemical preparations listed below.

(*a*) Saturated salt solutions

Salt solution	Relative humidity (%)
Potassium dichromate	98.0
Potassium nitrate	92.5
Barium chloride	90.2
Potassium chloride	84.3
Potassium bromide	80.7
Sodium chloride	75.3
Sodium nitrate	73.8
Strontium chloride	70.8
Sodium bromide	57.7
Magnesium nitrate	52.9
Lithium nitrate	47.1
Potassium carbonate	42.8
Magnesium chloride	33.0
Potassium acetate	22.5
Lithium chloride	11.1
Sodium hydroxide	7.0

(*b*) Glycerol/water mixtures

Glycerol/Water (%) (w/w)	Relative humidity (%)
33	90
51	80
64	70
72	60
79	50
84	40
89	30
92	20
95	10

Safety data (see *Safety in Science Laboratories* HMSO DES, No. 2)

(i) GENERAL LABORATORY SAFETY CHECKLIST

Safety awareness:
: Training of new employees.
  Safety meetings and contests.
  Training in first aid.
  Periodic inspections.
  Engineering standards.
  Safe working conditions.

Layout of rooms:
: Adequate exits, corridors, stairways, etc.
  Properly designed doors.
  Effective ventilation.
  Proper lighting.
  Correct storage facilities.
  Organized equipment arrangement.

Safety equipment:
: Safety showers and eye-baths.
  Fire extinguishing equipment.
  Personal protective equipment, safety glasses, face masks, gloves, aprons, respiratory equipment, etc.

Emergency facilities:
: First-aid kits and posted first-aid procedures for poisoning, burns, bleeding, and unconsciousness, etc.
  Telephone numbers posted for physician, ambulance, hospital, fire, and police.
  Rest beds and stretchers.
  Source of running fresh water.

(ii) SAFE PRACTICE CHECKLIST

General conduct:	Carry out only authorized procedures in work rooms
	Protect face, eyes, hands, body.
	Learn basic first-aid.
	Know where to get qualified help fast.
	Know the locality of telephone and emergency equipment.
	Report all unusual occurrences.
Handling chemicals and glassware:	Exercise care when attempting to insert glass tubing into a stopper or tubing.
	Protect the hands with gloves or cloths.
	Do not force glass tubing into holes.
	Pour reagents with care held away from the body.
	Read the labels first.
	Pour slowly down a glass rod or down the inner wall of the vessel.
	Use funnels where possible.
	Wipe up any spillage.
	Return bottles to safe storage area.
	Protect the eyes.
	Do not mix unknown chemicals.
	Careful with flames.
	Never heat a test tube with the opening towards anybody.
	Do not carry bottles by the neck.

(iii) FIRST-AID—OUTLINE PROCEDURES

The following is not a course in first-aid. This is an outline of the basics that should be included in a course as an obligatory part of any program for a laboratory worker.

An accident occurs—call the physician immediately

Until the doctor arrives administer the appropriate first-aid.

(a) *Remove the patient* from the accident area. Care with suspected bone fractures.

(b) *Eye injuries* due to chemicals need to be flushed with fresh water for 10–15 minutes. Never attempt any neutralization with another chemical.

(c) *Chemical burns* require to be flushed with flowing water for a long period. Remove any contaminated clothing.

(d) *Heat burns* require to be gently cleaned and covered with a dry sterile gauze. Look out for any shock reactions if the burns are extensive.

(e) *Bleeding* needs to be stopped without any contamination of the wound. A sterile gauze-pad can be pressed over the wound. It is not wise for an inexperienced first-aider to attempt tight tourniquets. Keep the wound uppermost. Relax the patient; loosen the clothing.

(*f*) *Bone fractures* must be treated with care so that the broken ends do not do more tissue damage. Keep the patient still. Do not move the patient if possible.

(*g*) *Poisoning by mouth* will generally require that the patient has the offending chemical diluted and that the patient vomits (unless this is specifically ill advised as for some poisonings). Give the patient copious water or milk to drink. Do not attempt this with an unconscious patient. In order to induce vomiting lukewarm salty water (2 tablespoons of salt in a glass of warm water) should be swallowed. The vomit should be clear when the stomach is flushed out. If vomiting cannot be induced, tell the doctor. The administering of antedotes is best left to a medically trained person.

(*h*) *Unconsciousness* for whatever reason necessitates that the person is laid face down so that the tongue cannot fall back into the throat and choke the patient. Never give anything by mouth (see "coma position" demonstrated).

If breathing has stopped apply artificial respiration. This technique must be learnt with an instructor. When the doctor arrives, give him all details of chemicals and the procedures already administered.

(iv) PLANTS POISONOUS TO ANIMALS

Popular name	Latin name	Toxic parts
Woodland Plants		
Wild Arum (Cuckoo pint)	*Arum maculatum*	All parts
Poison Ivy	*Rhus toxicodendron*	All parts
Mistletoe	*Viscum album*	Fruits
Oak	*Quercus species*	Fruits and leaves
Toadstools	*Amanita muscaria*	All parts
Shrubs and Trees		
Cherry laurel	*Prunus laurocerasus*	All parts
Laburnum (Golden Rain)	*Laburnum anagyroides*	All parts
Yew	*Taxus baccata*	All parts, seeds lethal
Broom	*Cytisus scoparius*	Seeds
Rhododendron	*Azalea* (American laurel) (Mountain laurel)	Leaves and flowers
Hedgerow Plants		
Buttercups	*Ranunculus species*	The sap
Deadly Nightshade	*Atropa belladonna*	All parts
Privet	*Ligustrum vulgaris*	The berries
Marshland Plants		
Hemlock	*Conium maculatum*	All parts
Marsh marigold	*Caltha palustris*	The sap
Garden Flowers		
Aconite (winter)	*Eranthis hyemalis*	All parts
Foxglove	*Digitalis purpurea*	All parts
Iris (Blue flag)	*Iris versicolor*	All parts
Larkspur	*Delphinium ajacis*	Leaves and seeds
Lily of the Valley	*Convallaria majalis*	All parts
Lupin	*Lupinus species*	All parts
Monkshood	*Aconitum anglicum*	All parts
Narcissus (daffodil)	*Narcissus species*	Bulbs

(iv) PLANTS POISONOUS TO ANIMALS (*continued*)

Popular name	Latin name	Toxic parts
Garden Vegetables		
Potato	*Solanum tuberosum*	Green sprouting tubers and the leaves
Rhubarb	*Rheum rhaponticum*	Leaves
House Plants		
Castor oil plant	*Ricinus communis*	Seeds
Hyacinth	*Hyacinthus species*	Bulbs
Poinsettia	*Euphorbis pulcherrima*	Leaves and flowers

When feeding fresh vegetation to animals it should be recognized that the above plants must be excluded, otherwise unpleasant symptoms may be induced.

8: *Self-testing Exercises*

Diagnostic tests—short answer questions

Students are encouraged to test their knowledge repeatedly. One way to monitor study progress is to attempt short answer questions. Do this factual testing before moving on to essay style questions.

Here are some short answer questions.

A. THE LAW AND LABORATORY ANIMALS

1. The main Act of Parliament regulating the use of laboratory animals is ....
2. This Act regulates the carrying out of experiments on ....
3. Which government official is responsible for the administration of this Act?
4. This Act requires that all personnel carying out experiments are holders of .....
5. A person who holds a ..... may be exempt from certain restrictions written into the issued ...... One such exemption may be ....
6. Which type of ..... must one hold in order to be able to inject or withdraw fluids from an animal without an anesthetic?
7. It is possible to delegate the experimental responsibilities to a *suitably qualified* person in another laboratory situation? Yes/No
8. What are *three* important clauses to be observed under the "pain condition"?
9. Which Committee (1963) made recommendations to update the Act previously mentioned?
10. Have the recommendations of this Committee been implemented yet? Yes/No
11. What does the Dog Act 1906 state?
12. What provisions are made in the Protection of Animal Act 1911?
13. Name *five* notifiable animal diseases?

B. GENETIC ASPECTS OF ANIMAL BREEDING

14. A sperm may be described genetically as haploid/diploid.
15. Homozygous/heterozygous animals are sometimes described as hybrids.
16. In mice agouti is dominant/recessive to albino.
17. The phenotypes of mice cc and +c will be ....

    (C = black   c = no color   + = agouti)

18. The "back-cross" is a useful method of checking the ....
19. Name two mouse homozygotes that are lethal?
20. Name two color mutant genes in mice.

21. Work out the mating between two mice as follows: (male) $+C^{ch} \times$ (female) $C^{ch}C$.

    ($+$ = agouti   C = black C^{ch} = chinchilla)

22. Some mutant strains cannot be bred in the usual manner. Name *one* example of a mutant strain that cannot be bred normally because the homozygous female cannot suckle her young.
23. If one carries out sibling (brother/sister) crosses for many generations what term is used to describe the type of offspring?
24. How many generations must one interbreed as brother/sister crosses in order to arrive at a pedigree animal?
25. If one has a stock lacking hybrid vigor (cc) what type of genotype would one introduce in order to freshen-up the stock?
26. Give *two* consequences of long term sibling breeding.

C. PRODUCTION BREEDING METHODS

27. What is "the index of productivity"?
28. What factor would one use to assess the "quality in productivity"?
29. State *two* factors which must be taken into account before selecting the species for a particular project?
30. Give two factors concerning the species selected for a project that will influence cost.
31. According to the World Health Organization specifications pre-clinical tests must be made for drugs in terms of their carcinogenicity and teratogenicity. What criteria have been defined for a pre-clinical test?
32. What three major points of information are given by the International Index of Laboratory Animals?
33. What criteria makes an animal breeder eligible for accreditation?
34. What is understood by the term—"the shelf-life of an animal"?
35. Under what circumstances is stock likely to be culled?
36. List *four* factors that are necessary before a breeding system is described as successful?
37. Name two animals that are best bred as monogamous pairs.

D. TRANSPORTATION

38. What two categories of animal are most disturbed by being transported?
39. List three important factors to be taken into account when considering transportation of animals.
40. Non-returnable, non-metal transportation containers should show at least the following *three* features. . . .
41. The packing density of animals for transportation is important. How many adult rats can be transported in a container of $100\,cm^2$ floor area?
42. Write out a formula for constructing a suitable container for transporting animals.
43. Where is the best place to locate ventilation vents on a transportation container?
44. What are the recommended shapes of the vents?
45. How would you ensure that animals do not suffer in transit through lack of water?

46. List some factors that may influence the accurate time and place of delivery of a consignment of animals.
47. What items of information should be included on a label attached to an animal container in transit?

E. HYGIENE

48. List four major agents of infection likely to be of concern in the animal house.
49. Which is more effective sterilization process, the dry-heat or wet-heat method?
50. At what temperature, and for how long must an item be left in a hot-air oven in order to be sterilized?
51. If ultraviolet lamps are used for sterilization purposes, what precautions must be taken?
52. Name a sterilization gas used in fumigation.
53. List five possible entry points of infectious agents to the animal house.
54. How are carcasses and waste materials best disposed of?
55. Name one infectious agent that may be transmitted to man from animals.
56. What contribution has a hysterectomy technique to the hygiene of a colony?
57. What is the contribution of Brown's tubes in the sterilization process?

F. S.P.F. UNITS

58. Name a virus that may pass over the placental barrier.
59. Why is diathermy used in hysterectomy techniques for caesarian delivery?
60. What does a dunk tank contain?
61. What precautions must S.P.F. personnel take before entering an S.P.F. unit?
62. Apart from the dunk tank what other sterilization methods are employed in the S.P.F. unit?
63. How would one sterilize books and papers entering the S.P.F. unit?
64. Why is the previously mentioned sterilizing agent to be used with care?
65. How does one sterilize food entering the S.P.F. unit?
66. To what use may an S.P.F. animal be put?

G. GNOTOBIOTIC ANIMALS

67. What is a gnotobiotic animal?
68. How are gnotobiotics obtained initially for rearing?
69. Under what circumstances are gnotobiotic animals reared?
70. Is there any recognisable internal difference between the conventional and gnotobiote rodent?
71. What kind of monitoring procedures are necessary with gnotobiotes?
72. To what use can the gnotobiotic animal be put in the laboratory investigations?
73. What action have gut microorganisms within the animal?
74. What major difficulty becomes apparent when using gnotobiotic animals?
75. What dietary precautions are necessary with gnotobiotic animals considering the absence of bacterial flora and the fact that feed has to be sterilized?

H. ANESTHESIA

76. What by definition is a local anesthetic?
77. Name two drugs used as local anesthetics.
78. How may local anesthetics be applied?
79. What is the value of premedication prior to anesthesia?
80. How would general anesthesia be described?
81. Name a non-volatile anesthetic used in non-recovery cases.
82. What is the most widely used intravenous anesthetic for short action?
83. Name three inhalational anesthetics used with the vaporizer.
84. What accompanying gas must be supplied whilst anesthesia is being administered?
85. State briefly the physiological basis of a muscle relaxant.
86. Give one danger or precaution to be considered when using anesthesia.

I. HUMANE KILLING (EUTHANASIA)

87. What does the term euthanasia mean?
88. What animals are best killed by a physical method?
89. What is the use of a captive bolt?
90. Name animals most suited to carbon-dioxide poisoning?
91. What drug is frequently employed in euthanasia?

9: *Bibliography*

ADAMS, C. C., The Rabbit, *The UFAW Handbook*.

American Anti-Vivisection Society, 1903 Chestnut Street, Philadelphia, Pennsylvania 19103.

American Association for Laboratory Animal Science (1967). *Manual for Laboratory Animal Technicians* (AALAS, 2317 W. Jefferson Street, Suite 208, Joliet, Illinois 60435).

American Biology Teacher, **22**, 479–83 (Animal Welfare Institute, 1960). Abuse of Animals in the Classroom and how it can be avoided.

ANDREWES, C. H., WALTON, J. R. (1977). *Viral and Bacterial Zoonoses* (Baillière-Tindall).

ARCHENHOLD, W. F., JENKINS, E. W., WOOD-ROBINSON, C. (1978). *School Science Laboratories*; a handbook of design, management, and organization (John Murray).

ASHBY, G. J. (1972). Earthworms; Locusts; The Housefly (*The UFAW Handbook*—Churchill-Livingstone).

BALL, D. J. (1969). Handling Reptiles (*Journal IAT*, **19**, 4); (1969) Housing Reptiles (*Journal IAT*, **20**, 4).

BALL, D. J., BELLAIRS, A. A. (1972). Reptiles (*The UFAW Handbook*—Churchill-Livingstone).

BARBER, B. R. (1972) Experiences in Large Scale Rat and Mouse Breeding (*Journal IAT*, **23**, 4).

BARNETT, S. A. (1967). *A Study in Behaviour* (Methuen).

BAUER, J. D., ACKERMANN, P. G., TORO, G. (1974). *Clinical Laboratory Methods* (C. V. Mosby Company).

BEANLAND, G. D., LAWRENCE, M. C. (1973). A Small Experimental Isolator of a Simple and Rigid Construction (*Journal IAT*, **24**, 3).

"Beauty without Cruelty", 175 West 12th Street, New York, N.Y. 10011.

BENIRSCHKE, K., GARNER, F. M., JONES, T. C. (editors, 1978). *Pathology of Laboratory Animals*, vol. i and ii (Springer-Verlag, New York).

BERCI, G. T. (1970). The Problems of a Supervisor in the Management of an Animal House (*Journal IAT*, **21**, 2).

Biological Sciences Curriculum Study: Biology Teacher's Handbook (1963) (Wiley, New York).

BLEBY, J., PORTER, G. (1969). Timed Assessment of the Duties of Animal Technicians (*Journal IAT*, **20**, 1).

BOTERENBROOD, E. C. 1972). Urodeles (*The UFAW Handbook*—Churchill-Livingstone).

BOVÉ, F. J. M.S.-222 Sandoz (*Anesthetic and Tranquilizer*) (Commercial pamphlets—Basle/Switzerland).

BOWMAN, C. J. (1974). *An Introduction to Animal Breeding* (Arnold).

BRANDER, G. C. ELLIS, P. R. (1977). *The Control of Disease* (Baillière-Tindall).

BROADFOOT, J. (1969). Hand Rearing Rabbits (*Journal IAT*, **20**, 3).

BUSH, B. M. 1975). *Veterinary Laboratory Manual* (Heinemann, Medical).

Carolina Bioreview Sheets:

    No. 42.1610. *Animal Meiosis.*

    No. 42.1620. *Animal Mitosis.*

    No. 42.3950. *Animal Cell.*

    No. 42.1256. *Invertebrate Life Cycles.*

    No. 42.1342. *Rabbit Anatomy.*

(Carolina Biological Supply Company, Burlington, North Carolina 27215).

Catalogue of Uniform Strains of Laboratory Animals Maintained in Great Britain (Laboratory Animals Centre, Carshalton, Surrey, England).

CHANDLER, A. C. (1955). *Introduction to Parasitology* (Wiley).

CHARLES, R. T. (1972). Physical Hazards in the Laboratory Animal House (*Journal IAT*, **23**, 4).

CLOUGH, G., GAMBLE, M. R. (1976). *Laboratory Animal Houses*—a guide to the design and planning of animal facilities (Medical Research Council, Carshalton, Surrey, England).

COLE, J. H. (1972). Beetles (*The UFAW Handbook*—Churchill-Livingstone).

COMBER, L. C. (1976). *Organisms for Genetics* (Hodder & Stoughton).

CRESS, S. J., WILDER, R. G. (1972). Patterns and Problems in an Experimental Rodent Unit (*Journal IAT*, **23**, 2).

CRUICKSHANKS, J., WOOD, L. C. (1970). Urine Collection from Individual Rats (*Journal IAT*, **21**, 1).

DENMAN, R. F., EASTCOTT, A. (1972). A Versatile Collapsible Metabolism Cage (*Journal IAT*, **23**, 1).

DROWER, J. D. L. (1974). Staff Management in Animal Units (*Journal IAT*, **25**, 1).

ELVIDGE, H. (1973). Hair Loss in Guinea Pigs (*Journal IAT*, **24**, 3).

EVANS, S. M. (1970). *The Behavior of Birds, Mammals and Fish* (Heinemann Educational).

FALCONER, D. S. (1972). Genetic Aspects of Breeding Methods (*The UFAW Handbook*—Churchill-Livingstone).

FARRIS, E. J., GRIFFITHS, J. Q. (1962). *The Rat in Laboratory Investigation* (Hafner, New York).

FESTING, M. F. W. (1976). Genetics and the Animal Technician; past, present and future (*Journal IAT*, **27**, 1).

FESTING, M. F. W. (1978). Increasing Use of Inbred Strains in Biomedical Research (*Journal IAT*, **29**, 2).

FESTING, M. F. W., ATWOOD, S. (1970). The Maintenance of Inbred Strains of Laboratory Animals (*Journal IAT*, **21**, 2).

FIENNES T-W., R. N. (1972). Primates (*The UFAW Handbook*—Churchill-Livingstone).

FORD, D. J. (1977). Future Developments in Laboratory Animal Nutrition (*Journal IAT*, **28**, 2).

FRANCIS, R. A. (1970). Tyzzer's Disease in Laboratory Animals (*Journal IAT*, **21**, 4).

FRAZER, J. F. D. (1972). Anura (Frogs and Toads) (*The UFAW Handbook*—Churchill-Livingstone).

FULLER, J. L., BARNAWELL, E. B. (1964). *The Laboratory Mouse and the Use of Inbred Strains for Genetics Research* (American Cancer Society, 219 E. 42nd Street, New York, N.Y. 10017).

Fund for the Replacement of Animals in Medical Experiments (312a Worple Road, London SW20 8QV, England).

FUSSEY, J. H. (1971). The Migratory Locust (*Journal IAT*, **22**, 1).

GAMBLE, M. R. (1976). The Importance of Environmental Control in Animal Houses (*Journal IAT*, **27**, 1).

GRAHAM-JONES, O. (1964). *Small Animal Anesthesia* (Pergamon Press).

Griffen Biology Experimental Notes (1972): *Keeping Guppies for Genetics Experiments*; *Keeping Mice for Genetics* (Ala); *Genetics Experiments with Mice* (Alb) (Griffen and George Ltd.).

GRINHAM, W. E. (1952). The Management of a Breeding Colony of Ferrets (*Journal Animal Technicians Association*, **2** (4) 3–6).

GOODWIN, D. (1978). Automatic Drinking Systems for Laboratory Animals (*Journal IAT*, **29**, 1).

HAFEZ, E. S. E. (1970). *Reproduction and Breeding Techniques for Laboratory Animals* (Lea & Febiger, Philadelphia).

HAMMOND, J. (1969). The Ferret as a Research Animal (*Carnivore Genetics Newsletter* **6**, pp. 126–7).

HAMMOND, J., CHESTERMAN, F. C. (1972). The Ferret (*The UFAW Handbook*—Churchill-Livingstone).

Handbook of Laboratory Animals (Institute of Animal Resources (NAS-NRC), Washington, DC).

HARRIS, R. J. C. (edited 1962). *The Problems of Laboratory Animal Disease* (Academic Press, London).

HARRY, E. G., COOPER, D. M. (1972). The Fowl (*The UFAW Handbook*—Churchill-Livingstone).

HEATH, J. S. (1978). *Aids to Nursing Small Animals and Birds* (Baillière Tindall, London).

HEGAN, M. A. (1976). Simple Cage Labelling and Record Keeping for Inbred and Outbred Strains of Rats and Mice. (*Journal IAT*, **27** (1)).

HEND, R. W. (1978). The Concept of Specific Pathogen Free and its Influence on Laboratory Animal Research (*Journal IAT*, **29** (1)).

HERVEY, G. F., HEMS, J. (1968). *The Goldfish* (Faber, London).

HIME, J. M. 1972). The Dog (*The UFAW Handbook*—Churchill-Livingstone).

H.M.S.O. (London). *Manual of Nutrition* (Her Majesty's Stationery Office).

H.M.S.O. (London). *Safety in Science Laboratories* (DES, No. 2).

HOLT, S. (1973). Continuous Culture of Hydra (*School Science Review*, 187, **54**, pp. 309–10).

HUBBLE, D. R. (1971). Maintenance of "Tropical" Fishes in the Laboratory (*Journal IAT*, **22**, 3).

INNES, W. T. (1966). *Exotic Aquarium Fishes* (Metaframe Corporation, Maywood, New Jersey, USA).

Institute of Laboratory Animal Resources (National Institutes of Health, Bethesda, Maryland 20014).

Institute of Laboratory Animal Resources (USA, 1974). *Guide for the Care and Use of Laboratory Animals* (ILAR, 2101 Constitution Avenue N.W., Washington, DC 20418).

KANABLE, A. (1977). *Raising Rabbits* (Rodale Press, Emmaus, Pennsylvania).

KELLY, P. J., WRAY, J. D. (1975). *The Educational Use of Living Organisms: A Source Book* (English Universities Press).

KIMURA, M., CROW, J. F. (1963). On the Maximum Avoidance of Inbreeding (*Genetics Research*, **4**, pp. 399–415).

KING, J. O. L. (1970). Transport of Animals (*The UFAW Handbook*—Churchill-Livingstone).

KORNFELD, F. (1966). *Biocidal Ampholytic Surfactants* (account of "Tego" in "Food Manufacture").

Lab-Aids Inc., 130 Wilbur Place, Bohemia, New York 11716.

Laboratory Animal Symposia (1968). *The Design and Function of Laboratory Animal Houses* (Laboratory Animals Ltd., Laboratory Animals Centre, Medical Research Council, Woodmansterne Road, Carshalton, Surrey, UK).

LANDAUER, W. (1961). The Hatchability of Chicken Eggs as Influenced by Environment and Heredity (Monograph I of the Agricultural Experimental Station, University of Connecticut).

LANE, D. R. (1980). *Jones's Animal Nursing*, 3rd edn., Pergamon Press.

LANE-PETTER, W. (edited 1963). *Breeding and Management of Some Common Laboratory Animals* (Academic Press, London).

LANE-PETTER, W. (1972). The Economics of Purchasing and Breeding Laboratory Animals (*Journals IAT*, **23**, 1).

LANE-PETTER, W. (1975). Presentation of Compound Diets (*Journal IAT*, **26**, 2).

LANE-PETTER, W., and CHATTELL, M. (1969). The Production of Standardized Litters of Mice (*Journal IAT*, **20**, 3).

LANE-PETTER, W., and LANE-PETTER, M. E. (1970). Dried Sugar Beet Pulp as Bedding (*Journal IAT*, **21**, 2).

LAPAGE, G. (1962). *Monnig's Veterinary Helminthology and Entomology* (Williams & Wilkings, Baltimore, Maryland).

LATIMER-SAYER, D. (1973). *Indoor Aquaria* (English Universities Press).

LEY, F. J. (1975). Radiation Sterilization of Diets (*Journal IAT*, **26**, 2).

MAHONEY, R. (1966). *Laboratory Techniques in Zoology* (Butterworth, London).

MANSFIELD, K. J. (1971). Installing a Cage-washer and Auxilliary Plant as Means of Saving Cost (*Journal IAT*, **22**, 3).

MAY, D. (1969). Synchronization of Oestrus in the Rat (*Journal IAT*, **20**, 4).

MAY, D., and GORDON, S. J. (1970). Effect of Early Mating on Conception in Rats (*Journal IAT*, **21**, 2).

MAY, D., and SIMPSON, K. B. (1971). An Improved Method of Synchronizing Oestrus in the Rat (*Journal IAT*, **22**, 3).

McGINNIS, T. (1974). *The Well Dog Book* (Wildwood House, Bookworks Book, USA).

Medical Research Council (1977). *Manual Series No. 2, Standardized Laboratory Animals* (Laboratory Animals Center).

Medical Research Council (1976). Genetics Monitoring (GM) Scheme (Laboratory Animals Center).

Medical Research Council (1974). *Manual Series No. 1, The Accreditation and Recognition Schemes for Suppliers of Laboratory Animals* (Laboratory Animals Center).

Medical Research Council. *Microbiological Examination of Laboratory Animals for Purposes of Accreditation* (Laboratory Animals Center).

MORGAN, D. R. (1977). Observations on the Breeding and Maintenance of Athymic Nude Mice (*Journal IAT*, **28**, 2).

MORHOLT, E., BRANDWEIN, P. F., JOSEPH, A. (1966). *A Sourcebook for the Biological Sciences* (Harcourt, Brace and World).

MOWLEM, A., DUTHIE, I. F. (1971). Lord Rank Research Center (*Journal IAT*, **23**, 1).

NEEDHAM, J. G. (1959). *Culture Methods for Invertebrate Animals* (Dover).

NEEDHAM, J. R. (1978). The Control of Mange Mites in a Conventional Mouse Colony (*Journal IAT*, **29**, 1).

NICHOLLS, P. J. (1970). Trouble-free Intravenous Injection of Rabbits (*Journal IAT*, **21**, 1).

NOSSAL, G. J. V. (1969). *Antibodies and Immunity* (Pelican).

Nuffield Foundation Science Teaching Project. Advanced Science: Biological Science (1970) (all texts in the Penguin Publications).

OLSEN, L. D. (1969). Chronic Middle Ear Infection in the Laboratory Rat (*Journal IAT*, **20**, 1).

PARROTT, R. F., EVELEIGH, J. R. (1970). Method of Planning Hysterectomies and its Use in Establishing a Specific Pathogen Free Colony of Inbred Mice and Rats (*Journal IAT*, **21**, 4).

PAYNE, L. N. (1972). The Chick Embryo (*The UFAW Handbook*—Churchill-Livingstone).

PERKINS, F. T., DARLOW, H. M., SHORT, D. J. (June 1967). Further Experience with Tego as a Disinfectant in the Animal House (*Journal IAT*).

POCKSON, A. P. (1956). The Breeding and Management of a Small Colony of Ferrets (*Journal IAT*, **7**, 7–9).

PORTER, G., FESTING, M. (1970). A Note on Breeding Performance of Mice in Four Different Types of Plastic Cage (*Journal IAT*, **21**, 3).

PORTER, G., and LANE-PETTER, W. (edited 1962). *Notes for Breeders of Common Laboratory Animals* (Academic Press, London).

PROSSER, L. (1960). Code for Use of Animals in High School Biology (*American Biology Teacher* **22** (8), pp. 478).

RAINIS, K. G. (1976). Enteric Protozoa in Frogs (*Carolina Tips*, vol. 39). Carolina Biological Supply Company, Burlington, North Carolina 27215.

Rabies (Importation of Mammals) Order 1971, H.M.S.O.

READ, A. T. (1976). A Modified Powder Feed Hopper for Rats (*Journal IAT*, **26**, 1).

Report of the Departmental Committee on Experiments on Animals (1965, *The Littlewood Report*), H.M.S.O. (2641).

Research Defence Society (1969). Notes on the Laws Relating to Experiments on Animals in Great Britain; the act of 1876.

RICHARDS, J. R., BROWN, J. J., DRURY, J. K. (1976). A Simple Method for Weighing Rats (*Journal IAT*, **27**, 1).

RITCHIE, D. H., HUMPHREY, J. K. (1970). Some Observations on the Mating of Rats (*Journal IAT*, **21**, 3).
ROBINSON, R. (1971). *Genetics for Cat Breeders* (Pergamon Press).
ROCHFORD, P. J. (1972). The American Cockroach (*The UFAW Handbook*—Churchill-Livingstone).
RUDDOCK, P. A., RUFFLE, W. G. (1972). The Care and Maintenance of a Breeding Xenopus Colony (*Journal IAT*, **23**, 2).
RUDOLPH, J. S., SMOLEN, V. F. (1970). Oral Dosing of Rabbits with Dry Solids (*Journal IAT*, **21**, 1).
RUSSELL, W. M. S., BURCH, R. L. (1959). *The Principles of Humane Experimental Technique* (Methuen).
SANDFORD, J. C. (1979). *The Domestic Rabbit* (3rd edition, Crosby Lockwood Staples, Granada Publishing).
SCOTT, P. P. (1972). The Cat (*The UFAW Handbook*—Churchill-Livingstone).
SEARLE, A. G. (1968). *Genetics Experiments with Mice* (Biology Experimental Notes) (Griffen and George Ltd., Frederick Street, Birmingham B1 3HT, UK).
SHARP, J. A. (1977). *An Introduction to Animal Tissue Culture* (Edward Arnold).
SIMPSON, K. B., MAY, D. (1973). Some Effects of the Oestrous Cycle in the Female Rat (*Journal IAT*, **24**, 1).
SIROCKIN, G., CULLIMORE, S. (1969). *Practical Microbiology* (McGraw Hill).
SMYTH, D. H. (1978). *Alternatives to Animal Experiments* (London, Scolar Press with Research Defence League).
SMYTH, D. H. (1976). *Introduction to Animal Parasitology* (2nd edition, Hodder & Stoughton).
SOFTLY, A., GRIFFITHS, D. (1972). Animal Technician Training in Western Australia (*Journal IAT*, **23**, 2).
SOLLEVELD, H. (1972). Simple Method to obtain Germfree Mice (*Journal IAT*, **23**, 4).
STEVENS, C. (1970). Attitude Towards Animals (*American Biology Teacher*, **32** (2), 77–9).
STRICKBERGER, M. W. (1962). *Experiments in Genetics with Drosophila* (Wiley).
SULU, R. (1976). *The Training of Animal Technicians in North America* (The Winston Churchill Memorial Trust).
SWEETNAM, G. W. (1970). Animal Food Handling and Storage (*Journal IAT*, **21**, 2).
SZLAUER, K. (1972). Non-spill Food Hoppers for Mice and Rats (*Journal IAT*, **23**, 1).
TOBIN, J.O'H. (1968). Viruses Transmissible from Laboratory Animals to Man (*Laboratory Animals*, **2** (1), 19–28).
THOMAS, J. L. (1969). Diseases of Animals Law (6th edition, Police Review Publishing Co. Ltd., London).
THOMPSON, A. (1970). Rat Metabolism Cage (*Journal IAT*, **21**, 1).
THOMPSON, R. (1971). The Water Consumption and Drinking Habits of a Few Species and Strains of Laboratory Animals (*Journal IAT*, **23**, 1).
TUCKER, D. K., PEACOCK, W. (1974). A Comparison of Two Rat Breeding Systems (*Journal IAT*, **25**, 1).
Turtox Service Leaflets (CCM, General Biological Inc., 8200 South Hoyne Avenue, Chicago, Illinois 60620):
    No. 4.   *The Care of Protozoan Culture in the Laboratory.*
    No. 5.   *Starting and Maintaining a Fresh Water Aquarium.*
    No. 7.   *The Care of Frogs and Other Amphibians.*
    No. 8.   *How to Prepare Microscope Slides of Simple Objects.*
    No. 10.   *The School Terrarium.*
    No. 13.   *Rearing the Silkworm Moth.*
    No. 15.   *The Culture of Drosophila Flies and their Use in Demonstrating Mendal's Laws of Heredity.*
    No. 16.   *The Culture of Planaria and its Use in Regeneration Experiments.*
    No. 17.   *Incubation, Fixation and Mounting of Chick Embryos.*
    No. 21.   *The Embalming and Injection of the Cat and the Laboratory Care of Embalmed Specimens.*
    No. 23.   *Feeding Aquarium and Terrarium Animals.*
    No. 28.   *Reptiles in the School Laboratory.*
    No. 34.   *The Care of Living Insects in the School Laboratory.*
    No. 36.   *Practical Microscopy.*
    No. 40.   *The Care of Rats, Mice, Hamsters and Guinea Pigs.*
    No. 48.   *Aquarium Troubles; their prevention and remedies.*
TV Vet (1974). *The TV Dog Book* (Farming Press Ltd.).
TV Vet (1977). *Cats; their Health and Care* (Farming Press Ltd.).
Universities Federation for Animal Welfare (1976). *The UFAW Handbook on the Care and Management of Laboratory Animals* (5th edition, Livingstone).
VEVERS, H. G. (1972). Freshwater Fish (*The UFAW Handbook*—Churchill-Livingstone).
VEVERS, H. G. (1972). Marine Aquaria (*The UFAW Handbook*—Churchill-Livingstone).
WALKER, G. (1976). Planning an Animal House (*Journal IAT*, **27**, 2).
WALLACE, M. E. (1971). *Learning Genetics with Mice* (Heinemann Educational Books, London).
WEESNER, F. M. (1970). *General Zoological Microtechniques*, p. 114. (Robert E. Krieger Publishing Co., Huntingdon, New York).
WESTON, R. (1975). Endoparasites in Dogs Supplied for Laboratory Use (*Journal IAT*, **26**, 2).
WHEELER, M. R. (1972). Fruit Flies (*The UFAW Handbook*—Churchill-Livingstone).

WHITTINGHAM, R. A. (1971). Training Animal Technicians (in Kenya) (*Journal IAT*, **23**, 1).
WILLS, J. E., SUTHERLAND, S. D. (1970). Increased Productivity in a Large Guinea Pig Colony (*Journal IAT*, **21**, 4).
WRAY, J. D. (1974) *Animal Accommodation for Schools* (Schools Council—English Universities Press).
WRAY, J. D. (1974). *Small Mammals* (Schools Council—Hodder & Stoughton).
WRAY, J. D. (1974). *Recommended Practice for Schools Relating to the Use of Living Organisms and Material of Living Origin* (English Universities Press).
YUNKER, C. E. (1964). Infections of Laboratory Animals Potentially Dangerous to Man (*Laboratory Animal Care*, **14** (5), 455–65).

10 : *Contact Addresses*

Animal Management Review, 36 Park Crescent, Hornchurch, Essex RM11 1BJ, UK.
American Association for Laboratory Animal Science, 2317 W. Jefferson Street, Suite 208, Joliet, Illinois 60435, USA (Publication: *Laboratory Animal Care*).
American Institute of Biological Sciences, 1401 Wilson Boulevard, Arlington, Virginia 22209, USA (Publication: *BioScience*).
Animal Protection Institute of America, Box 22505, Sacramento, California 95822, USA.
Fund for the Replacement of Animals in Medical Experiments (FRAME), 312A Worple Road, Wimbledon, London SW20 8QU, UK.
Institute of Animal Technicians, 5 South Parade, Summertown, Oxford OX2 7JL, UK (Publication: journals, bulletins).
Institute of Biology, 41 Queen's Gate, London SW7, UK.
Institute of Laboratory Animal Resources, National Academy of Sciences, National Institutes of Health, Bethesda, Maryland 20014, USA.
Laboratory Animal Science Association, c/o The Biochemical Society, 7 Warwick Court, London WC1R 5DP, UK (Journal: *Laboratory Animals*).
Medical Research Council, Laboratory Animals Center, M.R.C. Laboratories, Woodmansterne Road, Carshalton, Surrey SM5 4EF, UK.
National Association of Biology Teachers, 1420 N. Street, N.W., Washington, DC 20005, USA (Journal: *American Biology Teacher*).
Research Defence Society, 11 Chandos Street, London W1M 9DE, UK (Journal: *Conquest*).
Royal Society for the Protection of Birds, The Lodge, Sandy, Bedfordshire, SG19 2DL, UK.
School Natural Science Society (Publications), 44 Claremont Gardens, Upminster, Essex RM14 1DN, UK.
Universities Federation of Animal Welfare, 230 High Street, Potters Bar, Hertfordshire EN6 5BP, UK. (Publications: handbooks, leaflets).

Suppliers of Zoological Materials and Animal House Equipment (Cages, Washers, Hoppers, Bottles).

Ann Arbor Biological Center, 6780 Jackson Road, Ann Arbor, Michigan 48103, USA (313-761-8600).
Associated Grates Ltd., Coronation Street, Stockport, Cheshire SK5 7PL, UK (061-480-3016).
Bleak Hall Bird Farm (Luton) Ltd., Cresta House, Alma Street, Luton, Bedfordshire LU1 2PL, UK (0582-23730) (Birds and reptiles).
Carolina Biological Supply Company, Burlington, North Carolina 27215, USA (919-584-0381).
Culture Centre of Algae and Protozoa, 36 Storey's Way, Cambridge CB3 0DT, UK (0223-61378) (Protozoa).
Griffin Biological Laboratories Ltd., Gerrard House, Worthing Road, East Preston, West Sussex BN16 1AS, UK. (Rustington 090-62-72071/5).
Harris Biological Supplies Ltd., Oldmixon, Weston-super-Mare, Somerset, UK (0934-413063).
JG Animals, 19 Streatham Vale, London SW16 5SE, UK (01-764-4669) (Amphibia and reptiles).
Larujon Locust Suppliers, c/o Welsh Mountain Zoo, Colwyn Bay, Clwyd, UK (0492-2938) (Locusts).
Marine Biological Association, Supply Department, Citadel Hill, Plymouth, Devon PL1 2PB, UK (0752-21761) (Marine animals).
E. H. Taylor Ltd., Beehive Works, Welwyn Garden City, Hertfordshire AL6 OA2, UK (043871—4401) (Hive Bees).
Turtox, CCM: General Biological Inc., 8200 South Hoyne Avenue, Chicago, Illinois 60620, USA (800-621-8980).
Wellcome Reagents Ltd., Wellcome Research Laboratories, Beckenham, Kent BR3 3BS, UK (01-658-3541).
Worldwide Butterflies Ltd., Over Compton, Sherborne, Dorset DT9 4QN, UK (0935-4608).
Xenopus Ltd. (Biological Suppliers), 151 Frenches Road, Redhill, Surrey RH1 2HZ, UK (Redhill 0737-67224).

11: *Glossary and Index*

The glossary together with index will help the student do a little "self-teaching".
The glossary may be used as an aid to "self-testing" when the time for examination revision comes around.

Canine tooth. Conical or pointed flesh-tearing tooth. 149
Canis (dog) 165, 171, 267
Canker. A popular name for disorders of the skin resulting from infestations by mites. Usually in the region of the ears. 76, 103
Capillary. Smallest blood vessel passing through the tissues. From the capillary blood materials pass into the tissues (salts, glucose, water, amino-acids). 129, 243, 252
Carassius (**goldfish**). 196
Carausius (**stick insect**) 210, 277
Carbohydrates. Energy-producing food nutrients such as sugars and starches. 129
Carbol fuschin. Stain used with bacteria. 122
Carbon dioxide. Heavy colorless gas. A component of the atmosphere (0.03%). Poisonous in concentrations above 6% causes overstimulation of the heart and depression of reflexes. 248
Carbon monoxide. Poisonous gas found in exhaust fumes, coal gas, and many smokes. Combine with hemoglobin in the way in which oxygen does, therefore, causing asphyxia. 248
Carcass (disposal). Incinerators 14
Carcinogen. A substance that may cause cancer, i.e. tobacco smoke, tars. 2, 119
Cardiac. Of the heart. 243
Cardiac puncture. 243, 244
Cards (cage). 43
Carnivore. Flesh eating animal. 128, 149
Carotid arteries. Pair of arteries running up the neck to supply the head. 247
Cat. 5, 104, 265
  Data
  Mating
  Reproduction
  Weaning
Catabolism. The chemical breakdown of complex substances to simple ones in the body, i.e. digestion starch to glucose. The reverse of anabolism. 138
Categories (of animal). Star gradings of animal in terms of microorganisms. 120
Catalepsy. Paralysis posture. 210
Catalyst. A substance that speeds up or slows down a chemical reaction without itself being changed, i.e. enzymes are biological catalysts. 129, 130
Catarrhine. Anthropoid from the Old World, e.g. chimpanzee, baboon. 6
Caudal. Relating to the tail blood source 251
Cavia (**guinea-pig**). 4, 260
Cecum (caecum). A blind ended part of the digestive tract. 150
Ceepryn (USA). Quaternary ammonium detergent like, bactericidal agent. 88
Ceilings. Solid versus suspended. 15
Cell, The functional unit of tissues. Made up of cytoplasm and nucleus (usually) enveloped by a plasma membrane. 179
Cell division. Two types: mitosis, meiosis 181

Cellulose—Roughage component. 132
Ceratophyllum. Water plant—Hornwort. 29
Cerci. Paired, usually sensory, appendages at the rear end of the abdomen in many insects. 212
Cereal. 134, 137
Certificates. 237
Certrimide (UK). Quaternary ammonium detergent like, bactericidal agent. 88
Cervical. Relating to the neck. Mammals are characterized by having only seven cervical vertebrae in the neck.
  Cervical dislocation 246
Cestodes. Tapeworms, e.g. *Taenia solium* (pig tape). 73
Cetaceans. Whales, porpoises, dolphins. Aquatic placental mammals. 278
Chaga's disease. A trypanosomal infection 73
Chalkley's medium. Protozoan culture solution. 228
Chamber system. Anesthesia method. 241
Checks (environmental) 39
Chelonia. Tortoises, turtles. Reptiles enclosed in shells. Have no teeth. 250, 278
Chemical data. 289
Chirotera. Bats. Winged placental mammals. 278
Chickenpox. A very infectious and common virus disease. The same virus causes *herpes zoster* (shingles). 70
Chloral hydrate. A synthetic drug for promoting sleep. 240
Chloramines. Chlorine compounds used as disinfectants. 89
Chlorine. A gas used as a disinfectant for public water supplies. 88
Chloroform. Used as an anesthetic for short duration surgery. 249
Chlorohydra. Green hydra. 227
Cholera. An acute infection caused by a comma-shaped bacterium, *Vibrio cholerae.* Contracted from food or drinking water. 71
Cholesterol. A fat-like substance found in most tissues. Blood of man contains about 0.2%. 269
Chordates (animals with backbones) 278
Chorionic gonadotrophin. "Pregnyl". 205
Chorioptes cuuiculi. An ectoparasitic mite infesting rabbits, causing ear canker. 103
Chromatid. A strand which is the result of a duplication of a chromosome during cell division. 182
Chromosomes. Thread-like structures that usually become visible during mitosis and meiosis. They are the carriers of genetic material (DNA) within the cell nucleus. There are definite chromosome numbers for a species of animal. 180
Chronic disease. A disease that persists over a longer period of time. 95
Chrysalis. A pupa. A life stage of an insect. 222
Cimex **species.** *Hemipteran* (bugs), "Bedbugs". 77, 78

Dermanyssus. Blood-sucking "red mite" of fowls and pigeons. 75

Dermatitis. Inflammation of the skin. 110

Dermatophyte. Any microscopic fungus that infects the skin to cause ringworm and other diseases. 72, 82

Dermestes. Flesh-eating beetle. 83

Detergents. "Wetting agents" that lower the surface tension of water, e.g. soaps, synthetics, such as "Tego". 88

Development profile (mouse). 174

DDT (Dichloro-diphenyl-trichloroethane). A Contact insecticide. 82

Diabetes mellitus. "Sugar diabetes". A disorder resulting from a defect in the pancreatic insulin-secreting ability. Blood-sugar levels rise and glucose appears in the urine. 145

Diagnosis. The identification or recognition of diseases, employing various techniques. 95

Diagnostic aid. 95
Post-mortem.
Pregnancy.

Diarrhoea. Fluid feces usually resulting from intestinal inflammation. 95, 149

Diastema. 149

Diets.
Additional factors in. 138
Balanced. 129
Commercial. 136
Composition. 134

Diet 18. Proprietary diet for guinea-pigs and rabbits. 136

Diet PRM. 136

Diet RGP. 136

Diet SGI. Proprietary diet for guinea-pigs and rabbits. 136

Diet 41B. Proprietary diet for rats, mice, and monkeys. 136

Diet 86. Proprietary diet for rats and mice. 136

Diffusion. The movement of molecules in a fluid from an area of high concentration to an area of low concentration. 129, 167

Digestion. The breakdown of complex foodstuffs to simpler soluble substances that are absorbed through the intestine lining into the blood. 129, 144

Dihybrid inheritance. 184

Diphtheria. An acute infection generally caused by a bacillus (*Corynebacterium diphtheriae*). It affects the throat and some mucous membranes. 71

Diploid. The double or paired chromosome number of adult body cells. 182

Diptera. Two-winged insects, including gnats, mosquitoes, houseflies. 80

Dipylidium. A tapeworm that produces cysts (bladderworms) in dogs and cats. 73, 79

"Dirty" areas of an animal house. 11

Disaccharide. A compound sugar made up of two monosaccharides, i.e. sucrose, maltose, lactose. 129

Diseases (animal). 95, 107

Disease-free. 112

Disinfectant. A chemical agent that kills most growing forms of pathogen.
Characteristics of the ideal. 87

Dissection. 146

Distemper. Viral disease of dogs. 104

Diuretic. Any substance that promotes the flow of urine. 149, 242

Doe. Female rabbit. 163

Dog (*Canis familiaris*). 5, 104
Data. 266
Reproductive. 165

Dominant. A genetic term referring to a gene that shows its characteristics in both "pure" and "hybrid" situations, i.e. darker pigments. 182

Doors 15, 115

Dose (dosing). The required amount of an administered drug. Usually related to the body weight of the patient. 242

Down's syndrome. A genetic disorder "Mongolism". 180

Drainage (in animal rooms). 15, 85

Drosophilia fruit-flies. Drosophila melanogaster used in genetics experiments. 186
Data. 274
Rearing. 214

Drug. Any substance taken into the body to modify natural body processes, generally used to relieve symptoms or to combat disease. 242, 281

Drug injections. 242

Ductless gland. Endocrine glands. Those glands that secrete hormones directly into the bloodstream, not by way of ducts, i.e. thyroid gland, sex glands. 157

Duodenum. First part of the intestine following on from the stomach. 150

Dusting.
Wet and dry floor cleaning. 85
Insecticide dusting. 84

Duties (animal house). 38

Dysentery. An infective disorder of the large intestine caused either by a protozoan (see **Ameba**) or by a bacterium (**Shigella**). Diarrhoea is a common symptom. 72, 95

Ear. Organ of hearing and balance (see **Middle-ear disease**). 99

Ear and toe punches. 50, 51

Earthworms. (Lumbricus and others). 225, 277

Ecdysis. Molting. Occurs in the arthropoda. A vulnerable time in the life history. 208, 209

Echinococcus. A tapeworm which has cysts in dog liver and lungs. 73

Echinodermata. A phylum of animals with spiny skins, i.e. star fish, sea urchins. 278

Economic breeding life. The usefully productive breeding life span of an animal, i.e. rat 6 months. 161, 175, 288

Ectoparasites. Parasites living on the outside of the body. 68

Ectromelia (mouse pox). A virus infection.   96
Eelworms (anguillula).   144
Egg (fowl).   192
Eimeria perforans. A protozoan parasite causing intestinal coccidiosis in rabbit (also *Eimeria magna*).   101
Eimeria steidoe. Protozoan endoparasite in rabbit causing liver coccidiosis.   102
Eisenia. Small earthworms.   225
Electricity.   16, 56
Electrocution.   247
Elephantiasis. A disorder of the lymphatic system resulting from an accumulation of round worms (filaria).   75
Elodea (Canadian pond weed).   29
Embryo. The unborn offspring from the time of fertilization until such time as it is recognizably an animal of a particular species. From this time on it is known as a *fetus*.   167, 179
Emetic. A substance used to induce vomiting, i.e. strong common salt solution.   293
Enamel. Hard covering of the crown of the tooth.   149
Encephalitis. Brain inflammation.   107
Enchytraeus albidus. "White worms" Annelids used as fish food.   226
Endemic. A disease that is always present in a region or group of animals.   95, 107
Endocrine. Relates to "ductless" glands that secrete hormones directly into the blood, i.e. thyroid, pituitary.   131
Endometrium. The lining of the uterus.   157, 159, 160, 167
Endoparasites. Parasites living within the body.   68
Endoskeleton. Internal skeleton. Characteristic of the vertebrates.   278
Endotracheal tube. Tube inserted into trachea during anesthesia.   242
Energy. Provided by slowly burning foods releasing stored chemical energy.   138
Engineers room.   10, 12
Entamoeba. A protozoan. *E. histolytica* causes amoebic dysentery.   73, 95
Enteritis. Inflammation of the intestine. A symptom of a variety of diseases.   96
Entomology. Study of the insects.   77
Enzyme. A biological catalyst produced by living cells to promote chemical change.   129, 138
Eosin. An acidic, red dye used in biological staining.   65, 232
Eosinophil. A white blood cell which readily stains with eosin. Their numbers increase during allergic illness.   110
Ephestia species. Grain and flour moths.   83, 219
Epidermis. Outer layer of the skin.   82, 109, 242
Epididymis. Coiled tube attached to the mammal testis.   158
Epidural.   241, 244
Epiglottis. A membranous structure closing off the windpipe during swallowing.   147, 148
Epinephrine. Adrenaline (USA).   242
Epistaxis. Bleeding from the nose.   292

Epithelium. Lining tissue such as skin or mucous membrane.   82, 109, 232
Erythrocyte. Red blood cell. See Animal data charts.   255
Escherichia. Bacteria that are mostly harmless. *E. coli* normal intestine inhabitant. Used in water monitoring.   70, 117
Esophagus (see **Oesophagus**)   147, 150
Estrous cycle (see **Oestrous cycle**).   159, 160
Estrogen (see **Oestrogen**).   159, 160
Estrus (see **Oestrus**).   159
Estrus, post-partum (see **Oestrus**).   161
Ethanol. Ethyl alcohol C_2H_5OH.   89, 215, 250
Ether. Diethyl ether is a volatile anesthetic. It was one of the first general anesthetics to be used.   240, 249
Ethylene oxide. A toxic, inflammable gas used as a sterilizer.   90, 117
Ethyl chloride. Local anesthetic.   241
Excretion. The removal from the body of the by-products of body chemistry.   149
Exoskeleton. The skeleton covering the outside of the body. Characteristic of the arthropoda.   75
Experiments.
 Animal, return of   238
 Licences   237
Explosions.   55, 249
Extatosoma. Larger stick insect.   210
Extradural.   241, 244
Eye Socket (Orbitalsinus). Location for withdrawal of body fluids   245

F_1 **(first filial generation).** The offspring produced by the crossing of the P (parent) generation.   182
F_2 **(second filial generation).** The offspring resulting from the crossing of the F_2 members.   182
Factor (genetic). Gene.   180
Facilities (animal house).   9
Fallopian tube (oviduct). The tube that leads from the ovary to the uterus. Eggs fertilized by sperms here.   157
Farm animals.—see appropriate UFAW Handbook.   107
Fats. High-energy food nutrient stored in animals.   130, 146
Fauna. Animal populations as in aquaria and terraria.   31
Feces. The material that is indigestible plus intestinal bacteria, cells, and other substances. Expelled by way of the rectum and anus.   86, 107, 149
Fehling's test. Test for reducing sugars.   145
Felis species. The cats.   5, 104, 265
Femoral. Associated with the thigh. Collection site; intramuscular   243
Fercet. *Mustela furo*, a carnivorous mammal classified together with mink and weasel.   5, 164, 263
Fertility.   167
Fertilization. The fusion of male gamete (sperm)

with female gamete (ovum) to produce a zygote. 167

Fetus. The unborn animal following on from the embryo. 16, 179

Filters (animal room). Devices to eliminate particulate matter from the air entering the animal room (see *IAT Manual*, chap. 2; *UFAW Handbook*, chap. 6). 17, 117

Filters (aquaria).
Under gravel types. 25
Lift type filters. 26

First aid 292

Fish (*Pisces***)** 47, 105, 194

Fish foods.

Fish-meal. 137

Fixation. "Pickling" living cells or tissues to preserve their life-like condition. The first stage of microscopical specimen preparation. 232

Flaming. Dry heat sterilization. 89, 90, 94

Flatworms. Platyhelminthes, i.e. planarians, tapeworms, flukes. 73

Fleas. Siphonapterans. Blood-sucking insects; carriers of typhus and bubonic plague. 79

Flies. Dipterans. Two-winged insects. Include the mosquitoes, horse-flies and bot-flies. 80, 276

Floors.
Animal pen floors 14
Animal room floors 22
Metal-grid cage floors 34
Solid cage floors 34

Flour beetles. (Tribolium). 276

Fluke. Parasitic flatworm (Trematoda), i.e. Fasciola (liver fluke). 73

Food poisoning. Generally caused by bacteria of the species *Salmonella* which contaminate food (see botulism). 71, 96, 98, 107, 108

Foods. Digestible substances containing a variety of nutrients supplied in natural or commercially prepared form. 133

Foot and mouth disease. 107

Formaldehyde. A gas that dissolves in water to produce a powerful disinfectant (formalin). The gas is used in fumigation. 89

Formalin (formaldehyde). A solution of formaldehyde gas in water. A poisonous solution. 89

Fostering. Caesarian derived rats, mice, and rabbits are put with foster mothers who have been deprived of their litter. Scent spray over the nest, mother, and new litter may help disguise "foreign" odors. (See J. S. Paterson, *IAT Manual*, chap. 25). 114

Fowl (Gallus). 192, 270

Frogs. 272

Fumigation. Sterilization process involving the release of toxic fumes such as formaldehyde. 89

Fungus. Parasitic and saphrophylic plants, e.g. molds, yeasts. 72

Gall bladder. Stores bile juice. Usually within or near the liver. It contracts to expel bile juice when food (especially fatty food) is in the intestine. Not present in rat or horse. 130, 150

Gallus (fowl). 270

Gamete. A sex cell such as a sperm or ovum. 179

Gamma BHC. Insecticide dust. 84

Gamma radiation. Ionizing radiations given off from Cobalt 60 (isotope) and used in sterilizing diets and packaged goods. 94

Gas.
Anesthetic. 241
Cylinders. 290
Euthanasia. 248
Fumigant. 89

Gasterosteus (stickleback). 194

Gastric. Of the stomach. 151

Gastropoda. A class of mollusca including snails and slugs. 277

Gauge (hypodermic needles). 243, 251

Genes. The agents of heredity located on the chromosomes. Consists of Deoxyribonucleic acid (DNA) 180

Genital. Relating to the reproductive organs (genitalia) 157, 231

Genetics. The study of the behavior of heredity factors (genes). 179

Genotype. Describes the genetic constitution, in contrast to the physical characteristics (phenotype). 182

Genus. Biological category of closely related species, e.g. *Rattus rattus, Rattus norvegicus.* 277

Gerbil (Mongolian). *Meriones unguiculatus.* 4, 262

Germ cell. Gametes (sperm or eggs). 179

German measles. Rubella. A virus infection. Milder than true measles (rubeola). Virus affects the embryo in pregnant female. 270

Germicide. Popular name for chemical agents they destroy all "germs"—namely, bacteria, fungi, etc. 89

Gestation. The length of time from conception to birth in live bearing animals. 167

Gill. Female ferret. 164

Gills. Respiratory organs of aquatic organisms.
Internal, as in fish. 196
External, as in tadpoles. 200

Gill fungus (of fish). Saprolegnia. 72, 105

Gizzard. Muscular, forward part of the alimentary canal where food is ground or broken up. Found in birds. 148

Gland. An organ that produces a useful chemical secretion. 147
Exocrine—ducted glands, i.e. digestive.
Endocrine—ductless glands, i.e. sex glands.

Glanders. One of the zoonoses affecting horses. 108

Glomerulus. A projection of capillaries into Bowman's capsule of the vertebrate kidney. Ultra filtration of blood takes place across these capillaries. 151

Glossina species. Tsetse flies. 80

Glucose (dextrose). A hexose (6 carbon) sugar. A major energy source in cellular metabolism.

Basic unit which in combination form disaccharides and polysaccharides.   129, 145

Glycogen. Soluble polysaccharide energy store ("animal starch") found mostly in the liver and muscles. Made up of multiple glucose molecules.   129

Gnat (*Culex*).   80

Gnotobiotic animals. Animals living in association with *known microorganisms.*   118

Goldfish (*Carassius*).   196

Gonad. Animal organ which produces gametes (ovary, testis).   157

Gonadotropins. Hormones secreted by the anterior lobe of the pituitary gland of vertebrates. Affect sex rhythms and oestrous cycle.   159

Graafian follicle. A spherical, fluid-filled area in the mammal ovary. An ovum is released periodically (ovulation) from mature follicles.   157, 158

Grain mite (flour mite). Pest of stored cereals.   84

Grains.   83, 134, 137

Gram's stain. Used in bacteriology to distinguish: Gram-positive bacteria (stain violet), staphylococci, streptococcus; Gram-negative bacteria (stain red), typhoid bacteria.   125

Grasses   137, 139

Greens (vegetables).   137, 139

Griseofulvin. An antibiotic taken by mouth to combat the fungus causing "ringworm" or atheletes foot.   82

Guinea-pig (*Cavia porcellus*)
Data.   4, 259
Reproductive.   163

Gullet (see **esophagus**).   147

Guppies.   197

Hair. Dead keratin outgrowth from the skin of mammals.   180

Halogens. Iodine and chlorine disinfectants.   88

Halothane. Inhalation anesthetic.   241

Haltere. Modification of the hind wing of flies. Has sensory function.   80, 221

Hamster (Syrian). *Mesocricetus auratus.*   4, 164, 261

Handling (mammals).   229

Hand rearing. This may be necessary if the young are caesarian derived for SPF use. Rabbits and guinea-pigs—for techniques, see J. S. Paterson, *IAT Manual*, chap. 25.   114

Haploid. Half chromosome number, result of meiosis (reduction division).   180, 182

Hard pad. A virus infection of dogs associated with distemper.   104

"Hard" ticks. Ticks with a hard shell over their backs, e.g. *Ixodes* (castor-bean tick).   76

Harem. A mating method whereby one or more males are run with several females. The pregnant females are removed just previous to giving birth. (This is a sort of interrupted polygamy.)   178

Hay.   139

Hay infusion. Protozoan-rearing medium.   143, 228

Hazards (in the animal house).   55

Health (animal)   82

Heating.
Animal rooms.   17
Aquarium water.   25

Helminths ("worms"). A group name for the flukes and flatworms that are endoparasites of animals and man.   73

Hematology. The study of blood and its disorders.   269

Hematopinus species. Louse of pigs, horse, cattle. Anopluran (sucking louse).   78

Hematoxylin (Ehrlich's). Nuclear stain   232

Hemiptera (bugs). Insects with fluid-sucking mouth parts, e.g. *Cimex* (bed bug).   77

Hemoglobin. Respiratory pigment in the blood of vertebrates and some lower invertebrates. Carries oxygen in loose combination.   269

Hemolysis. Breakdown of red blood cells.   289

Hemophilia. A sex-linked recessive genetic disease due to the absence of one of the factors needed for normal blood clotting.   187

Hepatic. Of the liver.   97

Hepatitis. Inflammation of the liver. Two virus forms of an acute type—epidemic jaundice and serum hepatitis.   97, 104, 110, 270

Herbivore.   128, 149

Herpes. Virus infections of two types: Herpes simplex, "cold sore"; Herpes zoster, "shingles" (see chicken-pox). Herpes beta is a disease fatal to man. Affects monkeys.   70, 105

Heterosis (hybrid vigor). The increased vigor of fertility, resistance to disease and growth in a hybrid between two inbred animals.   178

Heterodont. Animals having different kinds of teeth, i.e. incisors, canines, molars.   149

Heterozygous. A pair of chromosomes having two different allelomorphs in the two corresponding loci.   181

Hibernation. Dormant state during cold season. Metabolism is slowed down and the body temperature sinks to that of the environment.   202

High power microscope.   65

Hirudinea. A class of annelid: the leeches.   278

Histamine. Found in damaged tissues. Causes inflammation responses. Symptoms of allergy caused by histamine. (See antihistamine.)   110

Histology. The study of tissues.   149, 245

Hob. Male ferret.   164

Holiday standby.   41

Homoiothermic. Able to maintain the body temperature above that of the surroundings.   162

Homo sapiens (Man).   269, 270

Homozygous. Chromosomes having identical genes in the two corresponding loci at pairing.   181

Hookworms. Nematode, roundworms that infest the alimentary canals of some animals.   74

Hoplopleura species. Rat louse USA.   78

Hoppers (food).   141, 142

Hoppers (locust). The immature stages in the life history of a locust.   209

Hormone. Chemical messengers. Produced in an endocrine gland and transported by way of the bloodstream to a "target" area where an effect is produced (i.e. growth), e.g. thyroxine, insulin. 158, 159

Host. An animal at whose expense a parasite lives. 74

Host (intermediary). An animal carrying some stage of the life history of an endoparasite, e.g. a snail in the life-cycle of liver fluke. 74

Hot-air oven. Dry heat sterilizer for glassware, etc., used at 160°C (320°F) over 2 hours. 90

"Hot spots". 16

Humidity regulating solutions. 290

Hybrid. A popular term referring to organisms of mixed genetic background (heterozygotes). 182

Hydra. A genus of small freshwater coelenterata. 227

Hydrochloric acid. Public water supplies may be acidified (pH 2.5) by adding hydrochloric acid as a precaution against contamination when used for animals. 117

Hydrophobia. Rabies. 70

Hygiene. Personnel hygiene in SPF units. 115

Hygrometers. Instruments used for measuring the atmospheric humidity. 59

Hypochlorites. Disinfectants such as calcium hypochlorite. They release free chlorine when dissolved in water. 88

Hypoderma **species.** Bot flies. 80

Hyla (tree frog). 198

Hypodermic. Under the skin. 243, 251

Hypertension. High blood pressure. Spontaneously hypertensive rats. See data capsules. 3

Hypertonic. A solution having a concentration that gains water by osmosis. 289

Hypotonic. A solution having a concentration that loses water by osmosis. 289

Ichthyophthirius. "White spot" disease of fish caused by this protozoan. 105

Identification techniques. Techniques of tattooing or ear-clipping individual animals. 48

Incineration. 57

Incubation. 192
  Fowl eggs
  Insect eggs

Incubator. 192

Infection. A disorder resulting from the parasitic existence of a virus or bacterium on the host animal tissues, i.e. a bacterial infection. 70

Infestation. The invasion of a host animal's body surface or living quarters by another dependent living organism, i.e. a mite infestation. 75

Infiltration anesthesia. Local injection technique, i.e. procaine. 241

Influenza. 70

Infusoria cultures. 196, 228

Inhalational anesthesia. i.e. diethyl ether, halothane, nitrous oxide. 248

Injecting. 242

Inspections. 238

Integument. An outer skin. 75, 244

Interferon. Protein that prevents growth of viruses. 110

Intradermal. 243, 244

Intrathoracic. 244

Ileum. The length of intestine prior to the large intestine. 150

Ill health (signs of). 106

Imago. Sexually mature adult insect. 214

Immunity. Natural resistance to infection. 111
  Active immunity.
  Passive immunity.

Inbreeding. This is the mating of close relatives (i.e. brothers and sisters) over twenty-plus generations resulting in "pure bred" animals. 175

Inbreeding, coefficient of. The degree to which animals are inbred. 175

Inbreeding, Depression. This is the reduced fertility and resistance to disease that results from continued inbreeding. 177

Incisors. Front wedge-shaped teeth on mammal jaws (cutting teeth). 149

Indigenous. Animals native, not introduced to the region. 4

Inflammation. A response to tissue damage. A protective mechanism. 110

Injections. The introduction of drugs beneath the skin by means of a hypodermic syringe, e.g. Xenopus hormone injection. 242, 251, 283

Insects. Land-dwelling (some return to water) arthropods with three pairs of legs and two pairs of wings, e.g. cockroach, fly. 77, 83

Insectivore. Mammal group feeding on insects, e.g. shrews, hedgehog, mole. 128

Instar. A stage between molts in the larval development of insects. 214

Intestine. The digestive tract or alimentary canal between the stomach and anus. 148, 150

Intramuscular. The injection of drugs into the skeletal muscles, e.g. local anesthesia.

Intraperitoneal. The injection of drugs into the abdominal cavity.

Intravenous. The injection of drugs into the venous bloodstream 242, 251

Invertebrates. Non-backboned animals 277

Iodine. A non-metal element found as iodides in water and some plant juices. Important in the diet for the manufacture of the thyroid hormone (thyroxin). 131

Iron. A necessary metal element for the manufacture of the respiratory pigment, hemoglobin. Found in liver, meat, eggs. 131

Isolators. Apparatus with sterile atmosphere in which gnotobiote (germ-free) animals are reared. 119

Isotonic. A solution having a concentration that neither gains nor loses water by osmosis. 281

Itch mite. Ectoparasitic arthropod (acarina) with four pairs of legs. Live on the skin or burrow in the skin, e.g. sarcoptes 75, 76

acterized by hair, milk secretion, and a diaphragm used in respiration.   278

Materials (cage).   34

Mating.   161

Maximum/minimum thermometer.   59

Mealworm (Tenebrio).   276

Measles (rubeola). A most contagious and widespread virus infection transmitted by air droplets (German measles—rubella).   70, 270

Mebendazole. An antihelminth preparation used to kill nematodes and tapeworms in dogs and cats.   82

Medication.   242

Meiosis. Cell division taking place in sex organs producing gametes that have half (haploid) the chromosome number of the parent cells.   181

Melanin. Dark brown pigment present in many animals which in different concentrations gives brown and yellow coloration.   184

Melopsittacus (budgerigar)   271

Mendelism. Study of the behavior of genes and inheritance.   182

Menstrual cycle. Modified oestrous cycle of primates such as monkeys, anthropoid apes, and man. No well-marked period of "heat".   159, 160

Menstruation. Period bleeding in some female primates of reproductive maturity.   159, 160

Menthol. Used to narcotize invertebrates.   250

Mesentery. Double layer of peritoneal membrane slinging the intestines, spleen, etc., to the dorsal wall of the abdominal cavity.   150

Mesocricetus **(hamster).**   261

Metabolism cages. Special apparatus used to determine food and water consumption by laboratory animals.   152

Metamorphosis.   200, 201

Methanal. Formalin.   89

Methylene blue.   105, 122

Methyl violet (gentian violet).   49

Micrometer (micron). Unit of length used in microscope work. It is 1/1000th of a millimeter.   279

Microorganisms. Microscope organisms including viruses, bacteria, protozoans, and fungi.   70

Microscopes.   63

Microworms. Minute nematodes used as fish food.   144

Micturition. Urination; act of passing urine.   149

Middle-ear disease. A bacterial disease of rats associated with *Streptobacillus moniliformis*.   99

Milk.   137
  Dried.
  Skimmed.
  Whole.

Millimicron. Nanometer (nm) 1/1000th of a micrometer.   279

Millon's test. A biochemical test for proteins.   146

Minimal inbreeding.   178

Mineral salts.   131

Minnow (Phoxinus).   194

Mites. Parasitic arthropods called acarina including the mange and canker mites.   75

Mitosis. Cell division taking place in growth areas producing genetically identical daughter cells.   181

Molars. Crushing, grinding teeth located at the back of mammal jaws.   149

Molecules. Smallest complete unit of a substance.   129

"Moldex". Mold inhibitor used in Drosophila medium.   215

Mollie (*Poecilia*** species).**   197, 274

Mollusca. Phylum of animals including snails, slugs, mussels, squids, octopus.   277

Monestrous. With one estrous cycle.   159

Monkey. See Rhesus.   171, 268

Monogamous. Male with single female.   179

Monohybrid inheritance.   182

Mononucleosis. Glandular fever. Involves the lymph glands.   110, 270

Monosaccharide. Simple hexose (6 carbon) sugar, e.g. glucose.   129

Mopping.   40

Mouse (*Mus musculus***).**   3, 161, 174, 255
  Data.
  Reproductive.

Mosquito. Dipteran insects.   73, 80, 108
  Anopheles—carrier of malaria.
  Aëdes—carrier of yellow fever.

Mucormycosis. "Moldy-hay disease" caused by a fungus Absidia ramosa.   72, 101

Mucous. Associated with *mucus*, e.g. mucous membranes.   109

Mucus. A thickish, clear-lubricating fluid protecting lining membranes.   109

Mud worms.   144

Mumps.   270

Murine typhus fever.   71, 96, 161

MS 222 (Sandoz). Narcotizing agent.   249

Mus **(mouse).**   3, 96, 108, 255

Musca domestica. The housefly, a two-winged insect (a dipteran).   220, 276

Muscle relaxants. Render muscles temporarily paralyzed during surgery.   240

Mustela **(ferret).**   164, 264

Mutant. A gene that has undergone mutation.   188

Mutation. A change in the structure of the gene, the chromosomal DNA, Can be caused by radiation or chemicals.   188

Mycelium. Fungal growth.   72, 101

Mycobacterium. Genus of bacteria resembling minute fungi, e.g. bacilli of tuberculosis and leprosy.   71, 105

Mycosis. Fungus caused by disease.   72

Myriophyllum. Water plant—Milfoil.   28

Myxomatosis. A virus disease of rabbits.   70, 102

Nanometer (nm). 1/1000th of a micrometer. Previously called a millimicron (10 Ängstrom units).   279

Narcosis. Anesthesia, dulling of consciousness.   249

Necrosis. Death of some portion of an organ, e.g. from damage to its blood supply.   95